The goal of this series is to provide concise but thorough introductory guides to various scientific techniques, aimed at both the non-expert researcher and novice scientist. Each book will highlight the advantages and limitations of the technique being covered, identify the experiments to which the technique is best suited, and include numerous figures to help better illustrate and explain the technique to the reader. Currently, there is an abundance of books and journals offering various scientific techniques to experts, but these resources, written in technical scientific jargon, can be difficult for the non-expert, whether an experienced scientist from a different discipline or a new researcher, to understand and follow. These techniques, however, may in fact be quite useful to the non-expert due to the interdisciplinary nature of numerous disciplines, and the lack of sufficient comprehensible guides to such techniques can and does slow down research and lead to employing inadequate techniques, resulting in inaccurate data. This series sets out to fill the gap in this much needed scientific resource.

More information about this series at http://www.springer.com/series/13601

Gerrit J. Gerwig

The Art of Carbohydrate Analysis

No Sugar, No Life !!

 Springer

Gerrit J. Gerwig
CarbExplore Research BV
Groningen, The Netherlands

ISSN 2367-1114 ISSN 2367-1122 (electronic)
Techniques in Life Science and Biomedicine for the Non-Expert
ISBN 978-3-030-77793-7 ISBN 978-3-030-77791-3 (eBook)
https://doi.org/10.1007/978-3-030-77791-3

This Springer imprint is published by the registered company Springer Nature Switzerland AG
The registered company address is: Gewerbestrasse 11, 6330 Cham, Switzerland

Key Features

- Introduction to carbohydrates
- Basic glycobiology summarized
- Description of protocols of various carbohydrate analysis techniques
- Step-by-step guidance and explanation

Preface

Carbohydrates (sugars) are the most significant actors in life science and biomedicine. Solitaire as well as attached to other molecules, they play key roles in many fundamental biological processes of life. Additionally, in biomedicine, carbohydrates are functional components of many drugs. Remarkably, carbohydrates are the most complex and diverse class of biological compounds found in nature.

The increasing notion of the tremendous importance of carbohydrates in physiological processes of health and disease has drastically raised the demand for detailed structural characterization of these biomolecules. The accessibility of precise sugar analysis methods is essential in carbohydrate research and in the modern biotechnological, pharmaceutical, and food industry. Until recently, carbohydrate analyses were typically available only through highly specialized laboratories.

This book introduces the reader to the world of carbohydrates and provides the principles and protocols of different carbohydrate analysis techniques. It should be noted that many methods for sugar analysis are existing in the scientific carbohydrate literature and several methods can be found on the Internet. The choice of methodology is highly dependent on the question of what kind of materials the analyst wants to investigate and the available amount and purity of the samples. Furthermore, the level of structural detail required is an item. Choosing the appropriate analysis method also depends on desired throughput, instrumental capabilities, available expertise and budget. In many cases, the full structural characterization of a carbohydrate sample is not straightforward and often requires significant time, resources, and skill.

Many natural carbohydrates, including the glycans of glycoproteins and glycolipids, display a broad structural diversity based on monosaccharide composition, linkage types, and branching. Since there is not a single technique that can cover all carbohydrate analysis needs, you have to employ several methods. This book summarizes several analytical protocols (chemical and physical techniques) in a simple, easy to read way for the non-expert researcher or novice scientist who starts to work with carbohydrates. This book does not claim to provide a comprehensive account of all analytical techniques in use.

An accurate carbohydrate analysis is also indispensable as quality control during production of recombinant therapeutic glycoproteins. At this moment, it is estimated that glycoprotein- and carbohydrate-based drugs represent a greater than $50 billion market, globally. Therapeutic glycoproteins are the fastest growing field in the pharmaceutical industry. More than 65% of the nearly 200 approved protein biopharmaceutical products today are glycoproteins. In 2018, out of the global top-ten best-selling pharmaceutical products, seven were glycosylated molecules. These glycosylated biotherapeutics comprise hormones, cytokines, growth and clotting factors, and monoclonal antibodies (mAbs). For instance, well known are erythropoietin (EPO), follicle-stimulating hormone (FSH), and thyroid-stimulating hormone (TSH). Systems to produce biotherapeutics include expression in bacteria, yeast, mammals, plants, insects, and transgenic organisms. Especially, during the preparation of recombinant glycoprotein therapeutics, detecting changes in glycosylation is important because this may affect the solubility, stability, functionality, efficacy, transport, and clearance of these biologic drugs, as well as their safety. Therefore, robust, high-throughput and quantitative analytical methods are required to characterize glycan structures in detail. Screening glycan profiles early in the clonal selection and cell culture optimization process of drug development can also provide better information for selecting the right clones. In fact, glycosylation is among the most crucial/critical quality attributes (CQA) and the most difficult parameter to control.

Current regulatory directives by the US Food and Drug Administration (FDA) and the European Medicines Agency (EMA) delineate increasingly stringent guidelines on the glycan analysis of recombinant glycoprotein therapeutics. This is expressed as follows: "For glycoproteins, the carbohydrate content (with particular attention to neutral sugars, aminosugars, and sialic acids) should be determined. In addition, the structure of the carbohydrate chains, the oligosaccharide pattern (antennary profile) and the glycosylation site(s) of the polypeptide chain should be analyzed, to the extent possible." Also, in the case of herbal medicines, carbohydrate analysis is very important for quality control and for understanding the carbohydrate's contribution to the holistic effects of herbs.

In general, the characterization of glycosylation in biological samples remains a considerable undertaking, requiring specialized analytical expertise, expensive equipment and a serious time commitment. Due to the large number of combinations in sugar types, the methods for carbohydrate analysis are far more diverse than those for protein or DNA analysis. Nevertheless, an increasing number of students and researchers are conducting experiments in carbohydrate-related studies, designated as *glycoscience* and *glycotechnology*. The interest in glycobiology studies is growing exponentially. It is getting distinct that glycosylation is the basis of most biological events. The analysis of carbohydrates is a challenging endeavor, due to their heterogeneity and abundance of functionally distinct isomers. Especially for beginners, the Materials and Methods or Experimental section in professional scientific journals may not provide sufficient information about the experimental procedures. Specific tricks and details are often lacking, because they are regarded as trivial or obvious for the carbohydrate expert. Moreover, many journals do not provide enough space for detailed description of the methods.

Although some background knowledge of carbohydrate chemistry of the reader is assumed, specific aspects necessary for the understanding of the described analytical techniques will be addressed. It is hoped that this volume in the series of "Techniques in Life Science and Biomedicine for the Non-Expert", entitled "*The Art of Carbohydrate Analysis*" is an easy starting reference for students and researchers, having a different background in the biological sciences, planning to enter the exciting field of glycobiology or glycobiotechnology. However, again, it is important to emphasize that the more than 50 protocols presented here are by no means comprehensive, but are restricted owing to the scope and limited size of the book. Therefore, references to original work and reviews in scientific journals are abundantly included and, where necessary, the reader is referred to other volumes covering more specific information and specialized methods.

Groningen, The Netherlands   Gerrit J. Gerwig

Acknowledgments

I thank Dr. A. Chrysovalantou Chatziioannou for assistance with the preparation of figures and tables. I am grateful to Dr. Stjepan K. Kračun, Dr. Martin Frank, and Dr. Sander S. van Leeuwen for critical reading of some chapters.

Disclaimer

The author declares that he has no conflicts of interest with the contents of this book. Any mention of commercial products is for information only; it does not imply recommendation or endorsement. Certain commercial equipment, instruments, or materials are identified in order to specify the experimental procedure adequately and not those that are necessarily the best available for the purpose.

No responsibility is assumed by the publisher/author for any injury and/or damage to persons or property as a matter of products liability, negligence, or otherwise, or from any use or operation of any methods, products, instructions, or ideas contained in the material of this book.

Contents

Appendix A

Appendix B

Appendix C

About the Author

**Gerrit J. Gerwig** earned his Ph.D. degree in Bio-Organic Chemistry from Utrecht University, The Netherlands. During 40 years, he was involved in Glycobiology and Carbohydrate Research at Utrecht University, together with Prof. Dr. JFG Vliegenthart and Prof. Dr. JP Kamerling. Thereafter, during 10 years, he had a guest position as carbohydrate researcher in the group of Prof. Dr. L Dijkhuizen at the University of Groningen, The Netherlands. At this moment, he is associated with CarbExplore Research BV, a glycobiotechnology company in Groningen. He has published over 150 peer-reviewed scientific papers in international journals and several chapters in different books on glycoscience.

Abbreviations

2-AA	2-aminobenzoic acid
2-AB	2-aminobenzamide
ANTS	8-aminonaphthalene-1,3,6-trisulfonic acid
APTS	8-aminopyrene-1,3,6-trisulfonic acid (9-aminopyrene-1,4-6-trisulfonate)
Ara	Arabinose
Asn	Asparagine
BGE	Background electrolyte (for CE)
CAD	Collision-activated decomposition
CAZY	Carbohydrate-active enzyme
CDG	Congenital disorders of glycosylation
CE	Capillary electrophoresis
CEA	Carcinogenic embryonic antigen
CFG	Consortium for Functional Glycomics
CID	Collision-induced dissociation
CMP	Cytidine monophosphate
CoV	Coronavirus
CZE	Capillary zone electrophoresis
DHB	2,5-dihydroxybenzoic acid
dHex	Deoxyhexose
DMSO	Dimethyl sulfoxide
DNA	Deoxyribonucleic acid
DP	Degree of polymerization
EI	Electron-impact/ionization
EPO	Erythropoietin
ER	Endoplasmic reticulum
ESI	Electrospray ionization
EVs	Extracellular vesicles
FACE	Fluorophore-assisted carbohydrate electrophoresis
FID	Flame ionization detector (free induction decay)
FLD	Fluorescence detection

Fru	Fructose
FSH	Follicle-stimulating hormone
Fuc	Fucose
GAG	Glycosaminoglycan
Gal	Galactose
GalNAc	*N*-acetylgalactosamine (2-acetamido-2-deoxygalactose)
GBP	Glycan-binding protein
GC	Gas chromatography
GDP	Guanosine diphosphate
GLC	Gas-liquid chromatography
Glc	Glucose
GlcA	Glucuronic acid
GlcN	Glucosamine
GlcNAc	*N*-acetylglucosamine (2-acetamido-2-deoxyglucose)
GPI	Glycosylphosphatidyl-inositol
GSL	Glycosphingolipid
GU	Glucose units
Hex	Hexose
HexA	Hexuronic acid
HexNAc	*N*-acetylhexosamine
HILIC	Hydrophilic interaction liquid chromatography
HMBC	Heteronuclear multiple quantum coherence spectroscopy
HPAEC	High pH (performance) anion-exchange chromatography
HMOs	Human milk oligosaccharides
HPLC	High-performance liquid chromatography
HS	Heparan sulfate
HSQC	Heteronuclear single quantum coherence spectroscopy
IdoA	Iduronic acid
IEC	Ion-exchange chromatography
LC	Liquid chromatography
LIF	Laser-induced fluorescence detection
m/z	Mass/charge
mAbs	Monoclonal antibodies
MALDI	Matrix-assisted laser desorption ionization
Man	Mannose
mRNA	Messenger RNA
MS	Mass spectrometry
MS/MS	Tandem MS
NCAM	Neural cell adhesion molecule
Neu5Ac	*N*-acetylneuraminic acid
Neu5Gc	*N*-glycolylneuraminic acid
NMR	Nuclear magnetic resonance
NOESY	Nuclear Overhauser effect spectroscopy
NP	Normal-phase
PAD	Pulsed amperometric detection

PAGE	Polyacrylamide gel electrophoresis
PCR	Polymerase chain reaction
PGC	Porous graphitized carbon
PMAA	Partially methylated alditol acetate
PNGase	Peptide-N^4-(N-acetyl-β-glucosaminyl) asparagine amidase
PSA	Prostate-specific antigen
PTM	Post-translational modification
Q-TOF	Quadrupole/time-of-flight
Rha	Rhamnose
RI	Refractive index
RNA	Ribonucleic acid
ROESY	Rotating frame nuclear Overhauser spectroscopy
RP	Reversed-phase
SARS	Severe acute respiratory syndrome
SDS	Sodium dodecyl sulfate
SEC	Size-exclusion chromatography
Ser	Serine
Siglec	Sialic acid binding-immunoglobulin lectin receptor
SPE	Solid phase extraction
TFA	Trifluoroacetic acid
Tg	Thyroglobulin
Thr	Threonine
TLC	Thin layer chromatography
TMS	Trimethylsilyl
TOCSY	Total correlation spectroscopy
TOF	Time-of-flight
tPA	Tissue plasminogen activator
TSH	Thyroid-stimulating hormone
UDP	Uridine diphosphate
UV	Ultra violet
WAX	Weak anion exchange
Xyl	Xylose
δ	Chemical shift (in ppm)

Chapter 1
The World of Carbohydrates

Abstract In nature, carbohydrates (sugars) are created in plants, where they function as building blocks and energy suppliers. Then, carbohydrates are one of the major components of our food. There is a difference between carbohydrates intended for consumption and carbohydrates as constituents of biomolecules. In the latter case, these carbohydrates (glycans) play most important roles in human life. The outer surface of nearly all cells in the human body is covered with carbohydrates linked to proteins and lipids present in the cell membrane. The carbohydrates are responsible for many biological events, such as cell–cell interactions and host–pathogen interactions. At the same time, most circulating proteins in the human body, like enzymes and hormones, are glycoconjugates, where the attached carbohydrates are often essential for functioning. This chapter shows some examples. It can be ascertained that sugars are involved in nearly all aspects of biochemistry of life.

Keywords Sugar · Biomass · Photosynthesis · Obesity · Diabetes · Human milk · Cell membrane · Glycocalyx · Fertilization · Zona pellucida · Glycocode · Glycobiology · Glycomics · Glycoproteomics

© Springer Nature Switzerland AG 2021
G. J. Gerwig, *The Art of Carbohydrate Analysis*, Techniques in Life Science and Biomedicine for the Non-Expert, https://doi.org/10.1007/978-3-030-77791-3_1

1.1 Carbohydrates Are All Around Us

Carbohydrates, also denoted as sugar(s) or saccharides, are widely distributed in nature. They represent, in the form of biomass, the most abundant organic substance on Earth. It has been suggested that half of the world's carbon exists as sugars. This enormous amount of carbohydrates is produced in green plants and algae from carbon dioxide (CO_2) and water (H_2O) through a process called photosynthesis using energy from sunlight. In this way, the basic sugar called glucose is created. Actually, photosynthesis consists of a series of chemical reactions, with each reaction being catalyzed by specific enzymes. In simple terms, the reactions of photosynthesis can be divided into two separate stages that take place in different locations within the chloroplast.

Stage 1: $12\ H_2O \rightarrow 24\ H^+ + 24\ e^- + 6\ O_2$
Stage 2: $6\ CO_2 + 24\ H^+ + 24\ e^- \rightarrow 6\ H_2O + C_6H_{12}O_6$

The oxygen produced during photosynthesis comes from water. The atmospheric CO_2 is converted into glucose ($C_6H_{12}O_6$). Much of the glucose produced by plants is immediately metabolized into different forms of energy that the plant uses for growth and germination. The portion of glucose that is not immediately used for energy is converted to complex sugar compounds, for instance, starch, where many glucose molecules are joined together chemically to form a polymer. Then, plants store starch for future energy needs or use it to build new tissues. The starch may be broken down and used as a substrate for the synthesis of a wide variety of saccharides, such as cellulose and sucrose. On our planet, plants produce ~10^{14} kg of cellulose per year. Obviously, useful industrial applications for carbohydrates are sought constantly. For many years, carbohydrates are abundantly used in the paper, plastics, textile, and cosmetic industry.

1.2 Carbohydrates in Food

Sugar is a hot topic nowadays. When the word *"sugar"* is dropped among a non-scientific community, people's first thought is about the sweet-flavored crystals (sucrose) that you add to coffee or tea. Additionally, severe worries are expressed about sugar that is excessively used in eatables, consequently providing high-calorie intake, causing obesity, diabetes, and cardiovascular ailments, increasing the risk of premature death. Over the past 10 years, several clinicians, professional organizations, and health charities have indeed waged war on sugars, thereby calling for dietary recommendations for soft drinks and sweet treats in an effort to reduce metabolic diseases. Adults and children are recommended to drastically reduce their daily intake of free sugar. However, it is important to stress that oils and fats are also a rich source of dietary energy and contain more than twice the caloric value of the equivalent amount of sugar. This means that obesity and diabetes mellitus type 2 are not exclusively carbohydrate diet-related diseases but are also caused by lack of physical activity. Nevertheless, overconsumption of refined sugar can be responsible for health problems, such as dental caries and inflammations of the intestine, liver, and kidneys [1]. On the other hand, carbohydrates are important to mental health. A strict non- or low-carbohydrate diet for a long time causes more anxiety, depression, and anger in people. Even the ability to memorize declines. Without sufficient glucose (hypoglycaemia), the central nervous system suffers, which may cause dizziness and mental weakness.

Source: Shutterstock

Nearly all living creatures derive their energy from the metabolism of glucose. In fact, glucose is the most important biological sugar, acting as the major cellular fuel for the fetus, the human brain, and muscles. Carbohydrates are involved in life-sustaining reactions. As an example, the importance of carbohydrates is beautifully demonstrated by saccharides in milk. Human milk usually contains, in addition to essential nutrients (lipids, proteins) and bioactive antibodies, a rich pool of carbohydrates. Lactose is the dominant saccharide, but more than 160 other saccharides (10–15 g/L) are present. Not all of these human milk oligosaccharides (HMOs) are digested by the infant. Many of them act as antimicrobial defense factors against pathogenic bacterial infections while the infant's immune system is developing and they act as prebiotics, stimulating the growth of beneficial bacteria in the gastrointestinal tract of the newborn [2, 3].

1.3   Carbohydrates in Life Science and Biomedicine

After DNA and proteins, carbohydrates are the third important class of biological molecules essential for life [4]. DNA (deoxyribonucleic acid), the most famous biological compound in all living organisms, already contains a sugar (deoxyribose) in its molecule, establishing the polymeric sugar–phosphate backbone. The study of carbohydrates has historically lagged behind that of the other major biomolecules. For a long time, carbohydrates have only been recognized as important structural components and compounds for energy storage, but now they are also recognized for their function as information carriers [5, 6]. The sugars that decorate the surface of cells represent a unique form of information used in the communication between cells. The enormous structural diversity of carbohydrate-containing molecules reflects their myriad biological functions in eukaryotic, but also in prokaryotic systems. In order to elucidate which kind of roles carbohydrates play in biological processes, methodologies that can analyze carbohydrate structures with increasing levels of sensitivity and simplicity are required [7]. There is no method for amplifying the amount of a carbohydrate analogous to overexpression for proteins or polymerase chain reaction (PCR) for nucleic acids, and so carbohydrate analysis is typically limited to what can be obtained from natural sources. The extreme complexity and diversity of glycoconjugate and polysaccharide structures continue to demand new processes for their elucidation. Another thing is that the production of recombinant glycoproteins and other carbohydrate-containing therapeutics regularly requires updates of techniques for their glycan analysis [8–12].

Due to the great diversity of carbohydrates, the capacity of information is much larger than in proteins. Carbohydrates contain a hidden code to biological recognition. Sometimes, this is regarded as "the sugar code of life" [13, 14]. This emphasizes that glycosylation of proteins is one of the most important posttranslational modifications. The majority of human proteins (~70%) are glycosylated. Glycosylation is predominantly found in cell–membrane proteins on the extracellular side. The overwhelming role of carbohydrates in living systems is still not completely understood as well as the importance of their three-dimensional structures.

A typical example for the importance of carbohydrate is demonstrated by the cell membrane [15]. On the outside of all animal/human cells are carbohydrates, which are chemically linked to proteins and lipids present in the phospholipid bilayer of the cell membrane (Fig. 1.1). These abundantly present glycans are composed of different monosaccharides that are linked together in a collection of structures, differing not only in their sugar constituents, but also in their configuration and spatial orientation in which the monomers are arranged. The thick layer of glycans is the first level of interaction of a cell with its environment. These glycans are the first cellular components encountered by approaching cells, pathogens, antibodies, or other molecules.

Intriguingly, each type of cell displays different carbohydrates at its cell surface and therefore the pool of surface carbohydrates, often termed the "glycocalyx," act

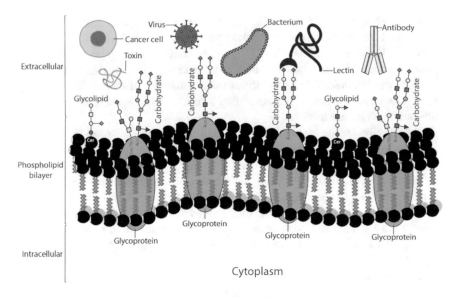

Fig. 1.1 The outer surface of all vertebrate/mammalian cells is decorated with a dense and complex array of carbohydrate chains (glycans) attached to proteins and lipids embedded in the cell membrane. These carbohydrates play important roles in cell–cell recognition. Pathogens, such as viruses and bacteria, have adapted to this and exploit surface sugars for adhesion and entry into the cell

as distinguishing markers for each cell. Thus, cell-surface glycans have specific recognition motifs that mediate the interactions between cells and the external environment and accordingly play a role in many important biological phenomena. Intercellular communication is essential to the correct functioning of multicellular organisms. This is also achieved through the secretion and uptake of extracellular vesicles (EVs). These are submicron particles that contain functional (glyco)proteins, RNA, and metabolites. They also have a lipid bilayer membrane equipped with external glycans [16–18].

Since the outer surface of mammalian cells is dominated by carbohydrates attached to proteins and lipids embodied in the cell membrane, it must be obvious that these carbohydrates play key roles in cell shapes and many fundamental biological processes. Among them are cell–cell signaling and recognition, due to specific recognition motifs of the glycans. Today, we are beginning to understand more and more the numerous critical functions of the carbohydrate chains (glycans) of cell surfaces.

The mammalian fertilization mechanism is an elegant example where contact between two cells is initiated via cell-surface glycans, in particular O-linked glycans. The sperm–egg interaction that leads to successful fertilization is mediated by the binding of saccharides present on the surface of the sperm cell with specific carbohydrate-binding proteins that are present on the egg's extracellular matrix, called the zona pellucida. Once the egg has been fertilized, it then releases an enzyme that cleaves the oligosaccharides from the surface of the sperm cells so that further fertilization is discouraged [19]. After fertilization, glycans and glycan-binding proteins (GBPs) are involved in the processes of implantation and placental functions in mammals. Obviously, the structural analysis of cell-surface carbohydrates is an important step in understanding a variety of biologically important protein–carbohydrate interactions.

Additionally, carbohydrates linked to (free) proteins are involved in many biological processes, such as blood clotting, wound healing, and pathogenesis. Overall, glycoproteins, including hormones and enzymes, play essential roles in the body. For instance, in the immune system, almost all of the key molecules acting in the immune response are glycoproteins [20]. Another example of the importance of carbohydrate is the fact that human blood groups are determined by specific carbohydrates (called antigens) on the surface of red blood cells (erythrocytes) and cells from various tissues. And furthermore, carbohydrate polymers lubricate all skeletal joints.

Inside the cell, varying among different cell types, the biosynthesis of glycans is extremely complex because the pathway is not under direct gene control (i.e., not template driven but secondary gene products). A great number of different enzymes and their substrates, transcriptional factors, and other proteins, together with environmental factors, determine the final glycan structure, which cannot be easily predicted from simple rules [21]. There is no sequence relationship between glycans and protein other than initial attachment sites. Ultimately, glycans are only related to genes via their biosynthetic enzymes and substrates. These glycan-related genes are named glycogenes. Mutations in glycogenes could cause defects of glycosylation. Defects in glycan synthesis cause a number of serious multisystemic diseases, known as congenital disorders of glycosylation, of which some can be lethal, in particular, those involving enzyme deficiencies [22, 23]. Furthermore, altered glycosylation also plays a pivotal role in the pathophysiological process of aging (see Chap. 3). Indeed, it is now recognized that the widely diverse class of carbohydrate molecules is functioning like a code that conveys information about a cell. In nature, this "glycocode" is deciphered by a host of glycan-binding proteins (GBPs). This is demonstrated by the preference viruses and bacteria show for infecting only specific cell types. In the lab, it is the chief objective of glycoscience to crack this "glycocode" and to elucidate how the different glycans displayed by cells contribute to the etiology of both healthy and diseased states [24].

Over the past three decades, researchers increasingly turned their attention toward understanding the role of carbohydrates and glycosylated proteins in cellular function. Recent years have witnessed an unprecedented growth of knowledge in biosciences. The study of cell–cell interactions in normal development, tissue morphogenesis, cellular homeostasis, embryogenesis, immune reactions, and pathological conditions such as tumor metastasis and inflammation, has made the study of carbohydrate structures inevitable. For instance, to understand how cancer cells change their immunological and adhesive properties to evade normal growth controls, it is necessary to determine their glycan expression at several stages [25]. It is observed that malignant transformation is associated with the appearance of abnormal carbohydrate determinants on the cell surface. It has been claimed that glycoconjugate glycans are directly involved in the pathophysiology of nearly every major disease. Changes in abundance and alterations in glycan profiles of serum proteins and cell-surface proteins have been shown to correlate with the progression of cancer and other disease states [26–28]. Consequently, the detection of specific alteration in glycosylation in bodily fluids, such as tears, saliva, urine, semen, or blood, might lead to the discovery of novel biomarkers for certain diseases. The US FDA has already approved nine cancer biomarkers and it is interesting to note that all nine biomarkers are glycosylated proteins, for example: α-fetoprotein (AFP-L3) in liver cancer, thyroglobulin (Tg) in thyroid cancer, prostate-specific antigen (PSA) in prostate cancer, and carcinogenic embryonic antigen (CEA), the latter is overexpressed in colon tumors, breast, and lung cancer. Carbohydrate structures found on bacteria are proving extremely valuable determinants for identification, serotyping, and classification. Research of carbohydrates from pathogenic bacteria will lead to the development of novel drugs and glycoconjugate vaccines.

The many carbohydrate-containing biomolecules, such as glycoproteins, proteoglycans, glycolipids, and carbohydrate polymers, form the fundamentals for a research field, called "*Glycobiology*," directed toward identifying the functions of carbohydrates in health and disease [15]. The study and characterization of the entire repertoire of complex carbohydrates in a living organism or cell ("the glycome"), along with identification of all the genes that encode glycoproteins and glycosyltransferases, are generally referred to as "*Glycomics*" [29]. Thus, functional glycomics is the study of structure–function relationships of glycans in a given biosystem during health and disease. "*Glycoproteomics*" is the protein-specific counterpart to glycomics, concerned with the analysis of the glycan attachment sites in the glycoproteins [30]. Glycoproteomics aims to determine the positions and identities of the complete repertoire of glycans and glycosylated proteins in a given cell or tissue. Over the last decades, many new methods and technologies, such as those based on high-performance liquid chromatography (HPLC), glycan microarray, mass spectrometry (MS), and LC-MS/MS have been developed to facilitate glycoscience study.

The knowledge about the important roles of carbohydrates in human life, medicine, agriculture, and bio-industry is growing fast. The past three decades have witnessed an expansion of interest in carbohydrate research among chemists and biologists, instigated, among others, by the discovery of novel carbohydrate-related human diseases. Early detection of such disease is undisputedly a critical factor for successful treatment and improved quality of life. Here, the search for (glyco)biomarkers plays an important role [31–34]. Glycomics has become one of the most important fields of life sciences [5, 35]. Moreover, the rapid growth of therapeutic recombinant glycoprotein pharmaceuticals has led glycoscience to the cornerstone of modern biomedical research. Many of the highest grossing protein therapeutics, including monoclonal antibodies (mAbs), Fc-fusion proteins, γ-interferon (FN-γ), erythropoietin (EPO), and tissue plasminogen activator (tPA), contain carbohydrates (glycans) covalently bound to their peptide backbone, being an inextricable part of their structure as well as their function. As glycosylation impacts the safety and efficacy of these protein therapeutics, worldwide regulatory agencies require a demonstration of consistency in glycosylation of manufactured glycoprotein drugs and biosimilars, during and after the bioproduction process [36].

There is a growing demand for diagnostic or prognostic biomarkers for disease and a vast and growing market of pharmaceuticals and vaccines. The observations of glycan aberrations of glycoproteins and/or glycolipids during some diseases indicate a potential of glycan profiles as biomarkers. Researchers have opened up new research fronts in terms of probing glycans as targets in the design and the development of novel drugs or therapies for infectious and neurological diseases, cancer, and metabolic disorders [37]. The current interest in carbohydrate research is also driven by the fact that carbohydrates/sugars are one of the most important components in many food products, where they contribute to the sweetness, appearance, textural characteristics, and, of course, to the caloric contents. Specific carbohydrate ingredients are often used for specific product applications. As mentioned, carbohydrates are extensively utilized as raw materials for several industries, e.g., paper, plastics, textiles, etc. All these aspects cause carbohydrate research an interesting scientific challenge.

It is evident that reliable carbohydrate analysis methods, suitable for a very diverse pallet of materials, are of utmost importance in life sciences, including glycobiology, biomedicine, and food technology. An accurate sugar analysis is the starting point for the study of structure–function relationships in many carbohydrate-involved biological/biochemical systems. However, multiple techniques are often needed to come to final conclusions. Because of the space limitations in a book such as this, the examples are illustrative rather than exhaustive. Some excellent scientific volumes dealing with glycobiology and carbohydrate analysis, have been published in the past [38–44]. Although many carbohydrate analysis methods and techniques were developed in the previous century (1960/1970/1980s), during the last two decades, the analytical methods have become more sophisticated and sensitive, due to great improvements in separation and detection techniques, advanced instrumentation, automated equipment, and computerized interpretation, including the use of online databases. This means that analyses can be performed more rapidly and with a greater degree of precision than ever before. To handle the large datasets generated by such high-throughput analyses, bioinformatics approaches have become important in many glycobiological studies.

Sugars are involved in nearly all aspects

of the biochemistry of life.

References

1. Preuss H, Bagchi D, editors. Dietary sugar, salt and fat in human health. Academic; 2020.
2. Bode L. Human milk oligosaccharides: every baby needs a sugar mama. Glycobiology. 2012;22:1147–62.
3. Van Leeuwen SS, Te Poele EM, Chatziioannou AC, Benjamins E, Haandrikman A, Dijkhuizen L. Goat milk oligosaccharides: their diversity, quantity, and functional properties in comparison to human milk oligosaccharides. J Agric Food Chem. 2020;68:13469–85.
4. Stick RV, Williams SJ. Carbohydrates: the essential molecules of life. 2nd ed. New York: Elsevier; 2009.
5. Hart GW, Copeland RJ. Glycomics hits the big time. Cell. 2010;143:672–6.
6. Taniguchi N, Endo T, Hart GW, Seeberger PH, Wong C-H, editors. Glycoscience: biology and medicine. Springer Reference; 2015.
7. Vliegenthart JFG. The impact of defining glycan structures. Persp Sci. 2017;11:3–10.
8. Li H, d'Anjou M. Pharmacological significance of glycosylation in therapeutic proteins. Curr Opin Biotechnol. 2009;20:678–84.
9. Mizukami A, Caron AL, Picanco-Castro V, Swiech K. Platforms for recombinant therapeutic glycoprotein production. Methods Mol Biol. 2018;1674:1–14.
10. Yang X, Bartlett MG. Glycan analysis for protein therapeutics. J Chromatogr B. 2019;1120:29–40.
11. Dalziel M, Crispin M, Scanlan CN, Zitzmann N, Dwek RA. Emerging principles for the therapeutic exploitation of glycosylation. Science. 2014;343:1235681.
12. Planinc A, Bones J, Dejaegher B, van Antwerpen P, Delporte C. Glycan characterization of biopharmaceuticals: updates and perspectives. Anal Chim Acta. 2016;921:13–27.
13. Gabius HJ, editor. The sugar code: fundamentals of glycosciences. Hoboken: Wiley-Blackwell; 2011.
14. Suzuki T, Ohtsubo K, Taniguchi N, editors. Sugar chains: decoding the functions of glycans. Springer; 2015
15. Varki A. Biological roles of glycans. Glycobiology. 2017;27:3–49.
16. Gerlach JQ, Griffin MD. Getting to know the extracellular vesicle glycome. Mol Biosyst. 2016;12:1071–81.
17. Costa J. Glycoconjugates from extracellular vesicles: structures, functions and emerging potential as cancer biomarkers. Biochim Biophys Acta. 2017;1868:157–66.
18. Williams C, Royo F, Aizpurua-Olaizola O, Pazos R, Boons G-J, Reichardt N-C, Falcon-Perez JM. Glycosylation of extracellular vesicles: current knowledge, tools and clinical perspectives. J Extracell Vesicles. 2018;7:1442985.
19. Gupta SK. The human egg's zona pellucida. Curr Top Dev Biol. 2018;130:379–411.
20. Kolarich D, Lepenies B, Seeberger PH. Glycomics, glycoproteomics and the immune system. Curr Opin Chem Biol. 2012;16:214–20.
21. Thaysen-Andersen M, Packer NH. Site-specific glycoproteomics confirms that protein structure dictates formation of N-glycan type, core fucosylation and branching. Glycobiology. 2012;22:1440–52.
22. Freeze HH. Understanding human glycosylation disorders: biochemistry leads the charge. J Biol Chem. 2013;288:6936–45.
23. Péanne R, de Lonlay P, Foulquier F, Kornak U, Lefeber DJ, Morave E, Pérez B, Seta N, Thiel C, van Schaftingen E, Mathijs G, Jaeken J. Congenital disorders of glycosylation (CDG): quo vadis? Eur J Med Genet. 2018;61:643–63.
24. Pilobello KT, Mahal LK. Deciphering the glycocode: the complexity and analytical challenge of glycomics. Curr Opin Chem Biol. 2007;11:300–5.
25. Mereiter S, Balmana M, Campos D, Gomes J, Reis CA. Glycosylation in the era of cancer-targeted therapy: where are we heading? Cancer Cell. 2019;36:6–16.
26. Gornik O, Lauc G. Glycosylation of serum proteins in inflammatory diseases. Dis Markers. 2008;25:267–78.

27. Schachter H, Freeze HH. Glycosylation diseases: quo vadis? Biochim Biophys Acta. 2009;1792:925–30.
28. Reily C, Stewart TJ, Renfrow MB, Novak J. Glycosylation in health and disease. Nat Rev Nephrol. 2019;15:346–66.
29. Cummings RD, Pierce JM. The challenge and promise of glycomics. Chem Biol. 2014;21:1–15.
30. Lazar IM, Lee W, Lazar AC. Glycoproteomics on the rise: established methods, advanced techniques, sophisticated biological applications. Electrophoresis. 2013;34:113–25.
31. Mariño K, Saldova R, Adamczyk B, Rudd PM. Changes in serum N-glycosylation profiles: functional significance and potential for diagnostics. Carbohydr Chem. 2012;37:57–93.
32. Ueda K. Glycoproteomic strategies: from discovery to clinical application of cancer carbohydrate biomarkers. Proteomics Clin Appl. 2013;7:607–17.
33. Etxebarria J, Reichardt N-C. Methods for the absolute quantification of N-glycan biomarkers. Biochim Biophys Acta. 2016;1860:1676–87.
34. Kailemia MJ, Xu G, Wong M, Li Q, Goonatilleke E, Leon F, Lebrilla CB. Recent advances in the mass spectrometry methods for glycomics and cancer. Anal Chem. 2018;90:208–24.
35. Springer SA, Gagneux P. Glycomics: revealing the dynamic ecology and evolution of sugar molecules. J Proteomics. 2016;135:90–100.
36. Zhang P, Woen S, Wang T, Liau B, Zhao S, Chen C, Yang Y, Song Z, Wormald MR, Yu C, Rudd PM. Challenges of glycosylation analysis and control: an integrated approach to producing optimal and consistent therapeutic drugs. Drug Discov Today. 2016;21:740–65.
37. Lebrilla CB, An HJ. The prospect of glycan biomarkers for the diagnosis of diseases. Mol Biosyst. 2009;5:17–20.
38. Brockhausen I, editor. Glycobiology protocols, Methods in molecular biology, vol. 347. Totowa: Humana Press; 2006.
39. Kamerling JP, Boons G-J, Lee Y, Suzuki A, Taniguchi N, Voragen AGJ, editors. Comprehensive glycoscience—from chemistry to systems biology. Amsterdam: Elsevier; 2007.
40. Cummings RD, Pierce JM, editors. Handbook of glycomics. Amsterdam: Elsevier; 2009.
41. Li J, editor. Functional glycomics: methods and protocols. Totowa: Humana Press; 2010.
42. Wang B, Boons G-J, editors. Carbohydrate recognition: biological problems, methods, and applications. Hoboken: Wiley; 2011.
43. Varki A, Cummings RD, Esko JD, Freeze HH, Stanley P, Hart G, Marth J, editors. Essentials of glycobiology. Cold Spring Harbor: CSHL Press; 2015.
44. Lauc G, Wuhrer M. In: Walker JM, editor. High throughput glycomics and glycoproteomics, methods and protocols, Methods in molecular biology, vol. 1503. Totowa: Humana Press; 2017.

Chapter 2
Basic Knowledge of Glycobiology

Abstract Free carbohydrates are divided into monosaccharides, oligosaccharides, and polysaccharides. This chapter gives the chemical structures of the most important monosaccharides. Molecular structural aspects will be conferred upon. Typical di- and polysaccharide structures are presented. Furthermore, the general composition of carbohydrate chains and their linkages to glycoconjugates, including glycoproteins, proteoglycans, and glycolipids, will be summarized. The participation of sugars in blood groups, in glyco(sphingo)lipids, and in GPI anchors will be discussed.

Keywords Glycobiology · Aldehyde · Ketone · Aldose · Ketose · Hemiacetal · Hemiketal · Pentose · Hexose · Epimer · Chiral · Stereoisomer · Fisher projection · Haworth projection · Pyranose · Furanose · Anomer · Enantiomer · Configuration · Conformation · Mutarotation · D/L-form · Nomenclature · Glycoconjugates

Glycobiology is defined as the study of biosynthesis, structure, function, and evolution of glycans distributed in nature, and the proteins that recognize them. Just to refresh the memory, here is a short summary of some basic carbohydrate chemistry. The generic term "carbohydrates" originates from the early thought that these compounds were "hydrates of carbon" ($C + H_2O$) on the basis of their composed chemical formula $C_n(H_2O)_n$, where n is 3–9, typically for single simple sugars. However, carbohydrates contain no water molecules as such. Chemically, carbohydrates are polyhydroxy carbonyls defined as polyhydroxy aldehydes (so-called aldoses) and polyhydroxy ketones (so-called ketoses), or substances that yield such compounds upon hydrolysis. Besides a large isomer diversity, carbohydrates in nature can display a great variety of functional groups (e.g., phosphates, acetates, and sulfates) as suited to the different biological roles that they play in the living organisms. Carbohydrates are classified according to their size (i.e., number of monomers) into three major classes: (1) monosaccharides, (2) oligosaccharides, and (3) polysaccharides.

G. J. Gerwig, *The Art of Carbohydrate Analysis*, Techniques in Life Science and Biomedicine for the Non-Expert, https://doi.org/10.1007/978-3-030-77791-3_2

2.1   Monosaccharides

Monosaccharides (Table 2.1) represent the simple sugar monomers, which are the building blocks to form larger carbohydrate structures. The most well-known sugar as monosaccharide is glucose (Glc). The word "glucose" comes from the Greek word γλυκός which means "sweet." It is the most abundant monosaccharide in nature. Its molecular structure was elucidated in the 1880s by the German chemist Emil Fischer (1852–1919). The diameter of a glucose molecule is about 1 nm, one millionth of a millimeter. Just to imagine, you can put about 20,000 glucose molecules next to each other on the diameter of an average human hair.

Glucose is important because it is involved in a number of downstream metabolic pathways, including glycolysis, and the production of other carbohydrates such as cellulose, starch, and sucrose in plants. Glucose also acts as the precursor for the biosynthesis of all the other carbohydrates in the human body, including the energy-storage polysaccharide, called glycogen.

Other monosaccharides differ from glucose in the different orientations of the hydroxyl (–OH) groups in the molecule (diastereomers or epimers), for instance, galactose is the C4 epimer of glucose and mannose is the C2 epimer of glucose (Table 2.1). Monosaccharides contain a number of stereogenic (chiral) carbon atoms, which means that simple monosaccharides with n chiral carbon atoms can exist as 2^n stereoisomers. For a hexose with four chiral carbons, glucose is one of the 16 possible stereoisomers. These monosaccharides differ only in their stereochemistry and consequently have identical atom composition and thus identical molecular masses. The monosaccharides have trivial (commonly used) names with historical origins from chemistry, medicine, and industry. Abbreviations and standard colored symbols (Table 2.1) are often used for graphical representations of the monosaccharides in complex carbohydrate structures.

2.1.1   The Molecular Structure

The majority of the natural monosaccharides have either five (pentoses, $C_5H_{10}O_5$) or six (hexoses, $C_6H_{12}O_6$) carbon atoms. Monosaccharides can exist as linear molecules in their open-chain forms, vertically drawn with carbon atom number 1 (C1) at the top (Fisher projection formula; Table 2.1). The presence of an aldehyde group (HC=O at C1) makes it an aldose and the presence of a ketone group (>C=O at C2) makes it a ketose (Fig. 2.1). The most important aldose is glucose (sometimes called dextrose or grape sugar) and the most important ketose is fructose (sometimes called levulose or fruit sugar). Both are known as the most common naturally occurring sugars.

In an aqueous solution, monosaccharides do not remain as aldehydes and ketones in a straight chain (Fischer projection), but they form, by intramolecular condensation, a hemiacetal bond between C1 and C5 for aldoses and a hemiketal bond between C2 and C5 for ketoses, resulting in cyclic (ring) forms, containing a C–O–C

Table 2.1 Structures of some naturally occurring important monosaccharides

Monosaccharide (Molecular mass)	Fischer projection formula	Haworth projection formula	Conformation representation

Tab. 2.1 (continued)

Monosaccharide (Molecular mass)	Fischer projection formula	Haworth projection formula	Conformation representation
L-Arabinose (Ara) ☆ (150)	HC=O H—C—OH HO—C—H HO—C—H H₂C—OH		
D-Ribose (Rib) ☆ (150)	HC=O H—C—OH H—C—OH H—C—OH H₂C—OH		
2deoxy-D-Ribose (dRib) ☆ (134)	HC=O H—C—H H—C—OH H—C—OH H₂C—OH		
L-Fucose (Fuc) ▲ (164)	HC=O HO—C—H H—C—OH H—C—OH HO—C—H CH₃		
L-Rhamnose (Rha) △ (164)	HC=O H—C—OH H—C—OH HO—C—H HO—C—H CH₃		
D-Glucuronic acid (GlcA) ◇ (194)	HC=O H—C—OH HO—C—H H—C—OH H—C—OH C=O OH		
D-Galacturonic acid (GalA) ◇ (194)	HC=O H—C—OH HO—C—H HO—C—H H—C—OH C=O OH		

bridge. Since the ring forms have lower energies, thereby more stable, they are present as such in intact oligosaccharides. The chemical schematic representations of the monosaccharides are usually drawn in their cyclic form (Haworth projection) or in a relatively energetically privileged chair conformation, where the relative positioning of the axial and equatorial hydroxyl groups can readily be visualized (Table 2.1). The most stable of these forms have the majority of the hydroxyl groups in equatorial positions, resulting in less steric hindrance.

The intramolecular hemiacetal/hemiketal formation during ring closure generates a new center of asymmetry at the carbon C1 (ring numbering is clockwise when viewed from above with C1 on the right side) and the rings are formed as pyranose (six-member ring α/β) or furanose (five-member ring α/β) structures. The α or β configuration (called anomers) of the monosaccharide (e.g., in D-pyranose anomers) is defined by the direction of the anomeric proton H-1 at the C1 atom (denoted anomeric carbon) compared to the direction of the proton H-5 at the C5 atom, which means for a D-hexopyranose, the same direction as C-H bond at C5 is β configuration. In the opposite plane to the C–H bond at C5 is the α configuration.

In an aqueous solution, there is a rapidly fluctuating equilibrium between the open-chain carbonyl form (aldehyde or ketone parent) and the cyclic hemiacetal/hemiketal form and interconversion of the α and β anomeric forms (called mutarotation). This is illustrated in Fig. 2.1 for glucose and fructose. When allowed to reach equilibrium at 25 °C and pH 7, approximately 65% of the glucose will adopt the β-glucopyranose conformation, with ~35% existing as the α-glucopyranose, and less than 1% existing as either the open-chain aldehyde or as α/β-glucofuranose. The equilibrium ratios may differ for other monosaccharides, but ring forms of most sugars predominate over open chains. For instance, the ketose sugar fructose in aqueous solution exists as approximately 68% β-pyranose, 3% α-pyranose, 22% β-furanose, 6% α-furanose, and less than 0.5% open chain form (Fig. 2.1).

It is remarkable that, although for all monosaccharides the open-chain form is present in such a small relative amount, it is actually this form that is responsible for most of the carbohydrate chemistry and chemical reactions.

2.1.2 D/L Forms of Monosaccharides

Glyceraldehyde is the simplest ("sugar") aldose molecule possessing three carbons, out of which the middle carbon is a stereo-center and therefore glyceraldehyde exists as a pair of two enantiomers, designated as D (dexter) and L (laevus). D/L-glyceraldehyde serves as the reference for the further configuration of the monosaccharides. The position of the hydroxyl (–OH) group of the highest-numbered asymmetric carbon atom (also denoted as the chiral carbon most remote from the carbonyl group C=O) dictates a D- or L-configuration in the Fischer projection formula (Rosanoff convention) (Fig. 2.2).

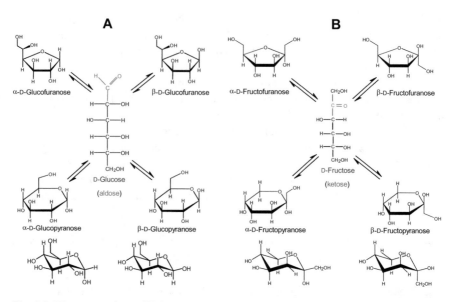

Fig. 2.1 The tautomeric equilibrium between the open-chain form (aldehyde or ketone parent) and the cyclic hemiacetal form and interconversion of the α and β anomeric forms (called mutarotation), illustrated for D-glucose (**A**) and D-fructose (**B**). The six-membered ring generally forms a chair of a fixed conformation providing the classification of protons as axial or equatorial

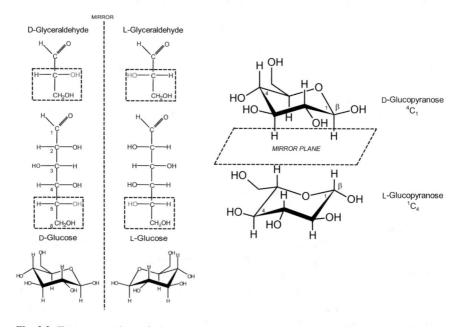

Fig. 2.2 Two presentations of D/L enantiomers (absolute configuration) of β-glucose, which can be viewed as mirror images. D- and L-glyceraldehyde is included for comparison. Most natural sugars belong to the D series

D/L isomers (also called enantiomers or optical antipodes) can be viewed as mirror images. Note that the orientation of all hydroxyl groups is reversed, not only the C5. They have identical chemical and physical properties. However, they have the property of being able to rotate plane-polarized light differently. Inevitably, the anomeric center (C1) results in diastereoisomeric α- or β-anomer.

Once again, the spatial orientation of the H and OH groups on the C atom adjacent to the terminal primary alcohol carbon (for hexose C5) determines whether the sugar is the D or L isomer. If the OH group is on the left side in the Fischer projection, the monosaccharide has L configuration. In living organisms, most sugars belong to the D-enantiomeric series, because, in living cells, they are biosynthetically produced from D-glyceraldehyde, via D-fructose/D-glucose, but the real reason why D-forms dominate in nature is unknown. It has to be noted that L-fucose can be present in glycans of human glycoproteins. For most sugars in aqueous solution, there is a predominance of the D-pyranose chair structure having $^4C_1$ conformation, whereas L-pyranoses (e.g., L-arabinose, L-fucose, and L-rhamnose) have $^1C_4$ conformation. This notation indicates the position of C4 with respect to C1 in the horizontal orientation of the monosaccharide molecule (Fig. 2.2). The ring geometry of the chair structure with the bulkiest groups in the equatorial position is the most favored. All monosaccharide building blocks found in mammalian glycoconjugates exhibit mainly a hexopyranose ring structure. For official "Nomenclature of Carbohydrates," the reader is referred to the IUPAC-IUBMB (International Union of Pure and Applied Chemistry—International Union of Biochemistry and Molecular Biology) rules: https://www.qmul.ac.uk/sbcs/iubmb/nomenclature/.

2.1.3 Specific Monosaccharides

Many modifications can occur on monosaccharide residues. The following are the most common ones:

Deoxysugars are monosaccharides, where a hydroxyl group (often at C6) is replaced by an H-atom (e.g., fucose is 6-deoxy-L-galactose, rhamnose is 6-deoxy-L-mannose). But also well-known is 2-deoxy-D-ribose as a constituent of the nucleotides of DNA.

Aminosugars are monosaccharides with an NH_2-group, usually at C2 (e.g., glucosamine). In nature, the amino group is generally acetylated, being an acetamido group. The most common N-acetylhexosamines found in biological materials are N-acetylglucosamine and N-acetylgalactosamine. Less common is N-acetylmannosamine, which occurs as a constituent of some polysaccharides.

Phosphorylated sugars (C6 or C1 phosphate ester) are intermediate metabolites, and key compounds in a variety of biological processes, such as the biosynthesis of oligosaccharides and nucleotides. In the majority of cases, the metabolic conversion of a sugar in cells begins by its conversion to a phosphoric ester. These compounds are also allosteric regulators for a whole range of enzymes and are involved in signal transduction and catabolic regulation. Mannose-6-phosphate, as a terminal monosaccharide of certain N-linked glycans in mammalian cells, targets these proteins to the lysosome.

Sulfated sugars, in which a hydroxyl is replaced by a sulfate group, are important components of surface-active polysaccharides and mucopolysaccharides, often occurring in algae and bacteria. Furthermore, sulfation can be abundantly present in mucin glycoproteins and proteoglycans.

Sugar acids (also called polyhydroxy carboxylic acids) are sugars with one or more carboxylic acid (COOH) function. Distinction is made between aldonic acids (the aldehyde group of an aldose is oxidized to a carboxy group, e.g., glucose gives gluconic acid) or (glycosyl)uronic acids (the primary hydroxyl group C6 of an aldose is oxidized, e.g., glucose gives glucuronic acid) or aldaric acids (both terminal ends are oxidized, e.g., glucose gives glucaric acid). Hexuronic acids are common constituents of proteoglycans, glycosaminoglycans, and glycoglycerolipids of animal tissue cells but are also found in polysaccharides of plants, fungal cell walls, and bacterial capsules. Glucuronic acid is the synthetic precursor of ascorbic acid (Vitamin C) in organisms. Uronic acids are commonly found as part of structural and/or extracellular polysaccharides of plants and bacteria. In uronic acids, the carboxylic group at C6 easily esterified intramolecularly with the hydroxyl group at C3, forming a lactone. Another very important class of sugar acids are the sialic acids (α-ketopolyhydroxyamino acids, e.g., *N*-acetylneuraminic acid) (see Chap. 9). Since they are abundantly present on the cell surface, they function as receptors for pathogens and toxins.

Sugar alcohols (also called polyols or alditols) are formed by reduction of the monosaccharide aldehyde group ($-CH=O$) to CH_2OH, giving a linear molecule, e.g., glucose gives glucitol (sorbitol); galactose gives galactitol; mannose gives mannitol; xylose gives xylitol. These compounds are widely used in food products as artificial sweeteners, because they carry fewer calories than sucrose and have a low glycemic index.

Iminosugars or azusugars are monosaccharides where the O atom in the ring has been replaced by a nitrogen (N) atom and *carbasugars* are compounds in which the ring oxygen has been replaced by a methylene moiety, but these compounds will not be discussed further.

Recapitulating, the most well-known "neutral" monosaccharides are glucose, galactose, and mannose (as aldohexoses), xylose, ribose, and arabinose (as aldopentoses), fucose and rhamnose (as 6-deoxysugars), and *N*-acetyl glucosamine and *N*-acetyl galactosamine (as aminosugars). Their chemical structures are depicted in Table 2.1. Together with some specific monosaccharides, such as sugar acids (glucuronic acid, galacturonic acid, and sialic acid), they occur most often in biological carbohydrate-containing compounds. In fact, there are more than 100 monosaccharide species if all variations are considered.

2.2   Oligosaccharides

Oligosaccharides are relatively low-molecular-weight polymers of linked monosac-charide units that are covalently bound through oxygen bridges, called glycosidic linkages. A glycosidic linkage is one of the most stable linkages among the different linkages occurring in bio(macro)molecules.

In general, oligosaccharides are defined as linear or branched combinations of 3–10 monosaccharides. There are many ways to link monosaccharides together, directed by linkage types (e.g., 1→2, 1→3, 1→4, 1→6) and the α or β anomeric configuration, depending on the orientation of the anomeric center (C1) of the monosaccharides involved. In the biological cell, the biosynthesis of oligosaccha-rides is governed by the coordinated action of a whole range of enzymes (glycosyl-transferases), which join monosaccharides together in regio- and stereo-selective reactions, and eventually transfer them to other biomolecules to afford glycoconjugates.

Disaccharides consist of two sugar monomers, linked by a glycosidic linkage, and are often classified outside the group of oligosaccharides (Fig. 2.3). After linkage of two monosaccharides forming a disaccharide, one of the monosac-charides usually still has a free (α/β) C1, which is called the reducing-end car-bon. Two important disaccharides without reducing-end are sucrose [α-D-glucopyranosyl-(1↔2)-β-D-fructofuranoside] and trehalose [α-D-glucopyranosyl-(1↔1)-α-D-glycopyranoside] (particularly found in fungi and insects), because both anomeric carbons (C1s) participate in the glycosidic

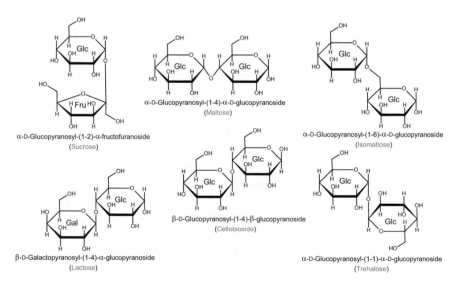

Fig. 2.3 Some biological important disaccharides

linkage. In the human body and in other animals, sucrose (also called saccharose, cane sugar, beet sugar, or table sugar) is digested and broken into its component simple sugars (glucose and fructose) for quick energy. Excess of sucrose can be converted into a lipid for storage as fat.

Lactose (milk sugar) is only found in mammalian milk, where it serves as one of the main energy sources for infants. Lactose, like sucrose, has a sweet flavour. As humans age, lactose becomes less tolerated for them because of the lack of the digestive enzyme lactase. People who suffer can take a lactase supplement to reduce symptoms of lactose intolerance: bloating, cramping, nausea, and diarrhea. Maltose (malt sugar) does not serve a specific purpose in the human body and it is only 50–60% digested and absorbed by the body. Cellobiose is the basic building block of cellulose, which is the main structural component of plant cell walls. Although cellulose is consumed in large quantities as part of plant-based foodstuffs, it cannot be degraded by humans, due to the lack of the enzyme cellulase. Trehalose has several medical and cosmetic applications.

Trisaccharides consist of three monomers, linked by glycosidic linkages, tetrasaccharides consist of four monosaccharides, etc. Thus, free oligosaccharides contain one terminal reducing sugar, called the reducing end of the oligosaccharide, typically drawn on the righthand side of the oligosaccharide structure. The opposite side of the oligosaccharide is called the nonreducing end. In the case of branching, there is more than one nonreducing end.

2.3   Polysaccharides

The majority of carbohydrates found in nature occur as polysaccharides in the plant kingdom, including seaweeds and fungi. Polysaccharides are macromolecules consisting of a large number of monosaccharide units connected by glycosidic linkages. They have a high molecular weight, usually ~10 kDa to ~1 MDa. Most natural polysaccharides contain somewhere between 100 and several thousands of monosaccharides, represented by aldoses, ketoses, anhydrosugars, aminosugars, and sugar acids. Unlike proteins, polysaccharides do not have defined molecular weights, as they are synthesized under secondary genetic control and not from a template like proteins. The biosynthesis of proteins follows a genetic code, but the biosynthesis of polysaccharides depends on the availability of enzymes, controlled by the genes expressing the particular glycosyltransferase and glycosidase enzymes. There is no specific stopping point for the enzymes involved in the biosynthesis of a polysaccharide, resulting in heterogeneity in molecular size of the polymer. Polysaccharides differ from each other in identity of the constituting monosaccharide residues, in types of bonds linking the monosaccharides, in lengths of their chains, and in the degree of branching. The special nature of the glycosidic bonds determines the shape of the polysaccharides, largely because some linkages inhibit the rotation of the residues toward each other.

Polysaccharides that contain the same type of monosaccharides are called *homopolysaccharides*. Most of these polysaccharides, consisting of glucose residues (glucans), serve for the storage of energy (e.g., starch in plants [abundant in potatoes] and glycogen in human and animals [abundant in liver and muscle cells]), and as structural components (e.g., cellulose in plants, chitin in fungi, arthropods, and invertebrates).

Polysaccharides that contain more than one type of monosaccharide are known as *heteropolysaccharides*; they are usually composed of a limited number of monosaccharides ordered in repetitive sequences (repeating units). Of note, glycosaminoglycans (GAGs), which are the carbohydrate components of proteoglycans (PGs), are often classified as heteropolysaccharides (see Sect. 2.4.2).

Many different monosaccharides, sometimes unusual types specific to only one species of organism, are found in bacterial polysaccharides. Although bacteria lack the glycosylation apparatus typical for eukaryotic cells, polysaccharides are produced in great variety by bacteria. A good example are lactic acid bacteria species [1].

Polysaccharides are usually considered as one of the pharmacological active compounds in medicinal plants and fungi. Several isolated/purified water-soluble polysaccharides are claimed to have important and significant pharmacological activities, such as anticancer, anti-inflammation, anti-oxidation, immune potentiation, and blood sugar reduction [2]. Several polysaccharides, which are biodegradable, have many uses as drug delivery vehicles, controlled drug carriers or scaffolds for tissue engineering, and as implantable biomaterials in medical science. Hence, polysaccharides are widely employed in the pharmaceutical and cosmetic industry.

The biological activities of polysaccharides are closely correlated to their physicochemical properties such as molecular size, types and ratios of constituent monosaccharides, and the glycosidic linkages (position, configuration). Their function in nature can be very diverse, e.g., energy sources for sustaining life and serving as structural elements providing shape and size. Detailed characterization of polysaccharides is necessary for ensuring their effectiveness and safety when they are used as medicine or used in beverage and food products. Polysaccharides as fiber are a vitally important dimension of nutrition health.

Some naturally occurring, important polysaccharides are:

Amylose (α1-4)-linked D-glucose residues

Amylopectin (α1-4)-linked D-glucose residues
with (α1-6)-branching

Starch consists of two glucose polymers, amylose (20–30% w/w) and amylo-pectin (70–80% w/w). The water-insoluble amylose (MW ~10^6 Da) consists of long linear chains of several thousand of (α1→4)-linked glucose units. It forms a loose random coil structure (helix; each coil contains 6 glucose residues). The water-soluble amylopectin (MW ~10^8 Da) is a branched macromolecule of which, on average, 1 in 20–25 (α1→4)-linked glucose units is branched by an (α1→6) glyco-sidic bond. Starch (MW ~10^6–10^9 Da) occurs as granules and is found in bulbs, seeds, and tubers of higher plants. It is the major energy source in the living world.

The breakdown of starch to low-molecular-weight products catalyzed by the enzyme α-amylase is one of the most important commercial enzyme processes. The obtained oligosaccharide products (maltodextrins or cyclodextrins) are widely used in the food, paper, and textile industry, together with cellulose.

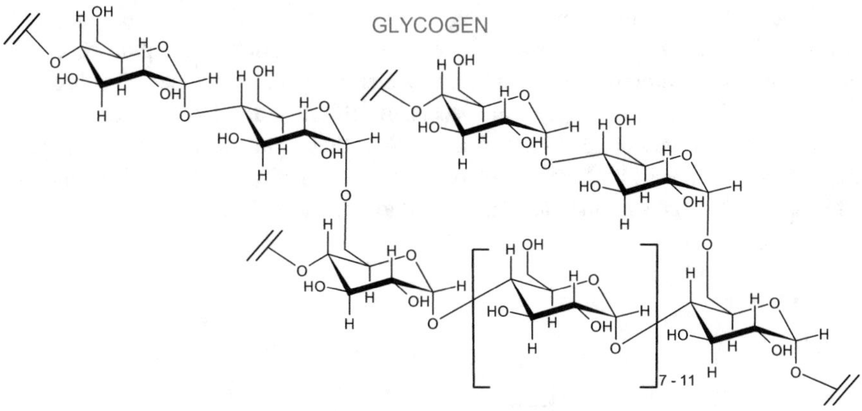

GLYCOGEN

Glycogen is also composed of ($\alpha1\rightarrow4$)-linked glucose residues, similar to amy-lopectin, but is highly ($\alpha1\rightarrow6$)-branched every ~10 residues. It is sometimes called "animal starch." Glycogen is stored in the cytoplasm of all animal/human cells but is mostly present in the cells of skeletal muscle and the liver. Glycogen is the storage form of sugar in the body and is used as an energy source.

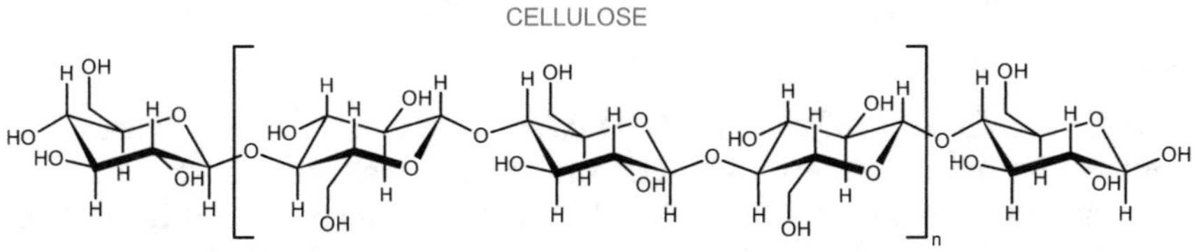

CELLULOSE

Cellulose, also a glucose polymer, is an unbranched ($\beta1\rightarrow4$)-linked polymer of ~2000–15,000 glucose residues, implying a molecular mass of over 1.5×10^6 Da and a chain length of about 5μm. Adjacent glucose residues alternate by 180° in orientation. Hydrogen bonding between cellulose chains affects compact cellulose fibers, comprising a densely packed structure of remarkable physicochemical strength. It is the main component of the cell wall of plants and the major constituent of the mass of wood. Cotton (*Gossypium*) is almost pure cellulose. Hemicellulose has all glucose residues substituted at C6 with a xylose residue.

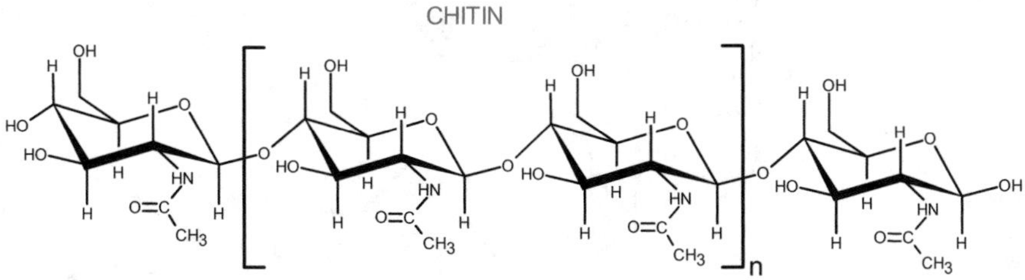

CHITIN

Chitin, probably the second most abundant polysaccharide after cellulose, is a linear homopolysaccharide consisting of ($\beta1\rightarrow4$)-linked *N*-acetylglucosamine

residues. Although not drawn in the structure above, also here the adjacent
N-acetylglucosamine residues alternate by 180° in orientation as in cellulose. Chitin
is the principal component of the hard exoskeletons of many species of arthropods
(e.g., crabs, lobsters, shrimps) and insects. In chitin, individual strands are held
together by hydrogen bonds as in cellulose. Note that cellulose and chitin contain β
linkages which are in both cases responsible for their physicochemical resilience.
Chitosan is the glucosamine variant, i.e., without the acetyl group.

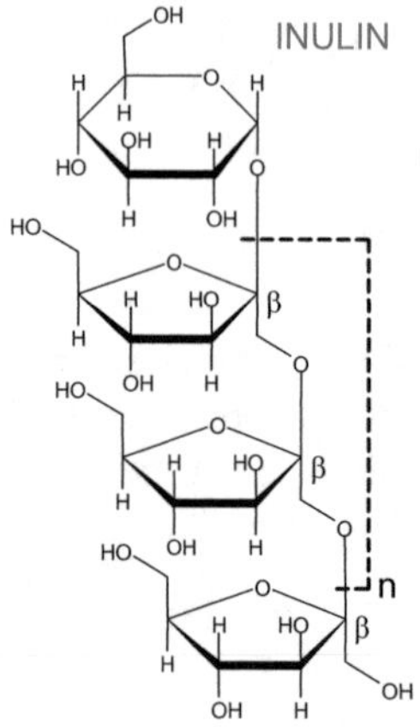

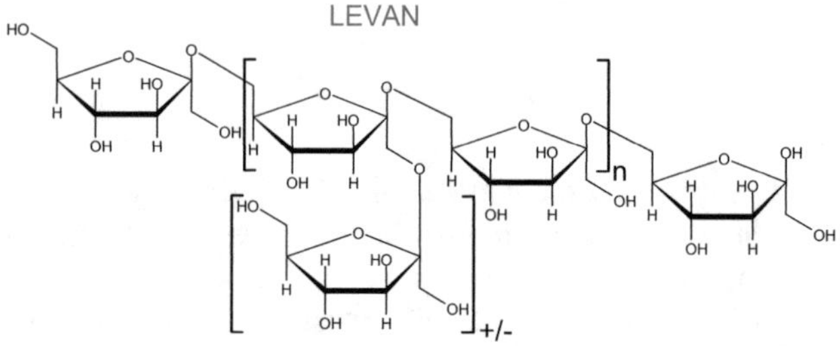

Inulin is a <10 kDa polymer of (β2→1)-linked D-fructose residues in furanose ring
forms with a terminal sucrose moiety. It is found in plants such as artichoke and chicory.

Levan is a bacterial polysaccharide (>100 kDa) made up of (β2→6)-linked fructose residues, linear and variable degrees of (β2→1)-branching. Both polymers are
classified as fructans.

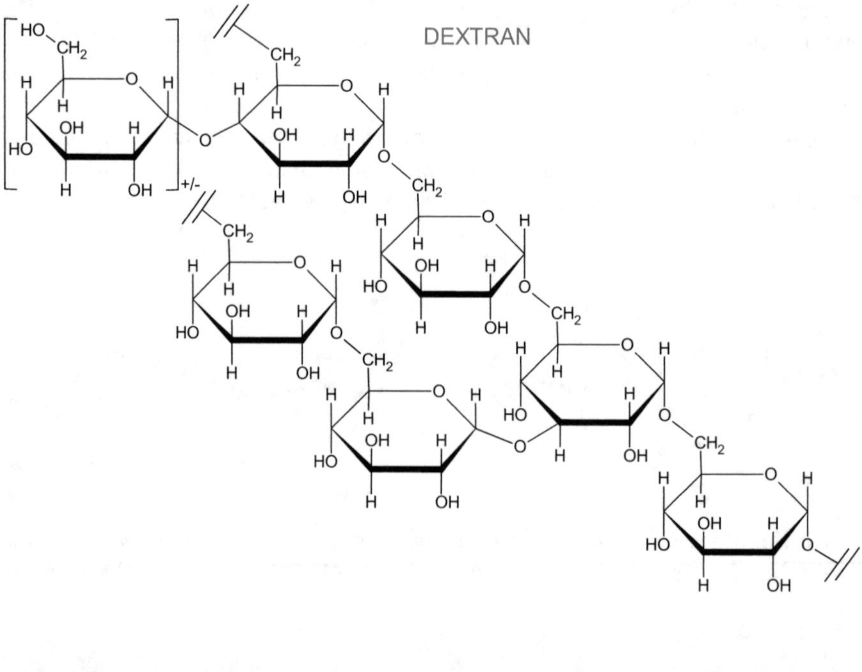

Dextrans are bacterial and yeast polysaccharides, consisting of $(\alpha1\rightarrow6)$-linked glucose polymer with $(\alpha1\rightarrow3)$ branches (~5%), and occasionally $(\alpha1\rightarrow2)$ or $(\alpha1\rightarrow4)$ branches. Dextran may consist of a mixture of different molecules, containing either short or long branched side chains. Dextrans with a molecular size of 40–70 kDa are used as blood plasma volume expander for emergency treatment of blood loss. Some bacteria growing on the surface of teeth are responsible for the forming of dental plaque, which is rich in dextran.

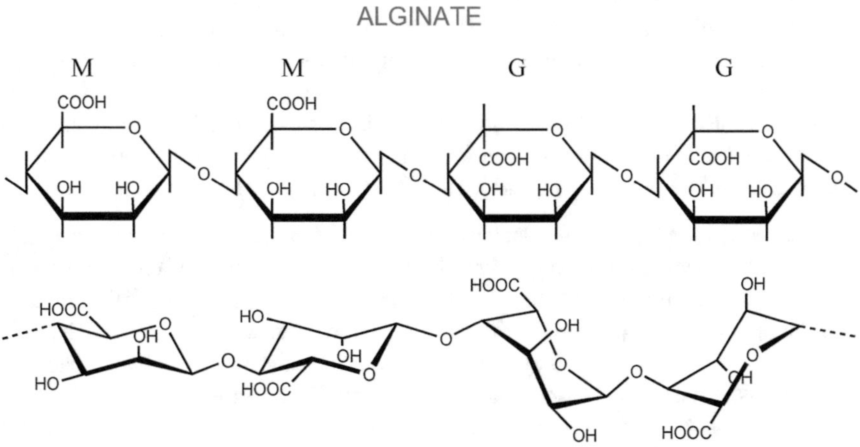

ALGINATE

M = β-D-mannuronic acid   G = α-L-guluronic acid

Alginate (alginic acid) is a polysaccharide widely distributed in cell walls of brown seaweed. The polymer contains D-mannuronic and L-guluronic acids with $(\beta1\rightarrow4)$- and $(\alpha1\rightarrow4)$-linkages, respectively. Thanks to hemostatic properties, alginate is used for wound treatment in various forms such as gel or sponge, but also as gelling and stabilizing agents in food.

Seaweed polysaccharide has emerged as a potential source of functional foods, having anticoagulant, anti-inflammatory, and antioxidant properties.

Further common polysaccharides are:

Pullulan is mainly a yeast glucose polymer composed of $(\alpha1\rightarrow4)$-maltotriose units linked together by $(\alpha1\rightarrow6)$ bonds.

Alternan is a linear glucose polymer with alternate $(\alpha1\rightarrow6)$ and $(\alpha1\rightarrow3)$ linkages.

Guar gum is a galactomannan comprising linear chains of $(\beta1\rightarrow4)$-linked mannose residues with side chains of single $(\alpha1\rightarrow6)$-linked galactose.

Curdlan is a water-insoluble linear $(\beta1\rightarrow3)$-glucan. Curdlan is a bacterial polysaccharide of significant interest due to its interesting and valuable rheological properties.

Xanthan (gum) is a high-molecular-mass heteropolymer with $(\beta1\rightarrow4)$-D-glucopyranosyl residues as main chain backbone, branched at C3 of every second unit with a trisaccharide D-Man$p(\beta1\rightarrow4)$-D-Glc$p$A$(\beta1\rightarrow2)$-D-Man$p(6O$Ac$)$ $(\alpha1\rightarrow3)$. Some of the nonreducing end units are terminated with a 4,6-O-pyruvyl cyclic acetal group. It is synthesized by *Xanthomonas* species and used as a thick-ener in food, pharmaceutical, and cosmetic industries.

Carrageenan is a galactose polymer (might be sulfated) distributed in the cell walls of red seaweeds (algae), containing ($\beta1\rightarrow3$) and ($\alpha1\rightarrow4$) linkages.

Gellan (gum) is a bacterial polymer with a tetrasaccharide repeating unit, which consists of two residues of D-glucose and one of each residue of L-rhamnose and D-glucuronic acid. The tetrasaccharide repeat has the following structure: $[\rightarrow3)$-D-Glc($\beta1\rightarrow4)$-D-GlcA($\beta1\rightarrow4)$-D-Glc($\beta1\rightarrow4)$-L-Rha($\alpha1\rightarrow]_n$.

Pectin mainly consists of ($\alpha1\rightarrow4$) polygalacturonic acid residues (with carboxyl groups partially methyl-esterified), interrupted by occasional ($\alpha1\rightarrow2$)-linked L-rhamnose residues (MW 20–400 kDa) or as repeating disaccharide [$\rightarrow4$) GalA($\alpha1\rightarrow2$)Rha($\alpha1\rightarrow$], but it is not branched. However, pectin can also contain minor amounts of other monosaccharides, such as galactose, xylose, and arabinose. Pectin is abundantly present in many fruits, particularly in orange and lemon rinds.

As already mentioned, polysaccharides are important in connection with bacteria [3]. Apart from extracellular production by several bacteria, polysaccharides consti-tute the major component of the complex structure of bacterial cell walls. In gen-eral, large amounts of polysaccharides occur on the surface of bacteria which serve a variety of purposes. They play critical roles in the interactions between bacteria and the host environments, and consequently contribute to the virulence of patho-genic bacteria. Bacterial polysaccharides are often immunogenic to humans and trigger the innate immune system. These polysaccharides represent a diverse range of macromolecules that include capsular and exopolysaccharides, lipopolysaccha-rides, and the prokaryotic peptidoglycan.

Bacteria are classified as Gram-positive and Gram-negative, depending on the structure of their outer membrane. Gram-positive bacteria possess, on the phospholipid-bilayer cytoplasmic membrane and periplasmic space, a thick cell wall containing multiple layers of peptidoglycans. These peptidoglycans contain polysaccharide consisting of alternating ($\beta1\rightarrow4$)-linked *N*-acetylglucosamine (GlcNAc) and *N*-acetylmuramic acids (MurNAc) residues, together with (lipo)tei-choic acids. In contrast, Gram-negative bacteria (e.g., *Escherichia coli* or *Salmonella typhimurium*) have a relatively thin cell wall consisting of a few layers of peptido-glycan surrounded by a thick second lipid membrane containing mainly lipopoly-saccharide (LPS), which contributes significantly to the cell structural integrity and pathogenicity. Most of these bacteria are encapsulated by a capsular polysaccharide (CPS), which plays a key role in the pathogenesis and manifestation of bacterial infection.

It should be noted that in prokaryotes, glycosylation leads to a greater diversity of glycan compositions and structures than found in eukaryotic cells. Prokaryotes produce a vast array of unusual monosaccharides and rare glycan structures on their glycoproteins, sometimes never observed before.

LPS consists of three structurally distinct regions: (1) the lipid A moiety (glucosamine-based phospholipid), (2) a core oligosaccharide, and (3) the O-antigen polysaccharide (also called the O-specific chain), containing multiple repeating units of oligosaccharides. The lipid A anchors the LPS in the outer membrane of the cell. Lipopolysaccharides form a class of compounds which present the widest variety of different monosaccharide residues in their structures both in the O-chain and in the core region (next to phosphatidylinositol), and are involved in the pathogenic virulence of these bacteria [4].

However, there are also many nonpathogenic bacteria producing extracellular polysaccharides. These polysaccharides are not attached to the bacterial cell wall and are useful in biotechnological applications, e.g., the food industry, in medicine, and renewable energy production. For instance, lactic acid bacteria (LAB) synthesize a wide variety of exopolysaccharides (EPS); these polysaccharides are enzymatically synthesized extracellularly from sucrose by glycansucrases (typically glucan- or fructansucrases) or intracellularly by glycosyltransferases. Several strains of LABs play a crucial role in various food fermentations, e.g., yoghurt, sauerkraut, and cheese [5]. Several microbial polysaccharides are used as emulsifiers, gelling, or thickening agents. The global production of bacterial polymers is increasing rapidly, caused by the growing demand for biobased polymers. The natural variety of EPS with specific properties has a huge potential for industrial utilization.

2.4 Glycoconjugates

Nature is full of carbohydrates, not only in free form but also associated or chemically bound (covalently linked) to other noncarbohydrate molecules, forming glycoconjugates. In general, the carbohydrate moieties (usually collectively called glycans) impart special properties to the (bio)molecules to which they are bound. Glycoconjugates, including glycoproteins, glycopeptides, glycolipids, peptidoglycans, and proteoglycans are expressed by eukaryotic as well as prokaryotic cells as secretory, intracellular, and membrane-bound components. But it is just 60 years ago that these bound sugars to protein and lipid attracted serious attention. The glycans that are expressed in glycoconjugates differ greatly between vertebrates and invertebrates. The exact role of most of the carbohydrate portions of glycoconjugates is still poorly understood.

The diverse biological functions of glycoconjugates are regulated by their complex oligosaccharide structures. The glycans can modulate the function of the biomolecules to which they are attached by the specific recognition of the glycan structure by carbohydrate-binding proteins.

To get insight into the structural aspects and characteristics of glycoconjugates, the unraveling of the function of the carbohydrate chains is an important step. As earlier mentioned, carbohydrate moieties on the surface of a cell play one of the most important roles in cellular recognition. In order to elucidate this phenomenon, the assessment of the structure of the glycans is indispensable. For instance, the glycan structures of the cell membrane glycoproteins are profoundly altered in cancer cells [6, 7].

2.4.1 Glycoproteins

Proteins to which carbohydrates are covalently attached are known as *glycoproteins*. Glycosylation is by far the most common posttranslational modification (PTM) of proteins. The occurrence of glycosylation of proteins is found in all three domains of life: Archaea, prokaryotes (bacteria), and eukaryotes. It is widely observed among the mammalian eukaryotic cell surface and extracellular secreted proteins. The human genome (~20,000 genes) seems to encode for about 40,000 different proteins and more than half of these proteins are glycosylated [8]. Many enzymes and hormones (e.g., pituitary gland hormones, EPO, hCG, FSH, luteinizing hormone) in the human body are glycoproteins or glycopeptides. Mammalian glycoproteins have an approximate molecular mass range of 20–200 kDa. The linked carbohydrate chains, making up 5–80% by weight, may consist of mono-, oligo-, or polysaccharides. As an example, EPO is a glycoprotein with approximately 40% of its weight due to glycosylation. Remarkably, ~70% of the proteins present in human breast milk are glycosylated.

In eukaryotes, glycosylation is performed co- or posttranslationally in the secretion machinery of the cell. Unlike the biosynthesis of DNA, RNA, and of proteins (where proteome is coded in the genome), the eukaryotic biosynthesis of glycans is not directly template-driven, but is directed by a large set of competing glycosylation enzymes (glycosyltransferases and glycosidases) with distinct substrate and linkage specificities [9]. In other words, the co/posttranslational modification of proteins is not directly encoded in the genome. However, a substantial proportion of mammalian genomes (~200 glycogenes) is dedicated to genes encoding proteins (i.e., enzymes) involved in glycosylation pathways, and these are highly conserved. This means that solid structural analytical data concerning glycan types and their distribution are the most important resources for the study of the meaning of these carbohydrates (glycans). The glycan structure is rarely predictable, determined not only by the nature of the protein it is bound to but also by the tissue or cell where it is produced. The same glycoprotein backbone can contain different glycans that differ in monosaccharide composition and linkages. Furthermore, the same glycan structure on different proteins may have different functions. The attachment of glycans affects the tertiary structure and physicochemical properties of the protein.

Conclusively, the human glycosyl-biosynthetic machinery within a cell is extraordinarily complex, involving a set of over 250 competing enzymes catalyzing various types of glycosylation, controlled by the genes expressing the particular glycosyltransferase and glycosidase enzymes.

There are two major categories of glycosylation, depending on whether the glycan is *N-linked* or *O-linked* to an amino acid in the polypeptide backbone. N-linked glycans are attached through an amide nitrogen linkage to the side-chain of an asparagine (Asn) residue. O-linked glycans can be attached in a more diverse manner, but most common to oxygen of the hydroxyl group on the side chain of serine (Ser) or threonine (Thr). Both types of glycans may be present in a single protein. The oligosaccharide chains of glycoproteins are fashioned by a series of enzymes acting in a specific sequence in different subcellular compartments, which means that the regulation of sugar chain biosynthesis is under the control of glycosyltransferases, the substrate specificities of the enzymes, and the localization of the enzymes in tissues and organelles.

One of the initial functions of N-glycosylation (in particular, the terminal Glc and Man residues) is to effect proper folding of the protein and regulation of the sorting (trafficking) of the glycoprotein. Generally speaking, the glycosylation of proteins in eukaryotes is a highly refined process [10]. Comparatively, few monosaccharide species (in particular, GlcNAc, GalNAc, Man, Gal, Glc, Xyl, Fuc, Sialic acid) are used to build the glycoprotein glycans, which in turn are relatively conserved compared to the huge number of potential structures that could be formed. Many proteins present several glycosylation sites and each site can be occupied by a variety of glycan structures. For instance, a protein containing three sites of glycosylation with ten different glycans in each site can result in about 1000 different glycoforms [11, 12].

It has been established that N-glycans are involved in:

1. the induction and maintenance of the protein conformation in a biologically active form,
2. the control of the life span of circulating glycoproteins in blood, and
3. protection of the peptide chain against proteolytic attack.

Also, O-glycosylation is found to play significant roles in the conformation, solubility, stability, and hydrophilicity of proteins. Finally, the glycans of glycoproteins in the cell membrane are of utmost importance for many biological features (vide infra). Moreover, aberrant glycosylation has been recognized as the trait of many mammalian diseases, including osteoarthritis, cystic fibrosis, and cancer (see Chap. 3).

2.4.1.1 N-Glycans

In eukaryotic cells, N-glycosylation happens almost exclusively on proteins that enter the secretory pathway. The biosynthetic process of *N-linked glycosylation* occurs in two steps when the translation of the nascent protein is almost complete. The first step occurs while the protein is being synthesized in the lumen of the endoplasmic reticulum (ER) of the cell [13]. A standard oligosaccharide ($Glc_3Man_9GlcNAc_2$) linked to a lipid polyisoprene carrier (dolichol-pyrophosphate, Dol-PP) is constructed in the cytosolic side and luminal side of ER, catalyzed by specific glycosyl transferases. Many different glycosyltransferases, as residents of the membranes lining the ER and Golgi apparatus (cis, medial, trans network), catalyze the transfer of a sugar residue from an activated donor, usually a nucleotide sugar (e.g., UDP-Gal, GDP-mannose, GDP-Fuc, UDP-GlcNAc, and CMP-sialic acid). The ER and Golgi membranes contain transporters for nucleotide sugars.

Then, catalyzed by an oligosaccharyltransferase (OST), the preassembled ($Glc_3Man_9GlcNAc_2$) oligosaccharide is en bloc transferred and covalently attached to the amide nitrogen of an asparagine (Asn) residue of the nascent protein back-bone. This is an L-asparagine in a particular tripeptide unit (sequon), Asn-Xxx-Ser/Thr, where Xxx is any amino acid except proline (Pro). But for reasons that are still not completely understood, not all such sequons are glycosylated. Only about 30% of the available N-glycosylation consensus sequons in the human proteome are actually occupied by N-glycans. Probably, the conformation of the protein back-bone in the vicinity of potential N-glycosylation sites may play a role. It has to be noted that on rare occasions, non-consensus motif N-glycosylation has also been observed, e.g., Asn-Xxx-Cys [14]. Since N-glycosylation occurs in the secretory machinery of the cell, only the proteins that contain a secretion signal peptide can be N-glycosylated.

Subsequently, after correct folding of the protein and enzymatic removal of the glucose residues, the glycoprotein is transferred, via vesicles, to the Golgi apparatus in the cell, where the second step of the N-glycosylation process occurs. While transiting through the cis-, medial- and trans-Golgi regions, the now protein-bound glycan is sequentially modified by several enzyme-catalyzed reactions, including trimming (removal of glucose and several mannose residues) and processing (attach-ments of galactose, N-acetylglucosamine, sialic acid, fucose) of sugar residues, resulting in a plethora of possible structures. Monosaccharides for the additions are provided by the abovementioned nucleotide sugar donors. To date, it is still impos-sible to predict the complete structure(s) of the glycosylation site chain(s) based on the protein sequence [15]. Various structures are synthesized via the actions of the various glycosyltransferases and availability of the sugar nucleotide donors during the cell cycle, and furthermore depending on the cell types, tissues or organisms, and different species [16].

The mature N-glycans can be crudely classified into three main types: (1) the high-mannose type, (2) the complex/lactosamine type, and (3) the hybrid type. They share a common pentasaccharide inner core structure ($Man_3GlcNAc_2$), presented as $Man(\alpha1\rightarrow3)[Man(\alpha1\rightarrow6)]Man(\beta1\rightarrow4)GlcNAc(\beta1\rightarrow4)GlcNAc(\beta1\rightarrow Asn$,   also called the tri-mannosyl-chitobiose core, but differ in their outer branches (antennae) (Fig. 2.4).

High-mannose type glycans (or oligomannose type) have two to six additional α-mannose residues linked to the core structure. Typical complex-type glycans con-tain two to four outer branches (typically denoted as tri-, tri-′ or tetra-antennae) with, typically, a sialyl-N-acetyllactosamine sequence. A variety of peripheral units can be present, such as sialic acid, Fuc, GalNAc, GlcA, sulfate, methyl, and phos-phate groups [12].

Hybrid-type glycans have the features of both oligo-mannose and complex type and can also contain an additional N-acetylglucosamine that is linked ($\beta1\rightarrow4$) to the β-linked mannose residue of the core (known as "bisecting GlcNAc"). It must be mentioned that complicated N-linked glycans may occur in many simple lower organisms [17].

The shape of a glycan may have a voluminous character due to the several differ-ent types of monosaccharides, their ring conformations, and their relative position-ing. Also, environmental parameters are of influence.

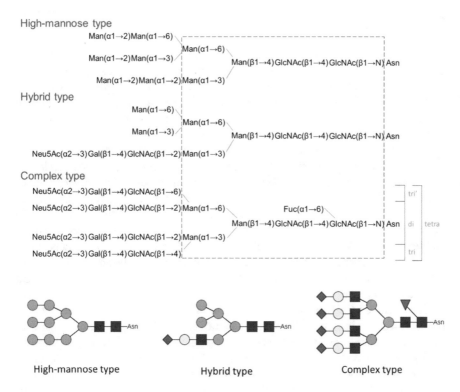

Fig. 2.4 Examples of N-glycans. The Man₃GlcNAc₂ core structure is indicated in the box. The di-, tri-, tri'- and tetra-antennary subtypes are indicated for the complex type structure. Below are the colored symbolic notations for the structures

Owing to the nature of their production, glycoproteins can be highly heterogeneous molecules with numerous structurally different glycan side chains present at a single site or at multiple glycosylation sites along a single polypeptide (denoted as micro- and macroheterogeneity). Complex-type N-glycans are predominantly present when cells are cancerous. In tumors, the changes in N- and O-glycan structures most often arise from disturbances in the expression and activity levels of the different glycosyltransferases and glycosidases along the secretory pathway, in the ER and Golgi of the cancer cells [18]. Notably, tumor cells increase the uptake of glucose. The most important feature is the altered, mostly elevated, sialylation of the surface of tumor cells, due to overexpression of sialyltransferases in the malignant cells. The aberrant sialylation is a vital way for tumor cells to escape immune surveillance and keep malignance. In humans, also genetic defects causing alterations in the biosynthetic pathways of N-linked glycoproteins may lead to severe medical consequences and these are referred to as congenital disorders of glycosylation [19] (see Chap. 3).

2.4.1.2 O-Glycans

O-linked glycosylation of proteins is found in all kingdoms of biology, including eukaryotes, Archaea, and a number of pathogenic bacteria. In eukaryotes, the biosynthetic process is supposed to occur entirely in the Golgi apparatus of the cell with no exact rules, after the protein has been synthesized. For the most common mucin-type O-glycans, the synthesis begins with the addition of N-acetyl-α-D-galactosamine (from UDP-GalNAc) to the hydroxyl oxygen of serine and/or threonine residues in the polypeptide backbone of the completely folded protein, at sequence stretches (coil/turn) rich in hydroxy-amino acids (mainly L-Ser and/or L-Thr in Pro-rich regions). There are often clusters of serine and threonine residues near sites of O-glycosylation. After this first step, numerous glycosyltransferase enzymes initiate the glycosylation process using nucleotide-activated sugars. The glycan is stepwise modified by a sequence of enzyme-catalyzed monosaccharide additions, in which the monosaccharides GalNAc, GlcNAc, Gal, Fuc, and Neu5Ac are usually involved. Similar to N-glycan chains, also O-glycan chains are frequently present on the outside surface of the cell membrane and they play important roles in the immune system.

The O-linked GalNAc-type glycosylation in the secretory pathway initiates in the cis Golgi and in some of the vesicles in trafficking ER-Golgi. The heterogeneity in monosaccharide attachment makes mammalian O-glycosylation mechanisms highly diverse and complex compared to N-glycosylation. The structures of O-glycans are very heterogeneous and may vary in length from monosaccharides to oligosaccharides containing more than 20 monomers, although O-glycans of membrane glycoproteins are much shorter and show less heterogeneity. The glycans O-linked via α-GalNAc to Ser/Thr lack a distinct common core structure as found with N-glycosylation, but are classified according to their root structures, designated as Core 1 to Core 8 (Fig. 2.5). Humans synthesize Core 1–4, although mainly Core 3 and a large variety of O-linked oligosaccharides may be built from these root structures depending on the different enzymes (glycosyltransferases) present in different cells. It has been observed that during cancerous transformation in the human large intestine, the Core 3 structure drastically decreases and the Core 1 structure increases.

Mucins are a special group of high-molecular-weight glycoproteins (1–20×10^6 Da), containing 70–85% carbohydrate in the form of highly clustered glycans O-linked to serine and threonine (Fig. 2.6). They are often highly negatively charged because of sialylation and/or sulfation [20]. In the peripheral parts of mucin-type O-glycans, in many cases, the blood group determinant is present (see Sect. 2.4.1.3).

Mucins form the major component of mucosal cells throughout the body (e.g., as glycoproteins in saliva, gastrointestinal, bronchial, nasopharyngeal, and cervical mucosae) and are known to play a physiological role as a barrier [21]. They protect the epithelial cells against noxious external influences like pH, and mechanical stress, but also from invasion of viruses, bacteria and chemical substances. Mucins aggregate as subunits linked through disulfide bridges and form gels on secretion at

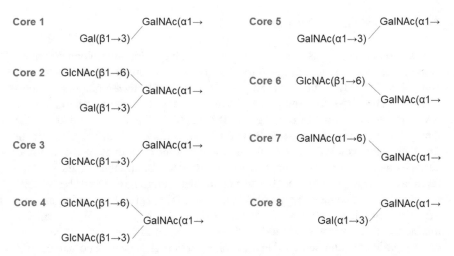

Fig. 2.5 Typical "core" structures of common mucin-type O-glycans. These structures are variably extended with Gal, GalNAc, Fuc, and sialic acid residues

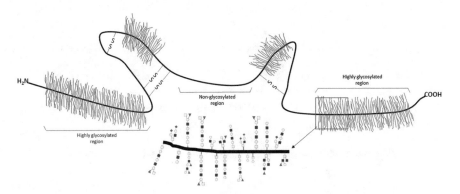

Fig. 2.6 Typical mucin glycoprotein, extensively decorated with O glycans. Intra- and intermolecular disulfide bridges may occur in cysteine-rich regions

the mucosal surface. The negatively charged mucins form a network of fibers and are the essential components of mucus which covers the entire epithelial lining of the gut protecting the intestinal cells from infections. Mucin glycoproteins can be important markers of normal and disease development (e.g., cystic fibrosis, inflammatory bowel disease, peptic ulceration, and intestinal cancer) [18]. Upon malignant transformation, mucins containing changes in O-glycans are secreted into the bloodstream of cancer patients.

2.4.1.3   Blood Group Determinants

The different blood types arise from differential glycosylation of blood cell membrane proteins and lipids. The AB0(H) family of blood group determinants (antigens) are assembled from D-galactose, *N*-acetyl-D-galactosamine, and L-fucose (Fig. 2.7). Blood group A individuals have the A determinant and a certain quantity of H, whereas those of blood group B have the B determinant and also a certain quantity of H. Blood group 0(H) individuals only have the H determinant. The determinants are present on the nonreducing terminal ends of the oligosaccharide chains of several glycoproteins and glycolipids, and thus on the cell surfaces in the majority of organs. In addition to these antigens on the red blood cells, the blood serum contains antibodies (IgG, IgA, IgM, IgE, IgD), which are proteins (immunoglobulins) that can identify and attack foreign cells. When transfusing blood, only compatible blood types must be used to avoid adverse reactions due to antibodies and some other factors. Interestingly, there is no anti-H antibody, so blood group 0(H) (rhesus factor must be negative) can always be used in case of emergency.

Blood group	Antigen on red blood cells	Antibody in serum	can receive blood types
0(H)	Gal(β1→3)GlcNAc(β1→ Fuc(α1→2)/	anti-A anti-B	0
A	GalNAc(α1→3)Gal(β1→3)GlcNAc(β1→ Fuc(α1→2)/	anti-B	A and 0
B	Gal(α1→3)Gal(β1→3)GlcNAc(β1→ Fuc(α1→2)/	anti-A	B and 0
AB	GalNAc(α1→3)Gal(β1→3)GlcNAc(β1→ Fuc(α1→2)/ Gal(α1→3)Gal(β1→3)GlcNAc(β1→ Fuc(α1→2)/	none	A, B, and 0

Fig. 2.7 Major blood group determinants on human erythrocytes. The types of oligosaccharides present on the surface of the red blood cells determine a person's blood type. It must be mentioned that additional blood subgroups exist, varying in glycans (not discussed here), indicated as Lewis types, e.g., Le^a, Le^x, Le^b, and Le^y

2.4.1.4 Different Types of Glycosylation

In contrast to the majority of the already mentioned N-glycosylation usually via GlcNAc($\beta1\rightarrow$N)Asn, and the O-glycosylation usually via GalNAc($\alpha1\rightarrow$O)Ser/Thr, several other types of glycosylation exist in nature, whereby different sugars as well as different amino acids are involved (Table 2.2).

A specific type of dynamic glycosylation, which is found on nuclear, cytoplasmic, and mitochondrial proteins, is the single β-GlcNAc O-linked to the hydroxyl group of Ser and/or Thr [22, 23]. This unique PTM, called O-GlcNAcylation (the GlcNAc is rapidly added and removed, and, may be phosphorylated), regulates the trafficking of proteins into and out of the nucleus, but does not occur on secretory glycoproteins. Hence, this nucleocytoplasmic modification is the dominant form of intracellular eukaryotic glycosylation. This type of glycosylation has also been found in viral proteins but not in prokaryotes.

In particular, O-glycosylation shows a great variety of linkages [24, 25]. For instance, GalNAc($\alpha1\rightarrow$O)Ser/Thr is the most common type of O-glycosylation and

Table 2.2 Examples of naturally occurring carbohydrate-amino acid linkage types

Class/Linkage	Class/Linkage
N-glycosylation	*O-glycosylation*
GlcNAc($\beta1\rightarrow$N)Asn	GalNAc($\alpha1\rightarrow$O)Ser/Thr
GalNAc($\beta1\rightarrow$N)Asn	GlcNAc($\beta1\rightarrow$O)Ser/Thr
Glc($\alpha/\beta1\rightarrow$N)Asn	Gal($\alpha1\rightarrow$O)Ser/Thr
L-Rha($\alpha1\rightarrow$N)Asn	Glc($\beta1\rightarrow$O)Ser/Thr
L-Rha($\alpha1\rightarrow$N)Arg	Man($\alpha1\rightarrow$O)Ser/Thr
GlcNAc($\beta1\rightarrow$N)Arg	Xyl($\beta1\rightarrow$O)Ser/Thr
Man($\alpha1\rightarrow$N)Trp	L-Fuc($\alpha1\rightarrow$O)Ser/Thr
	L-Ara($\beta1\rightarrow$O)Ser/Thr
C-glycosylation	Gal($\beta1\rightarrow$O)Hyl
Man($\alpha1\rightarrow$C)Trp	L-Ara($\beta1\rightarrow$O)Hyl
	L-Ara*f*($\beta1\rightarrow$O)Hyp
S-glycosylation	Gal($\beta1\rightarrow$O)Hyp
Gal($\beta1\rightarrow$S)Cys	Gal($\beta1\rightarrow$O)Tyr
Glc($\beta1\rightarrow$S)Cys	Glc($\alpha/\beta1\rightarrow$O)Tyr
GlcNAc($\beta1\rightarrow$S)Cys	
	Phosphoserineglycosylation
Glycation (Schiff base)	GlcNAc($\alpha1$-$P\rightarrow$O)Ser
Glc$\rightarrow$Lys	Man($\alpha1$-$P\rightarrow$O)Ser
Rib$\rightarrow$Lys	L-Fuc($\beta1$-$P\rightarrow$O)Ser
	FucNAc($\beta1$-$P\rightarrow$O)Ser
Glypiation (GPI anchors)	Xyl($\alpha1$-$P\rightarrow$O)Ser/Thr
Protein-CO-NH-(CH$_2$)$_2$-$P\rightarrow$6)Man	

is usually found in secreted and membrane mucin-type glycoproteins. Then, there is GlcNAc(β1-O)Ser/Thr, called the O-GlcNAc type of O-glycosylation, as already mentioned, found in human nuclear, mitochondrial, and cytoplasmic glycoproteins.

Abundantly found in proteoglycans and the extracellular matrix of human tissues, is the linkage Xyl(β1$\rightarrow$O)Ser/Thr, called proteoglycan-type of O-glycosylation. The linkage Gal$p$(β1$\rightarrow$O)Hyl (hydroxylysine) is mainly found in the fibrous protein Collagen (types IV and V) and thus called the collagen-type [26]. The linkages L-Ara$f$(β1$\rightarrow$O)Hyp (hydroxyproline) and Gal(α1$\rightarrow$O)Ser, called extensin-type glycosylation, are found in plant cell walls.

The O-glucosylation [27] and O-fucosylation are unusual modifications. In some cases, they can be elongated by the addition of Gal and sialic acid, for example, in Notch proteins. The O-mannosylation of Ser/Thr is mainly known from fungi and yeast proteins, but this linkage has also been found in vertebrate proteoglycans and glycopeptides of the brain. In proteoglycans, a low-molecular-mass glycosaminoglycan (GAG) can be attached to the protein via an O-linked mannose residue rather than the normal-linked xylose residue. O-mannosylation may function during muscle and brain development [28, 29]. O-mannosylation is initiated in the ER of the cell, but in the Golgi apparatus extension can occur with GlcNAc, Gal, Neu5Ac, ribitol phosphate, Xyl, and GlcA. The Man(α1$\rightarrow$O)Ser/Thr linkage is abundantly present in glycoproteins of the brain and nervous system (neuromuscular junctions). So far, this linkage has not been found in prokaryotes.

Notably, glycosylation on non-consensus amino acid sequons seems to be more frequent in rodents, where even glutamine-linked glycosylation has been observed. Furthermore, it is becoming clear that other unconventional N-glycan types, such as paucimannosidic (Man$_{1-3}$GlcNAc$_2$Fuc$_{0-1}$) and just the chitobiose core (GlcNAc$_2$Fuc$_{0-1}$), decorate some mammalian glycoproteins.

C-glycosylation is a rare form of glycosylation, featuring a single α-mannopyranosyl unit on a carbon (C2) of the indole side chain of a tryptophan (Trp) residue. The amino acid sequence for glycosylation usually comprises TrpXxxXxxTrp, wherein the first Trp carries the C-Man. This is found in animal cells, but not in yeast and bacteria. The function of C-mannosylation is still unknown. Such carbon–carbon linkages are not very common but can be found as C-glycosides, e.g., as flavonoids (flavones). The linkage cannot be hydrolyzed as the glycosidic carbon is not as conducive to protonation and subsequent cleavage as the glycosidic oxygen.

S-linked glycosylation refers to the attachment of oligosaccharides to the sulfur atom of a cysteine residue. Certain native glycopeptides contain S-linkages, such as galactosyl-cysteine, but S-glycosylation is extremely rare occurring in mammals, also in prokaryotes. There, this modification has been reported in *Bacillus subtilis* (*S*-glucosylation) and in *Lactobacillus plantarum* (*S*-GlcNAcylation).

S-glycosylation is mainly found in plant β-D-thioglucosides or glucosinolates, which contribute to the characteristic flavor of radish, cabbage, and mustard.

Glycation refers to the non-enzymatic attachment of reducing sugar to amino acids in proteins (both to N-terminus and to lysine and histidine side chains). A chemical process occurs when a protein or lipid covalently binds to a sugar, such as fructose, galactose or glucose, usually through overconcentration of the free sugar. It is often a haphazard process that is usually detrimental to the function of the protein and the organism as a whole. The reaction may be endogenously occurring in an organism, possibly in association with diabetes type 2 (haemoglobin glycation producing the HbA1C biomarker) and cardiovascular disease (formation of aneurysms), and can be detected in the blood.

Another type of exogenous glycation arises most commonly as a result of cooking of food. For example, at temperatures above 120 °C, glycation occurs readily. The browning of bread during toasting is a product of glycation by the so-called Maillard reaction, whereby a reactive sugar carbonyl group combines with a nucleophilic amine of an amino acid in a complex group of reactions, involving the formation of "Amadori compounds" and a high number of by-products. Many specific beverages such as coffee, beer, whisky, and tea owe their aroma properties to the Maillard reaction.

Finally, glypiation is the way to attach a protein to the outside of the cell via a glycosylphosphatidyl-inositol (GPI) anchor in the cell membrane. The C-terminus of the protein is linked to lipids through a glycan (see Sect. 2.4.4).

2.4.1.5 Glycosides

When the carbohydrate anomeric hydroxyl is linked to a small noncarbohydrate moiety (aglycone), which is not an amino acid, the compound is called a glycoside. In the simplest example, there is a monosaccharide with an O-methyl group at C1, which makes the monosaccharide a methyl glycoside (e.g., methyl glucoside or methyl galactoside).

The majority of secondary metabolites in plants are glycosides. In most natural glycosides, the sugar (mono- or small oligosaccharide) is O-linked to a great variety of aglycones (usually nonpolar, hydrophobic). Examples are glycoalkaloids, flavonoids, saponins, isoflavones, coumarins, and dihydrochalcones. Steviol glycosides are a nice example. Nowadays, these compounds (Fig. 2.8) are very popular as natural sweeteners, as replacement for sucrose as well as alternative for artificial synthetic sweeteners in food technology [30].

Finally, the sugar may also be linked through an amino or thiol group, giving N- or S-glycosides, respectively. Typical of the N-glycosides are the nucleosides, condensation products of ribose/deoxyribose and an amine. Adenine and guanine (purines) and cytosine, uracil, and thymine (pyrimidines) are the common bases in RNA and DNA. Here, the sugar linkages are always β.

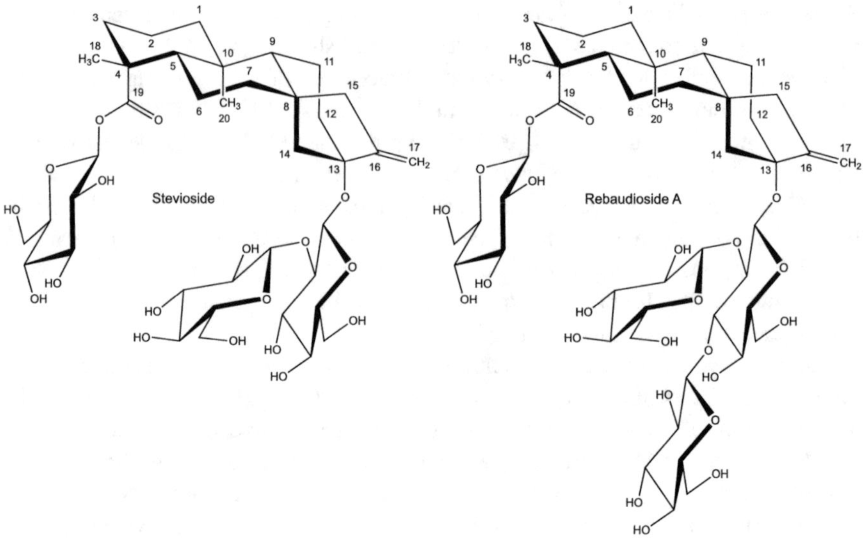

Fig. 2.8 The steviol glycosides Stevioside and Rebaudioside A

2.4.2 Proteoglycans and Glycosaminoglycans

Proteoglycans (PGs) represent a special class of heavily glycosylated proteins in eukaryotic species. PGs are composed of a variable, central core protein substituted with a high number of O-linked, long, unbranched carbohydrate polymers. These polymers are built up by disaccharide-repeating units, consisting of an amino sugar and a uronic acid or galactose. Mostly, they are highly charged by the carboxylic acids and *O*- or *N*-sulfate ester groups at different positions. PGs influence cell growth and they interact with a number of growth factors, chemokines, cytokines, and morphogens. Multiple species of proteoglycans exist in the central nervous system (CNS), and as such they are the major glycoconjugate in the microenvironment of neural cells. Neural proteoglycans are involved in a variety of cellular events including proliferation, adhesion, migration, differentiation, neurite elongation, and neuronal plasticity [31]. Furthermore, the cell-surface proteoglycans are important for the strength and flexibility of cartilage and tendons. The long, unbranched poly-disperse heteropolysaccharides, usually covalently O-linked to Ser/Thr, are called glycosaminoglycan (GAG), and they can be heterogeneously sulfated. The GAG portions have negative charge. In the past, they were denoted as (acid)mucopolysac-charides. It has to be noted that sulfated glycosaminoglycans are not found in pro-karyotes. GAGs are defined by the composition of their repeating disaccharide units and the chemical linkage between the constituting monosaccharides (Fig. 2.9).

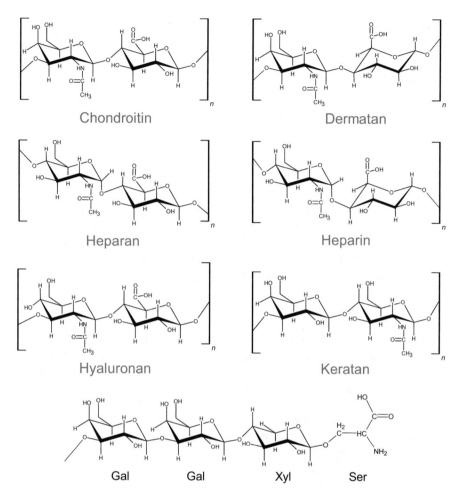

Fig. 2.9 The repeating disaccharide structures of glycosaminoglycans (GAGs), which can consist of up to 1000 repetitive units and can be variably sulfated (see text). Under is shown, the terminal sequence by which most of the GAGs are linked to Ser (or Thr) in the protein backbone

GAGs are divided in two main categories:

1. the *polyuronides*, which include chondroitin sulfate (CS) and heparan sulfate (HS). These are O-linked via a trisaccharide →3)Gal(β1→3)Gal(β1→4) Xyl(β1-O)- to certain Serine residues in the protein core, where the Ser residue is generally in the (often repeated) amino acid sequence-Ser-Gly-Xxx-Gly- (Xxx = any amino acid).
2. the *polylactosamines*, for instance keratan sulfates (KS I, KS II, and KS III). Corneal KS I is N-linked via *N*-acetylglucosamine in an alkali-stable bond to asparagine in the protein core and cartilage KS II is O-linked via *N*-acetylgalactosamine in an alkali-labile bond to Ser or Thr in the protein core (Fig. 2.10). Additionally, GAGs are also classified into four different families: (1) heparan/heparin sulfate (HP/HS), (2) chondroitin/dermatan sulfate (CS/DS), (3) keratan sulfate (KS), and (4) hyaluronan (HA).

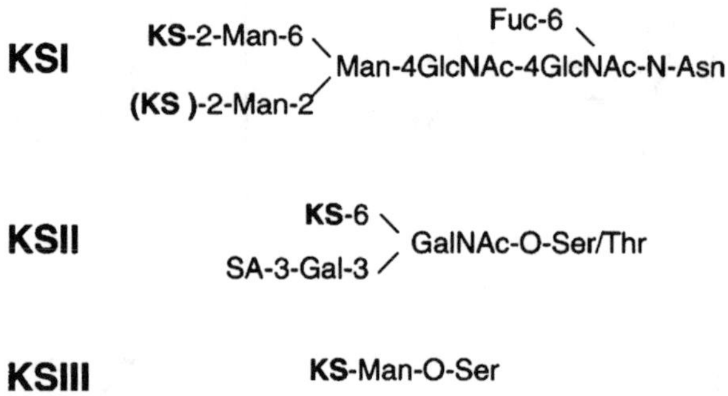

Fig. 2.10 Structures the
terminal sequence by
which the three classes of
Keratan sulfate (KS) are
linked to the protein core

KSI

KS-2-Man-6 ⟍
           ⟍ Man-4GlcNAc-4GlcNAc-N-Asn
(KS)-2-Man-2 ⟋

Fuc-6 ⟍

KSII

KS-6 ⟍
       ⟍ GalNAc-O-Ser/Thr
SA-3-Gal-3 ⟋

KSIII   KS-Man-O-Ser

Although proteoglycans are related to glycoproteins, the abundantly present gly-cans on the polypeptide backbone differ enormously from the N- and O-glycans of regular glycoproteins. The structural complexity is due to GAG polymerization and modifications, which varies with cell type and tissue source. The GAGs are pro-foundly expressed on the surface of cells as components of the extracellular matrix (ECM) in connective tissue, such as cartilage, to provide strength and elasticity. The ECM, as a gel-like material, fills the extracellular space between the cells and holds the cells together [32]. Furthermore, GAGs can be involved in processes such as cerebral development and wound healing. Specific biological functions are related to specific sequences within the carbohydrate chain. GAGs also play a role in viral invasion [33, 34] (see Chap. 3). A significantly reduced presence of proteoglycans is associated with diseases such as heart failure and skeletal development. Anomalies in GAGs are observed during osteoarthritis, angiogenesis, and cancer metastasis.

The disaccharide repeating units of the six main classes of proteoglycan glycos-aminoglycan, which are depicted in Fig. 2.9, can be described as follows:

Chondroitin sulfate (CS) is composed of repeating disaccharides consisting of N-acetyl-D-galactosamine linked to D-glucuronic acid: $[\rightarrow3]$GalNAc$(\beta1\rightarrow4)$ GlcA$(\beta1\rightarrow]_n$. GalNAc may be sulfated at C4 or C6 or both (denoted as CS-A and CS-C, respectively), but no N-sulfation. The molecular mass ranges from 5 to 50 kDa, depending on the tissue source. CS, as a proteoglycan, is present in extra-cellular matrices and on cell surfaces and is abundant (~40%) in cartilage, where it acts as a lubricant in skeletal joints by creating a gel-like medium that is robust to friction and shock. CS is also involved in the tensile strength of the walls of the aorta. Furthermore, chondroitin sulfate proteoglycans (CSPGs) play important roles in the development of the nervous system, including neuronal migration, and syn-apse formation. CSPGs are a key substance for injury repair of the central nerve system (CNS). Chondroitin sulfate is widely used as a treatment for osteoarthritis and cataracts, as it has anti-inflammatory and pain-reducing properties. CS is often found as a hybrid structure with dermatan sulfate.

Dermatan sulfate (DS), the primary GAG in the dermis, has a repeating disac-charide of N-acetyl-D-galactosamine linked to L-iduronic acid, $[\rightarrow3]$ GalNAc$(\beta1\rightarrow4)$-L-IdoA$(\alpha1\rightarrow]_n$ of which GalNAc is sulfated at C4 and C6 (and mainly C6 in adultery). The IdoA units may occasionally bear C2-sulfation. The number of disaccharides per chain differs from 30 to 130 (20–60 kDa). DS, also known as CS-B, contributes to the pliability of skin.

Heparan sulfate (HS) has a repeating disaccharide consisting of N-acetyl-D-glucosamine linked to D-glucuronic acid: $[\rightarrow 4)GlcNAc(\alpha 1 \rightarrow 4)GlcA(\beta 1 \rightarrow]_n$, where N-sulfation can occur. Additionally, clustered O-sulfation at C6, C3 of GlcNAc, and at C2 of GlcA is possible. D-Glucuronic acid (GlcA) can randomly be substituted by an α-L-iduronic acid residue (IdoA) to less extent. This makes the structure rather complex. The composition varies between different tissues. Typically, the HS chain has between 50 and 200 disaccharide units and is covalently O-linked to a serine residue of a protein via a GlcA-Gal-Gal-Xyl tetrasaccharide. Heparan sulfate proteoglycans (HSPGs) are present ubiquitously on the mammalian cell surface membrane and in the extracellular matrix (matrisome), including the basement membranes (heart, lung, liver, kidney, skin), and as such is a key mediator of the interaction of cells with their local microenvironment. HSPGs play important roles in a variety of developmental, morphogenic, and pathogenic processes. Many pathogens (viruses, parasites, and bacteria) use cell-surface HS and CS/DS with definite sulfation patterns and chain lengths for host cell attachment and invasion (see Chap. 3).

Heparin (HP), also called heparin sulfate, has a structure closely related to that of heparan sulfate. The structure consists of a repeating disaccharide consisting of N-acetyl-D-glucosamine linked mainly to L-iduronic acid: $[\rightarrow 4)GlcNAc(\alpha 1 \rightarrow 4)$-L-IdoA$(\alpha 1 \rightarrow]_n$ (generally 20–90 units; 5–15 kDa). Note that there are all α-linkages. L-IdoA can be replaced by D-GlcA for 10%. In general, a uniform (GlcNAc-6S/IdoA-2S) sulfation is found in heparin, which makes it the biomolecule with the highest negative charge density known. HP is restricted almost exclusively to intracellular granules of mature mast cells and is released as a mixture of free polysaccharide chains at points of injury. HP in the blood prevents blood coagulation (clotting) by binding to several plasma proteins, including antithrombin III. HP is added as anticoagulant to blood donated for transfusion. HP is, by far, the most common therapeutic carbohydrate with more than 100,000 kg produced annually worldwide. However, it should be noted that heparin does not dissolve blood fibrin clots, which causes thrombosis. Heparin sulfate has further anti-inflammatory and anticancer activities [35].

Keratan sulfate (KS) is mostly present in cartilage, the intervertebral disc, and the cornea. The repeating disaccharide (about 20–30 units, ~20 kDa) does not contain uronic acids but consists of D-galactose linked to N-acetyl-D-glucosamine: $[\rightarrow 3)Gal(\beta 1 \rightarrow 4)GlcNAc(\beta 1 \rightarrow]_n$ (called poly-N-acetyllactosamine), which has variable amounts of sulfate only at C6 of both residues. Furthermore, KS is classified by the core protein linkage [36]. KS I (corneal type) is linked to specific Asn residues via the fucosylated N-glycan core structure. K II (skeletal type) is attached to Ser/Thr through GalNAc, which can be extended with a Gal and SA residue. KS III, preferentially found in PG of brain and nervous tissue, is linked to C2 of Man-O-Ser (Fig. 2.10). KS is present in horny structures formed by dead cells, such as horn, hair, hoofs, nails, and claws but also on the surface of erythrocytes. Furthermore, keratan sulfates have associations with Alzheimer's disease and arthritis in general (see Chap. 3).

Hyaluronic acid (HA) (also known as **Hyaluronan** or **Hyaluronate**) is structurally the simplest member of the GAG family, but has the longest chain among all GAG types (>2500 disaccharide units). The repeating disaccharide consists of N-acetyl-D-glucosamine linked to D-glucuronic acid $[\rightarrow 3)\text{GlcNAc}(\beta 1 \rightarrow 4)$ $\text{GlcA}(\beta 1 \rightarrow]_n$ and it is the only non-sulfated GAG. The molecular mass of the linear polysaccharide is 5–500 kDa, depending on the tissue source. It should be noted that HA is not found as a real in vivo proteoglycan (PG), because it is actually not covalently attached to a protein core. However, HA is a major constituent of the extracellular matrix (ECM), linking proteoglycans and other binding molecules into macromolecular aggregates. HA is a major component of the synovial fluid of joints and is indispensable for the lubricating function [37]. It is also the major component of the vitreous fluid in the eye, maintaining the necessary spherical shape of the eye. Furthermore, HA is a major component of skin. Hyaluronic acids and chondroitin sulfate are widely utilized in medical treatments (drug delivery, eye surgery, wound healing) and in the cosmetic industry, due to their elastic and cushioning properties [38].

Source: Shutterstock

2.4.3 Glycolipids

There is another important class of glycoconjugates: *glycolipids*. A glycolipid consists of carbohydrate linked to a lipid molecule. The glycan structures of glycolipids differ from those of glycoproteins. Glycolipids mainly occur in diverse tissues of higher animals and humans but are also present in plant and microbial cells. They are present on all eukaryotic cells in the outer membrane, where they act as specific

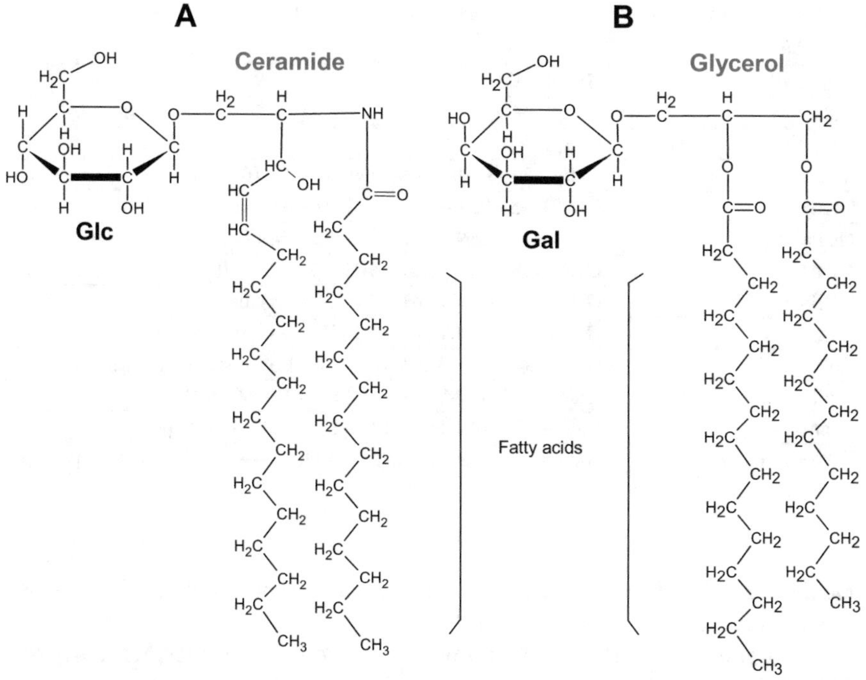

Fig. 2.11 Exemplary representation of two glycolipids. (**A**) Glucosyl-ceramide (cerebroside). The β-glucose residue is usually elongated at C4 with sugars forming glycosphingolipids (GSLs) (Table 2.3). (**B**) Galactosyl-diglyceride. β-Galactose is usually elongated at C6 with sugars. The fatty acids vary in chain length and degree of (un)saturation

sites for recognition by carbohydrate-binding proteins. They can also be present in the membranes of intracellular components like mitochondria, lysosomes, and the nucleus. The carbohydrate structure mostly confers the biological function of a particular glycolipid [39]. The amphipathic glycolipids molecules are unique because they display both polar hydrophilic (due to the high concentration of hydroxyl groups in the carbohydrate part) and nonpolar hydrophobic properties (due to the lipid component). The lipid moiety of glycolipids is generally buried in the cell membrane bilayer, leaving the oligosaccharide moieties exposed but in close proximity to the bilayer surface. Glycolipids represent more than 25% of the lipid content of myelin sheaths that insulate the axons of neurons in the nervous system. Glycolipids play an important role in immunogenicity.

Two glycolipid families are defined: (1) the *sphingosine* type and (2) the *glycerol* type (Fig. 2.11). The predominant class of most common glycolipids are *glycosphingolipids* (GSLs), where glucose, covalently bonded to ceramide (Cer), is extended with other carbohydrate residues, such as Gal, Man, GlcNAc, and/or GalNAc, resulting in large oligosaccharides. Ceramide is composed of (*N*-acyl)sphingosine substituted, through an amide linkage, with any of a variety of fatty acids. The carbon chain length may vary and rarely contains one or more

Table 2.3 Root structures of neutral glycosphingolipids (GSLs)

Type name	Structure
Glucosyl	Glc(β1→Cer
Galactosyl	Gal(β1→Cer
Lactosyl	Gal(β1→4)Glc(β1→Cer
Muco	Gal(β1→3)Gal(β1→4)gGal(β1→4)Glc(β1→Cer
Lacto	Gal(β1→3)GlcNAc(β1→3)Gal(β1→4)Glc(β1→Cer
Neolacto	Gal(β1→4)GlcNAc(β1→3)Gal(β1→4)Glc(β1→Cer
Globo	GalNAc(β1→3)Gal(α1→4)Gal(β1→4)Glc(β1→Cer
Isoglobo	GalNAc(β1→3)Gal(α1→3)Gal(β1→4)Glc(β1→Cer
Ganglio	Gal(β1→3)GalNAc(β1→4)Gal(β1→4)Glc(β1→Cer
Gala	GalNAc(α1→3)GlcNAc(β1→3)Gal(α1→4)Gal(α1→Cer
Mullo	GlcNAc(β1→2)Man(α1→3)Man(β1→4)Glc(β1→Cer
Arthro	GalNAc(β1→4)GlcNAc(β1→3)Man(β1→4)Glc(β1→Cer
Forssman GSL	GalNAc(α1→3)GalNAc(β1→3)Gal(α1→3)Gal(β1→4)Glc(β1→Cer

double bonds. Cerebrosides are composed of a sphingosine, a fatty acid, and galactose or glucose.

Glycosphingolipids (GSLs) are primarily located (but not homogeneously distributed) in the plasma membrane of animal/human cells, in the following way: a lipophilic part acting as anchor (ceramide portion) embedded in the phospholipid-based bilayer and the hydrophilic carbohydrate moiety, which protrudes from the cell surface. Glycosphingolipids participate in adhesion or adjoining cells via the carbohydrate moieties present on the outer layer of the cell membrane. The biological significance of variation in the lipid portion of GSLs is still not fully understood [40]. Depending on the monosaccharide linked to the ceramide lipid, two groups are indicated: (1) Glucosphingolipids and (2) Galactosphingolipids. These GSLs appear as (1) neutral, monohexosyl to (branched) polyglycosyl ceramides and (2) acidic negatively charged by sialic acid and sulfate. Typically, GSLs are divided into structural family types according to their root structure, using trivial names (Table 2.3).

As previously mentioned, the oligosaccharides of GSLs can be very large, for instance, [→3)Gal(β1→4)GlcNAc(β1]$_n$→3)Gal(β1→4)Glc(β1→Cer, where $n > 10$.

The *ganglio*-GSLs usually contain sialic acids and are called sialoglycosphingolipids or *gangliosides*. Because of considerable variation in their sugar components, more than 130 varieties of gangliosides exist. They are abundantly found in central and peripheral neuronal tissues, where they contribute to the stability of par anodal junctions and ion-channel clusters in myelinated nerve fibers. Gangliosides have neurotrophic activity. Common examples are monosialoganglioside (GM1 and GM3) Neu5Ac(α2→3)Gal(β1→4)Glc(β→Cer and disialoganglioside (GD2 and GD3) Neu5Ac(α2→8)Neu5Ac(α2→3)Gal(β1→4)Glc(β→Cer. They dominate the glycome of the brain early in development. Further acidic GSLs can contain sulfate and phosphate groups. Distributed on the cell-surface membrane and anchored in the external leaflet of the lipid bilayer by the ceramide moiety, they play important roles in the development and differentiation of nervous systems in vertebrates. Tumor cells express aberrant glycosylation in GSLs, as such GD2 and GD3 are almost exclusively expressed in tumor cells, and might show increased fucosylation.

The glycoglycerolipids or glyceroglycolipids are glycolipids of the glycerol type (Fig. 2.11). The oligosaccharide is linked by a glycosidic bond to the primary C1 of a glycerol molecule in the *sn* configuration. C2 and C3 are esterified by fatty acid molecules. These glycolipids predominantly occur in plants (in the membranes of chloroplasts), algae and cyanobacteria.

2.4.4 Glycosylphosphatidyl-Inositol Membrane Anchors

The cell membranes can contain another kind of glycolipid structure, called the glycosylphosphatidyl-inositol (GPI) anchors, by which proteins may be attached to the plasma membrane. GPIs are composed of phosphatidyl-inositol linked glycosidically to a tetrasaccharide (Man-Man-Man-GlcNH$_2$) that terminates with a phosphor-ethanolamine moiety, which is linked via an amide to the α-carboxyl group (C-terminal) of a protein (Fig. 2.12). In this way, the GPI-linked proteins (GPI-APs) occur on the outer leaf of the cell membrane lipid bilayer, where they play critical roles in numerous biological processes, such as cell recognition and interaction. Parasite cells (i.e., *Trypanosoma* or *Leishmania sp.*) are especially rich in GPI-APs. GPI anchors are found throughout eukaryotes, including protozoa, plants, fungi, invertebrates, and mammals.

GPI anchors contain a fairly conserved common structural core motif: ethanolamine-PO$_4 \rightarrow$6)Man(α1$\rightarrow$2)Man(α1$\rightarrow$6)Man(α1$\rightarrow$4)GlcNH$_2$(α1$\rightarrow$6)*myo*-inositol-1-PO$_4$-lipid. Notably, the glucosamine is not *N*-acetylated. In different organisms, GPIs may differ in their acyl/alkyl substituents in the phospholipid tail, extra ethanolamine phosphate groups on the core glycan structure, having additional sugar moieties on the mannose residues, and an acyl substituent on inositol.

The GPI is embedded in the lipid bilayer of the cell by two fatty acid chains, and the attached protein is exposed on the extracellular side of the plasma membrane. GPIs are structurally and functionally very diverse. The fatty acid residues can vary significantly. The lipid moiety of the GPI anchor can be a 1-alkyl-2-acyl phosphatidylinositol, diacyl phosphatidylinositol, or inositol-phosphoceramide. Different GPI anchors arise through substitutions to the core structure. For instance, Gal and GalNAc substituents can be present on the Man residue (α1$\rightarrow$4)-linked to glucosamine or an extra Man residue (α1$\rightarrow$2)-linked to the terminal mannose. An additional fatty acid chain can be attached to the myo-inositol at the 2-position. Some GPI anchors can contain palmitic acid, hydroxyester-linked to the inositol. The attached (glyco)proteins (GPI-APs) are very diverse, including size. In humans, on blood cells, already ~20 different GPI-anchored proteins occur.

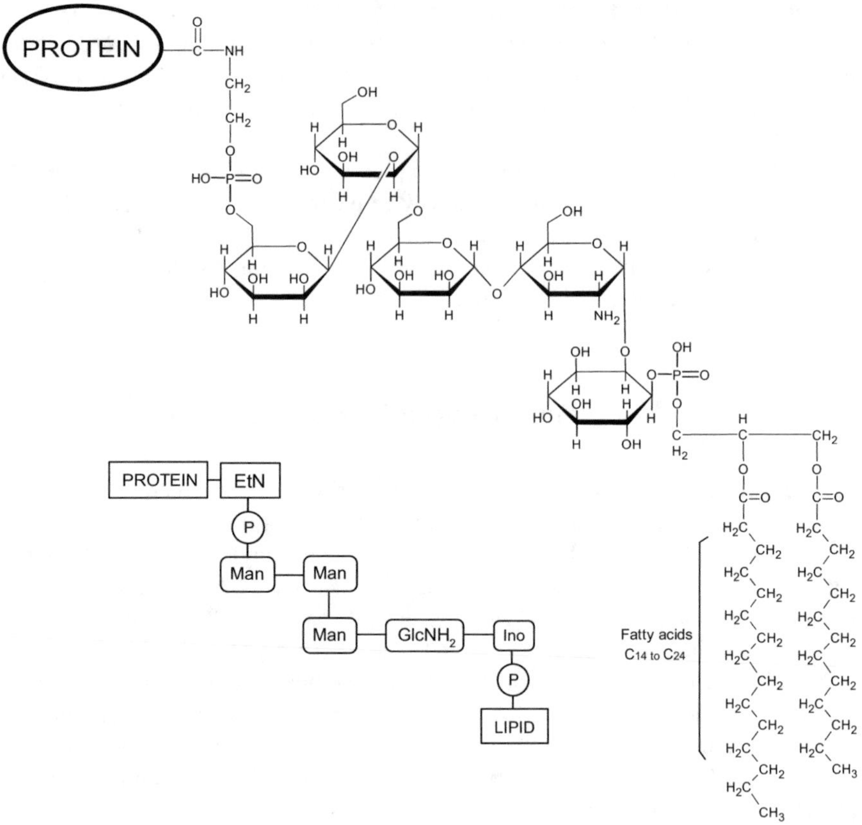

Fig. 2.12 Two schematic representations of the evolutionarily conserved structure of a typical glycosylphosphatidyl-inositol (GPI) anchor

GPIs can also be present without a linked protein, particularly in protozoa. The reader, interested in more information about the biosynthesis and biology of GPIs, is encouraged to consult these excellent reviews [41, 42]. Many GPIs have been structurally characterized but the comprehension of their biological functions, beyond the simple physical anchoring, remains largely speculative. The work on functional elucidation at a molecular level is still limited.

2.5 Intrinsic Complexity of Carbohydrates

It is clear that the biomolecular group comprising carbohydrates (including those of glycoproteins, glycolipids, polysaccharides) is prominently represented in life science and biomedicine. Carbohydrates are very heterogeneous, showing differences in primary structure (ring size and shape), degree of polymerization (mono vs. oligo vs. polysaccharides), macromolecular characteristics (linear structure vs. branched structure), linkage types (α or β glycosidic linkage, linkage position), and charge. Furthermore, the type and progress of glycosylation in each glycoconjugate is time-, tissue-, organism-, interaction-, and disease-dependent. The physical and chemical differences give rise to a large variety of properties, including solubilities, reactivities, and, for instance, susceptibility to digestive enzymes. Detailed characterization of the glycan moieties is often crucial for a molecular understanding of various biological processes.

Since carbohydrates occur as simple and/or complex mixtures in a large variety of compounds, it must be evident that their determination and characterization are often time-consuming, needing pretreatments, and an extensive arsenal of analytical methodologies. The large number of possible carbohydrate structures (high structural complexity) makes their identification a challenging process, requiring sensitive methods that can determine monosaccharide composition, sequence, anomericity, linkage positions, and branching. Despite the importance of carbohydrates but due to their complexity, research on their functions has remained, for a long time, in the shadows of genomic and proteomic studies. Another complicated factor is that many glycans have more than one disparate function.

Carbohydrate research has a reputation, not unmerited, for being technically difficult and rather impenetrable. The complexity of carbohydrate chemistry is often demonstrated in comparison to protein chemistry. For instance, only a single peptide can be formed from two identical amino acids, and three different amino acids can only be combined into six distinct peptide sequences. However, two identical monosaccharides (as α/β-hexopyranoses) can already generate 19 isomeric disaccharides, calculated only by the (α/β1→1,2,3,4,6)-linkages, while three identical hexoses can form 448 isomers. Three different hexoses can combine to form over a 1000 distinct structures. A hexasaccharide can, in theory, manifest about 10^{12} structural isomers [43]. It is now apparent that glycans are the most structurally diverse biopolymers formed in nature. Thus, the analysis of carbohydrates is very challenging. Luckily for the glycoscientist, not all the sugar structures possible are actually used in biological molecules, but the number of isomers to be considered in practical analysis is still quite large.

This volume of *Techniques in Life Science and Biomedicine for the Non-expert* only touches the basics of carbohydrate chemistry. For more extended information, including structures and terminology, the reader is referred to the following books on carbohydrate chemistry and glycobiology: [44–49]. The nonexpert in glycoscience could also consider the European Training Course on Carbohydrates, a 4-day course that takes place every 2 years at University of Wageningen, The Netherlands (http://www.vlaggraduateschool.nl/glycosciences/).

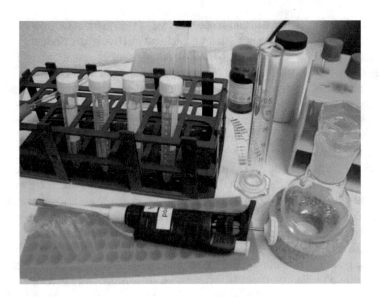

2.6   Working in a Laboratory

During working in a laboratory, serious safety issues have to be taken into account
due to the handling of toxic, corrosive, hot, and inflammable materials. This will be
evident to a trained analytical chemist. Always wear a lab coat, safety eyeglasses,
and non-powdered gloves, when working with chemicals. Care should be taken
while handling caustic acids like concentrated sulfuric acid (H_2SO_4), nitric acid
(HNO_3), hydrochloric acid (HCl). Sodium hydroxide (NaOH) and potassium
hydroxide (KOH) solutions are caustic too, so handle with care. Some chemicals
can be carcinogenic. Work in a laboratory fume cupboard or under a hood funnel.
Always use clean test tubes and glassware. Chemicals must be adequately dis-
carded. At all times, do work according to good laboratory practices outlined by
environmental health and safety protocols. This book will provide a list of the
reagents and equipment required for several analytical methods and give a description of how to carry out the protocols safely. To get reliable results for carbohydrate
analyses, always perform experiments at least in duplicate (optimally triplicate) and
include well-characterized reference standards. Methods using radioactive labeling
will not be discussed in this book.

References

1. Gerwig GJ. Structural analysis of exopolysaccharides from lactic acid bacteria. In: Kanauchi
   M, editor. Lactic acid bacteria: methods and protocols, Methods in molecular biology, vol.
   1887. Springer; 2019.
2. Popa V. Polysaccharides in medicinal and pharmaceutical applications. Smithers Rapra; 2011.

3. Özlem AD, editor. Microbial exopolysaccharides: current research and developments. Norfolk: Caister Academic Press; 2019.
4. Reid CW, Twine SM, Reid AN, editors. Bacterial glycomics: current research, technology and applications. Norfolk: Caister Academic Press; 2012.
5. Kanauchi M, editor. Lactic acid bacteria, Methods in molecular biology, vol. 1887. Totowa: Humana Press; 2019.
6. Pinho SS, Reis CA. Glycosylation in cancer: mechanisms and clinical implications. Nat Rev Cancer. 2015;15:540–55.
7. Pan S, Brentnall TA, Chen R. Glycoproteins and glycoproteomics in pancreatic cancer. World J Gastroenterol. 2016;22:9288–99.
8. Venter JC, Adams MD, Myers EW, et al. The sequence of the human genome. Science. 2001;291:1304–51.
9. Schmaltz RM, Hanson SR, Wong C-H. Enzymes in the synthesis of glycoconjugates. Chem Rev. 2011;111:4259–307.
10. Lombard J. The multiple evolutionary origins of the eukaryotic N-glycosylation pathway. Biol Direct. 2016;11:1–31.
11. An HJ, Froehlich JW, Lebrilla CB. Determination of glycosylation sites and site-specific heterogeneity in glycoproteins. Curr Opin Chem Biol. 2009;13:421–6.
12. Moremen KW, Tiemeyer M, Nairn AV. Vertebrate protein glycosylation: diversity, synthesis and function. Nat Rev Mol Cell Biol. 2012;13:448–62.
13. Aebi M. N-linked protein glycosylation in the ER. Biochim Biophys Acta. 2013;1833:2430–7.
14. Asperger A, Marx K, Albers C, Molin L, Pinato O. Low abundant N-linked glycosylation in hen egg white lysozyme is localized at nonconsensus sites. J Proteome Res. 2015;14:2633–41.
15. Thaysen-Andersen M, Packer NH. Site-specific glycoproteomics confirms that protein structure dictates formation of N-glycan type, core fucosylation and branching. Glycobiology 22 (2012) 1440-1452
16. Suzuki N. A bird's-eye view of glycan diversity. Trends Glycosci Glycotechnol. 2020;32:E1–6.
17. Schiller B, Hykollari A, Yan S, Paschinger K, Wilson IB. Complicated N-linked glycans in simple organisms. Biol Chem. 2012;393:661–73.
18. Adamczyk B, Tharmalingam T, Rudd PM. Glycans as cancer biomarkers. Biochim Biophys Acta. 2012;1820:1347–53.
19. Jaeken J, Matthijs G. Congenital disorders of glycosylation: a rapidly expanding disease family. Annu Rev Genomics Hum Genet. 2007;8:261–78.
20. Jensen PH, Kolarich D, Packer NH. Mucin-type O-glycosylation—putting the pieces together. FEBS J. 2010;277:81–94.
21. Tian E, Ten Hagen KG. Recent insights into the biological roles of mucin-type O-glycosylation. Glycoconj J. 2009;26:325–34.
22. Ma J, Hart GW. Protein O-GlcNAcylation in diabetes and diabetic complications. Expert Rev Proteomics. 2013;10:365–80.
23. Ogawa M, Okajima T. Structure and function of extracellular O-GlcNAc. Curr Opin Struct Biol. 2019;56:72–7.
24. Larsen ISB, Narimatsu Y, Clausen H, Joshi HJ, Halim A. Multiple distinct O-mannosylation pathways in eukaryotes. Curr Opin Struct Biol. 2019;56:171–8.
25. Holdener BC, Haltiwanger RS. Protein O-fucosylation: structure and function. Curr Opin Struct Biol. 2019;56:78–86.
26. Hennet T. Collagen glycosylation. Curr Opin Struct Biol. 2019;56:131–8.
27. Yu H, Takeuchi H. Protein O-glucosylation: another essential role of glucose in biology. Curr Opin Struct Biol. 2019;56:64–71.
28. Wells L. The O-mannosylation pathway: glycosyltransferases and proteins implicated in congenital muscular dystrophy. J Biol Chem. 2013;288:6930–5.
29. Endo T. Mammalian O-mannosyl glycans. Proc Jpn Acad Ser B. 2019;95:39–51.
30. Gerwig GJ, Te Poele EM, Dijkhuizen L, Kamerling JP. Stevia glycosides: chemical and enzymatic modifications of their carbohydrate moieties to improve the sweet-tasting quality. Adv Carbohydr Chem Biochem. 2016;73:1–72.

31. Oohira A. Multiple species and functions of proteoglycans in the central nervous system. In: Kamerling JP, editor. Comprehensive glycoscience: from chemistry to system biology, vol. 4. Oxford: Elsevier; 2007. p. 297–322.
32. Frantz C, Stewart KM, Weaver VM. The extracellular matrix at a glance. J Cell Sci. 2010;123:4195–200.
33. Afratis N, Gialeli C, Nikitovic D, Tsegenidis T, Karousou E, Theocharis AD, Pavão MS, Tzanakakis GN, Karamanos NK. Glycosaminoglycans: key players in cancer cell biology and treatment. FEBS J. 2012;279:1177–97.
34. Aquino RS, Park PW. Glycosaminoglycans and infection. Front Biosci. 2016;21:1260–77.
35. Fu L, Suflita M, Linhardt RJ. Bioengineered heparins and heparan sulfates. Adv Drug Deliv Rev. 2016;97:237–49.
36. Uchimura K. Keratan sulfate: biosynthesis, structures, and biological functions. Methods Mol Biol. 2015;1229:389–400.
37. Tamer TM. Hyaluronan and synovial joint: function, distribution and healing. Interdiscip Toxicol. 2013;6:111–25.
38. Köwitsch A, Zhou G, Groth T. Medical application of glycosaminoglycans: a review. J Tissue Eng Regen Med. 2018;12:e23–41.
39. Malhotra R. Membrane glycolipids: functional heterogeneity: a review. Biochem Anal Biochem. 2012;1:108.
40. D'Angelo G, Capasso S, Sticco L, Russo D. Glycosphingolipids: synthesis and functions. FEBS J. 2013;280:6338–53.
41. Paulick MG, Bertozzi CR. The glycosylphosphatidylinositol anchor: a complex membrane-anchoring structure for proteins. Biochemistry. 2008;47:6991–7000.
42. Kinoshita T. Biosynthesis and biology of mammalian GPI-anchored proteins. Open Biol. 2020;10:190290.
43. Cummings RD, Pierce JM, editors. Handbook of glycomics. Amsterdam: Elsevier; 2009.
44. Kamerling JP, Boons G-J, Lee Y, Suzuki A, Taniguchi N, Voragen AGJ, editors. Comprehensive glycoscience—from chemistry to systems biology. Amsterdam: Elsevier; 2007.
45. Varki A, Cummings RD, Esko JD, Freeze HH, Stanley P, Hart G, Marth J, editors. Essentials of glycobiology. Cold Spring Harbor: CSHL Press; 2015.
46. Lauc G, Wuhrer M. In: Walker JM, editor. High throughput glycomics and glycoproteomics, methods and protocols, Methods in molecular biology, vol. 1503. Totowa: Humana Press; 2017.
47. Lindhorst TK. Essentials of carbohydrate chemistry and biochemistry. Weinheim: Wiley-VCH; 2007.
48. Sinnott M. Carbohydrate chemistry and biochemistry: structure and mechanism. Cambridge: RSC Publishing; 2013.
49. Cipolla L, editor. Carbohydrate chemistry: state of the art and challenges for drug development. London: Imperial College Press; 2016.

Chapter 3
Carbohydrates Involved in Diseases

Abstract This chapter briefly looks at the involvement of carbohydrates in diseases. Glycosylation of proteins is one of the most common posttranslational modifications in eukaryotic cells. Subsequently, many glycoproteins specifically decorate the cells with their glycans. Pathogenic bacteria and viruses benefit from the exuberant presence of carbohydrates on the cell surface. Their interaction with specific glycan epitopes leads to infections and inflammations. Furthermore, significant alterations of cell glycans occur during many diseases, including cancer and congenital disorders of glycosylation (CDG). Monitoring changes in glycosylation (e.g., in human serum) reveals a specific and sensitive approach to biomarker discovery and disease diagnosis.

Keywords Diabetes · Bacteria · Viruses · Rheumatoid arthritis · COVID-19 · HIV · AIDS · Malaria · Inflammatory bowel disease (IBD) · Crohn's disease · Pompe's disease · Hurler's disease · Sandhoff disease · Tay-Sachs disease · Cancer

In view of the concept of this book series "Techniques in life science and biomedicine," it is relevant to briefly say something more about the participation of carbohydrates in human diseases. Currently, some far-reaching biomedical consequences of changes in the structure and metabolism of glycans have been discovered. As indicated a few times in previous sections, carbohydrates play crucial roles in infections by pathogenic bacteria, viruses, and parasites. These pathogens bind to specific glycan elements on host cells. Infections occur when several pathogenic microbes breach the protective barriers of the host, enter the body, multiply, damage cells, and disrupt normal tissue functions. Many infections start by specific interaction with host-native glycans (in particular, glycosaminoglycans), followed by actual penetration of the cellular membrane by the pathogen. It is important to note that many pathogens have evolved highly specific carbohydrate-binding proteins that can recognize aspects of the cell surface glycans they encounter in host species. Pathogens can elaborate highly specific enzymes, exo- and endoglycosidases, that could degrade host glycans. The density of oligosaccharides at the surface of cells is often altered in diseased states. Altered glycosylation has been observed as a universal feature of Alzheimer disease [1, 2], cystic fibrosis [3], and multiple sclerosis [4].

An altered glycome can lead to autoimmune diseases. Furthermore, altered gly-cosylation patterns have been observed for diseases and physiological states, such as congenital diseases, inflammatory diseases such as rheumatoid and juvenile arthritis, as well as during pregnancy and ageing [5–7]. Glycoforms containing bisecting GlcNAc increase and core-fucosylation decrease with age, observed most evident for females. The level of galactosylation and sialylation on immunoglobulin G (IgG) also changes with increasing age [8]. It is clear that the biological function of the human immunoglobulins is strongly related with their glycosylation [9]. Immunoglobulins are heavily glycosylated with N-glycans on sites located in the Fc and Fab region, as well as O-glycans in the Fab region of the molecules. Glycosylation is necessary for the maintenance of an open conformation of the two heavy chains.

In bipolar disorders and schizophrenia patients, alterations in sulfation patterns of chondroitin sulfate (CS) are observed. Obviously, the disruptions in the biosyn-thesis or catabolism of glycosylated biomolecules lead to severe diseases. An important aspect is to figure out glycans alterations as possible biomarkers of a disease. For excellent reviews on the glycosylation mechanisms and contributions to human disease, the reader is referred to [10–13].

In the first instance, when people are asked about diseases related to sugar, they say "diabetes." **Diabetes mellitus** is a metabolic syndrome characterized by chronic high blood sugar value (hyperglycaemia). The patient is unable to remove excess glucose from the bloodstream, which may result in damage to vital organs. Normally, the pancreas releases insulin to increase the conversion of glucose into glycogen. Malfunction of this important process is caused by absolute or relative insufficiency of insulin and/or decrements in sensitivity to insulin. Diabetes type 1 is instigated by an autoimmune mechanism and errors in ganglioside metabolism. The most common diabetes type 2 is caused by environmental and lifestyle factors, including the consumption of too much refined sugar. Significant changes in N-glycan com-position of serum proteins have been found in diabetes patients [14].

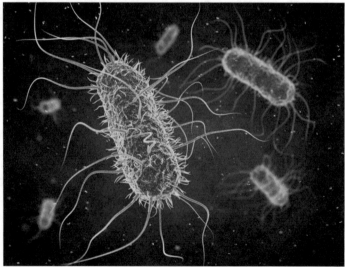

Source: Shutterstock

Many **bacterial infections** are mediated by the interaction of proteins (adhesins, fimbriae) of the bacterial cell wall with glycoconjugates (glycolipids and glycoproteins) present on the target human cell surface [15]. The pathogenic properties are caused by the initial binding of the bacterium to host cell surface glycans through specific glycan–protein interactions. Several pathogenic bacteria attach to the epithelial surface of the respiratory, genitourinary, or gastrointestinal tracts through specific recognition of oligosaccharide structures there present. For instance, the Gal($\alpha1\rightarrow4$) Gal($\beta1\rightarrow$epitope in glycans present on the urothelium is involved in the pathogenesis of bladder infections by *Escherichia coli*. The Gram-negative bacterium *Helicobacter pylori*, involved in peptic ulcers and chronic gastritis, recognizes through its flagella specific sialic acids of glycoconjugates on the surface mucous cells of the gastric mucosa (epithelial cells lining the stomach). The flagellin of many bacterial pathogens is an important factor that confers motility and allows colonization of host cells. Furthermore, it is known that some prokaryotes also have the disposal of glycoproteins which may play a role in infection and pathogenesis, and interference with host inflammatory immune response. Additionally, many pathogens express highly specific glycans on their own surfaces, which seems to modulate their antigenicity and/or their susceptibility to bacteriophages.

The Gram-negative bacteria *Pseudomonas aeruginosa* and *Staphylococcus aureus*, and their toxins, recognize epithelial glycans and are the cause of severe respiratory tract infections.

Other bacteria, such as *Vibrio cholerae* and *Clostridium tetani* use toxins, which bind to carbohydrate residues (usually sialic acids) of the gastrointestinal epithelium, leading to severe diseases such as cholera and tetanus, respectively. Blocking the attachment of the bacterial adhesins by competitive inhibition using oligosaccharides of identical structure that these toxins recognize, could be a solution to prevent infection.

The capsular polysaccharides of many bacteria are pathogenic to humans. Structural parts (conjugated with protein) of capsular polysaccharides from these bacteria (e.g., *Streptococcus pneumoniae* and *Neisseria meningitidis*) have been used as antibacterial infection vaccines to stimulate immunity [16]. Some polysaccharide-protein conjugate vaccines are among the safest and most successful carbohydrate-based vaccines formulations developed during the last 30 years [17]. However, the increased development of antibiotic resistance by pathogenic bacteria becomes a serious threat for the coming years. Many antibiotics, such as streptomycin, tobramycin, vancomycin and saccharomycin are glycoconjugates. Altering their glycosylation could be a way of creating new antibiotics to overcome the antibiotic resistance.

Comparable to bacterial infections, also **viral infections** start by viruses recognizing host cell's oligosaccharides (in particular terminal sialic acid residues of proteoglycan glycosaminoglycans), using coat proteins (haemagglutinins) for binding and to gain entry into host cells [18, 19]. The human influenza virus binds to the carbohydrate structures containing sialyl lactosamine, Neu5Ac($\alpha2\rightarrow6$)Gal($\beta1\rightarrow4$)GlcNAc($\beta1\rightarrow$, on the host cell membrane, while the receptor for the Avian virus is Neu5Ac($\alpha2\rightarrow3$)Gal($\beta1\rightarrow4$) GlcNAc($\beta1\rightarrow$. These glycans form part of the ACE2 protein on the host cell surface. SARS-CoV-2 is the Corona virus responsible for the recent global COVID-19 pandemic and seems to recognize specifically the (9-*O*-acetyl-)Neu5Ac($\alpha2\rightarrow3$)Gal sequence of glycoconjugates in the host cell membrane. The cell surface heparan sulfate (HS) interacts with the receptor-binding domain of the virus to facilitate binding and subsequent viral uptake in epithelia of the human respiratory tract [20].

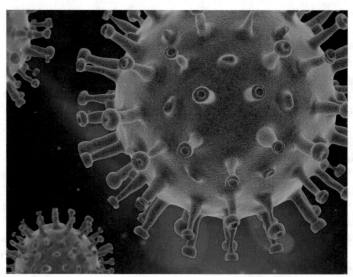

Source: Shutterstock

Viruses, as well as respiratory bacterial pathogens, exploit the host-cell glycos-aminoglycans (GAGs) of proteoglycans in the cell membrane for their infection. The trimeric spike (S) proteins and hemagglutinin esterase on the outer surface of viruses are essential in viral entry and cell membrane fusion [21, 22]. Additionally, the unique N- and O-linked glycosylation sites of the heavily glycosylated spike (S) proteins may protect the virus from antibody neutralization by the host immune system and may activate infectivity and pathogenicity. Subsequently, most viruses utilize the host glycosylation machinery to install glycans that mask and protect the virus from immune destruction. The virus uses an enzyme, called sialidase or neuraminidase, to release newly formed virus particles [23, 24]. The viral spike (S) protein is a key target in vaccine design efforts [25]. At the moment, the most used vaccine against COVID-19 is based on the viral mRNA encoding the spike (S) protein.

The human papilloma virus (HPV) infects mucosal epithelial cells to cause cervical tumors. HPV also uses cell surface proteoglycan heparan sulfate (HS) as an attachment receptor. Also, the *Herpes simplex* virus uses the host heparan sulfates to bind.

The initial step in HIV (human immunodeficiency virus) infection is the binding of the virus to cell surface endothelial and epithelial cells. This is initiated by the highly glycosylated protein gp120 [rich in N-glycans containing Neu5Ac($\alpha2\rightarrow6$) GalNAc] present in the virus envelope. In particular, binding to the host T-lymphocyte membrane (CD4 receptor). Also, another glycoprotein, called gp41, is part of the HIV viral coat and is active in the infection process. Once inside the T-cell, the virus replicates with no possibility of attack by the host's immune system, leading to AIDS. Therapeutically agents, preventing fusion of the virus and host cell by inhibiting the N-glycosylation of these viral coat glycoproteins, have shown reduced infectivity of the HIV.

Also, viral sialidases play a role in infection. A sialidase is an enzyme that cata-lyzes the release of a sialic acid residue from a glycan chain. They improve access of the virus to membrane-bound sialic acid residues by desialylating soluble mucins. An antiviral agent has shown potency by using a sialic acid analogue that is recog-nized by the virus sialidase and inactivates the virus through binding.

In many **parasitic infections**, it is the intense human immune response, raised to the antigenic oligosaccharide structures (carbohydrate epitopes) of the parasite and its toxins, that is often responsible for tissue damage. Detailed analysis of the spe-cialized glycoconjugates produced by parasites can help the development of diag-nostic tests and effective vaccines.

The most well-known parasitic infections are malaria caused by *Plasmodium falciparum*, Chagas disease caused by *Trypanosoma cruzi*, African sleeping sick-ness caused by *Trypanosoma brucei*, toxoplasmosis caused by *Toxoplasma sp.*, and schistosomiasis caused by *Schistosoma mansoni*.

For instance, *S. mansoni* recognizes ($\alpha 1 \rightarrow 3$)-fucosylated glycan structures. In the case of malaria, during the parasite invasion, an adhesin protein binds heparan sulfate on the host hepatocytes. Meanwhile, the parasite uses adhesins, which are specific for the Neu5Ac($\alpha 2 \rightarrow 3$)Gal sequence on O-linked glycans of the transmem-brane glycoprotein, glycophorin, on red blood cells. Furthermore, GPI-anchored proteins are abundant constituents of the cell membranes in parasites.

Glycosaminoglycans (GAGs) are used by pathogens at virtually every major portal of entry to promote their attachment and invasion of host cells, movement from one cell to another, and to protect themselves from immune attack [26].

During **inflammations** (severe damage of tissues), leukocytes of the immune system in the bloodstream are recruited and exit the blood vein at the site of inflam-mation. The leukocytes (neutrophils) adhere to endothelial cells of the venule medi-ated by selectins and their carbohydrate ligands (sialylated and fucosylated O-glycans). By using integrins (adhesion molecules), the neutrophils bind firmly to the endothelium and then migrate through it to invade the local tissue (inflammatory response). Alterations of glycosylation are observed in a number of inflammatory diseases [27, 28]. The chronic, systemic inflammatory disorder rheumatoid arthritis (assigned as an autoimmune disease) is characterized by excessive leukocyte recruitment [29]. In affected patients, the disease severity is correlated with the degree of circulating IgG molecules having decreased levels of galactosylation of their N-glycans [30, 31]. It has been noticed that altered glycosylation is involved in a number of other autoimmune disorders.

Chronic intestinal inflammatory disorders such as Crohn's disease and ulcerative colitis are associated with significant changes in sulfation and sialylation of glycans of gut mucins.

Inflammatory bowel disease (IBD) or irritable bowel syndrome (IBS) is charac-terized by a chronically inflamed mucosa of the gastrointestinal tract, caused by an underlying immune imbalance and triggered by luminal substances, including bac-teria. Carbohydrates play a crucial role in maintaining microbial communities and are the main modulators of the gut microbiota structure and function. Mucus forms a gel layer covering the gastrointestinal tract, acting as a semipermeable barrier between the lumen and the epithelium. Mucins, the building blocks of the mucus gel, determine the thickness and properties of mucus. In human IBD, alterations in both membrane-bound and secretory mucins have been observed, involving degree of glycosylation, sialylation, sulfation, and degradation of mucins. Typically, ($\alpha 2 \rightarrow 6$)-linked sialic acid increases and ($\alpha 2 \rightarrow 3$)-linked sialic acid decreases with progressive severity of inflammation. Also, here and with Crohn's disease, a shift to less-galactosyl glycoforms of IgG (the most abundant immunoglobulin in serum) is observed.

Many diseases, however, do not always originate from the penetration of external pathogens. Instead, the regulation and control mechanisms within the particular organism may be disturbed by various internal causes. For instance, cystic fibrosis is a genetic disorder causing altered glycosylation of glycoproteins (mucins) in the lung epithelium. The N-glycome of sputum from several inflamed and bacterial-infected cystic fibrosis patients is comprised of so-called paucimannosidic N-glycans [32, 33]. Altered terminal glycosylation, with increased fucosylation and decreased sialylation is a characteristic of the cystic fibrosis glycosylation phenotype.

In the case of certain **cancer** diseases, multiple and complex alterations in glyco-sylation of glycoconjugates are a general feature of cancer cells [34–36]. Glycosylation plays a major role in a number of cellular processes of key impor-tance for carcinogenesis and tumor progression [37]. An altered, mostly elevated, sialylation of glycans on the surface of tumor cells is often observed, showing an increase in sialylated lactosamine [Neu5Ac($\alpha 1 \rightarrow 3/6$)Gal($\beta 1 \rightarrow 4$)GlcNAc] sequences, caused by overexpression of sialyltransferases in the malignant cells. The expression of sialyl-glycans is positively correlated with aggressiveness and metastasis in many cancers. Higher sialylation increases the resistance of tumor cells to apoptosis [38]. Remarkably, the occurrence of N-glycolylneuraminic acid (Neu5Gc) has been observed for various human tumors, such as breast carcinomas. Additionally, ($\alpha 1 \rightarrow 2/3$)-fucosylation at terminal and subterminal regions and ($\beta 1 \rightarrow 6$)GlcNAc branching of N- and O-linked glycans occurs, but on the other hand, also truncated glycans have been reported in various cancer disorders. For instance, single GalNAc $\alpha 1 \rightarrow O$-linked to Ser/Thr (Tn epitope) is not found on nor-mal cells but is aberrant present in breast, prostate, lung, and pancreatic carcinomas [39]. A vaccine composed of sialyl-Tn carbohydrate linked to a protein carrier has been proved effective against colon, breast, and pancreas cancer.

The many changes in glycosylation in malignancy contribute to metastasis of cancer cells through blood and/or lymphatic vessels. The altered glycans can be of prognostic and clinical significance. Highly O-glycosylated proteins, i.e., mucins, are frequently detectable in the serum of human cancer patients. The presence of the glycotope Gal($\beta 1 \rightarrow 4$)GlcNAc($\beta 1 \rightarrow 6$)[Gal($\beta 1 \rightarrow 4$)GlcNAc($\beta 1 \rightarrow 2$)]Man($\alpha 1 \rightarrow$is associated with metastasis and indicates a poor prognosis for the cancer patient. Also, the presence of glycans bearing terminal GalNAc is associated with the ability of cancer cells to metastasize. Extensive research is in progress with glycosphingo-lipid oligosaccharide derivatives or monoclonal antibodies to carbohydrates to inhibit metastasis [40]. Pectin and hyaluronic acid may also have potential as antimetastatic agents. Several naturally occurring polysaccharides [e.g., branched (β1→3)-D-glucans] exert antitumor properties.

Finally, there are inherited carbohydrate-related diseases in humans. They are designated as **CDG**, which stands for congenital disorders of glycosylation. CDGs are defined as a group of several rare autosomal recessive inborn errors of metabolism (IEM) that affect the glycosylation of many proteins and/or lipids. They are often embryonic lethal. Their discovery accumulated from 2 in 1990 to over 120 at this moment [41, 42]. They represent a very heterogeneous family of autosomal recessive genetic defects in the biosynthesis and processing of N- and/or O-glycans. For decisive diagnosis, structural analysis of glycoproteins and their glycans is required. The importance of glycosylation is clearly revealed by the severe clinical symptoms of the patients, such as dysmorphia, brain abnormalities, malformation of several organs, ataxia, failure to thrive, severe mental retardation, epilepsy, recurrent infections, thrombosis, and often death within the first few years of life. Most of these diseases are caused by enzyme deficiencies (glycosidases and glycosyltransferases) in the biosynthetic pathway.

Oligosaccharides that are attached to proteins may determine their cellular destinations. Improper glycosylation can result in the failure of proteins to reach the correct cellular compartment.

There are at least 45 inborn errors in carbohydrate catabolism detected, leading to **lysosomal storage diseases**. Cellular lysosomes contain enzymes, such as exo- and endoglycosidases, necessary for the catabolic breakdown of glycoconjugates. Again, inherited deficiencies in these enzymes caused by gene defects are responsible. The lysosomal diseases are characterized by the accumulation of undegraded glycoconjugates in the lysosomes, leading to neurological symptoms. For instance, the enzyme α-glucosidase for breakdown of glycogen is deficient in Pompe's disease, α-L-iduronidase for breakdown of GAGs is deficient in Hurler's disease, or β-hexosaminidase for breakdown of glycolipids in Sandhoff disease and breakdown of glycoproteins in Tay-Sachs disease. The group of diseases known as sphingolipidoses, caused by incorrect breakdown of the GSLs, are often characterized by neurodegeneration and developmental disabilities.

A few CDGs can sometimes successfully be treated by administering the missing carbohydrate and by enzyme replacement therapy with recombinant lysosomal enzymes. Some congenital muscular dystrophies are due to glycosylation defects. Furthermore, it has to be mentioned that complex carbohydrate epitopes play a role in immune reactions during allergies.

References

1. Schedin-Weiss S, Winblad B, Tjenberg LO. The role of protein glycosylation in Alzheimer disease. FEBS J. 2014;281:46–62.
2. Haukedal H, Freude KK. Implications of glycosylation in Alzheimer's disease. Front Neurosci. 2020;14:625348.

3. Xia B, Royall JA, Damera G, Sachdev GP, Cummings RD. Altered O-glycosylation and sulfation of airway mucins associated with cystic fibrosis. Glycobiology. 2005;15:747–75.
4. Cvetko A, Kifer D, Gornik O, Klarić L, Visser E, Lauc G, Wilson JF, Štambuk T. Glycosylation alterations in multiple sclerosis show increased proinflammatory potential. Biomedicines. 2020;8:410.
5. Ruhaak LR, Uh HW, Deelder AM, et al. Total plasma N-glycome changes during pregnancy. J Proteome Res. 2014;13:1657–68.
6. Vanhooren V, Laroy W, Libert C, et al. N-Glycan profiling in the study of human ageing. Biogerontology. 2008;9:351–6.
7. Miura Y, Endo T. Glycomics and glycoproteomics focused on ageing and age-related diseases—glycans as a potential biomarker for physiological alterations. Biochim Biophys Acta. 2016;1860(8):1608–14. S0304-4165 (2016) (16)00022-2
8. Cobb BA. The history of IgG glycosylation and where we are now. Glycobiology. 2020;30:202–313.
9. Arnold JN, Wormald MR, Sim RB, Rudd PM, Dwek RA. The impact of glycosylation on the biological function and structure of human immunoglobulins. Annu Rev Immunol. 2007;25:21–50.
10. Planinc A, Bones J, Dejaegher B, van Antwerpen P, Delporte C. Glycan characterization of biopharmaceuticals: updates and perspectives. Anal Chim Acta. 2016;921:13–27.
11. Gerlach JQ, Griffin MD. Getting to know the extracellular vesicle glycome. Mol Biosyst. 2016;12:1071–81.
12. Mariño K, Saldova R, Adamczyk B, Rudd PM. Changes in serum N-glycosylation profiles: functional significance and potential for diagnostics. Carbohydr Chem. 2012;37:57–93.
13. Dennis JW, Nabi IR, Demetriou M. Metabolism, cell surface organization, and disease. Cell. 2009;139:1229–41.
14. Rudman N, Gornik O, Lauc G. Altered N-glycosylation profiles as potential biomarkers and drug targets in diabetes. FEBS Lett. 2019;593:1598–615.
15. Imberty A, Varrot A. Microbial recognition of human cell surface glycoconjugates. Curr Opin Struct Biol. 2008;18:567–76.
16. Vliegenthart JFG. Carbohydrate based vaccines. FEBS Lett. 2006;580:2945–50.
17. Hutter J, Lepenies B. Carbohydrate-based vaccines: an overview. Methods Mol Biol. 2015;1331:1–10.
18. Van Breedam W, Pöhlmann S, Favoreel HW, de Groot RJ, Nauwynck HJ. Bitter-sweet symphony: glycan-lectin interaction in virus biology. FEMS Microbiol Rev. 2014;38:598–632.
19. Thompson AJ, De Vries RP, Paulson JC. Virus recognition of glycan receptors. Curr Opin Virol. 2019;34:117–29.
20. Kim SY, Jin W, Sood A, Montgomery DW, Grant OC, Fuster MM, Fu L, Dordick JS, Woods RJ, Zhang F, Linhardt RJ. Characterization of heparin and severe acute respiratory syndrome-related coronavirus (SARS-CoV-2) spike glycoprotein binding interactions. Antiviral Res. 2020;181:104873.
21. Shajahan A, Supekar N, Gleinich AS, Azadi P. Deducing the N- and O-glycosylation profile of the spike protein of novel coronavirus SARS-CoV-2. Glycobiology. 2020;30(12):981–8.
22. Walls AC, Park Y-J, Tortorici MA, Wall A, McGuire AT, Veesler D. Structure, function, and antigenicity of the SARS-CoV-2 spike glycoprotein. Cell. 2020;181:281–92.
23. Vankadari N, Wilce JA. Emerging WuHan (COVID-19) coronavirus: glycan shield and structure prediction of spike glycoprotein and its interaction with human CD26. J Emerg Microbes Infect. 2020;9:601–4.
24. Watanabe Y, Allen JD, Wrapp D, McLellan JS, Crispin M. Site-specific glycan analysis of the SARS-CoV-2 spike. Science. 2020;369:330–3.
25. Shang J, Ye G, Shi K, Wan Y, Luo C, Aihara H, Geng Q, Auerbach A, Li F. Structural basis of receptor recognition by SARS-CoV-2. Nature. 2020;581(7807):221–4.
26. Aquino RS, Park PW. Glycosaminoglycans and infection. Front Biosci. 2016;21:1260–77.

27. Schnaar RL. Glycobiology simplified: diverse roles of glycan recognition in inflammation. J Leukoc Biol. 2016;99:825–38.
28. Groux-Degroote S, Cavdarli S, Uchimura K, Allain F, Delannoy P. Glycosylation changes in inflammatory diseases. Adv Protein Chem Struct Biol. 2020;119:111–56.
29. Bhusal RP, Foster SR, Stone MJ. Structural basis of chemokine and receptor interactions: Key regulators of leukocyte recruitment in inflammatory responses. Protein Sci. 2020;29:420–32.
30. Albrecht S, Unwin L, Muniyappa M, Rudd PM. Glycosylation as a marker for inflammatory arthritis. Cancer Biomarkers. 2014;14:17–28.
31. Gudelj I, Lauc G, Pezer M. Immunoglobulin G glycosylation in ageing and disease. Cell Immunol. 2018;333:65–79.
32. Venkatakrishnan V, Thaysen-Andersen M, Chen SC, Nevalainen H, Packer NH. Cystic fibrosis and bacterial colonization define the sputum N-glycosylation phenotype. Glycobiology. 2015;25:88–100.
33. Takakura D, Tada M, Kawasaki N. Membrane glycoproteomics of fetal lung fibroblasts using LC/MS. Proteomics. 2016;16:47–59.
34. Christiansen MN, Chik J, Lee L, Anugraham M, Abrahams JL, Packer NH. Cell surface protein glycosylation in cancer. Proteomics. 2013;14:525–46.
35. Mehta A, Herrera H, Block TM. Glycosylation and liver cancer. Adv Cancer Res. 2015;126:257–79.
36. Peixoto A, Relvas-Santos M, Azevedo R, Santos LL, Ferreira JA. Protein glycosylation and tumour microenvironment alterations driving cancer hallmarks. Front Oncol. 2019;9:280.
37. Dempsey E, Rudd PM. Acute phase glycoproteins: bystanders or participants in carcinogenesis? Ann N Y Acad Sci. 2012;1253:122–32.
38. Rodrigues E, Macauley MS. Hypersialylation in cancer: modulation of inflammation and therapeutic opportunities. Cancers. 2018;10:1–19.
39. Pan S, Brentnall TA, Chen R. Glycoproteins and glycoproteomics in pancreatic cancer. World J Gastroenterol. 2016;22:9288–99.
40. Magalhães AM, Duarte H, Reis CA. Aberrant glycosylation in cancer: a novel molecular mechanism controlling metastasis. Cancer Cell. 2017;31:733–5.
41. Schachter H, Freeze HH. Glycosylation diseases: quo vadis? Biochim Biophys Acta. 2009;1792:925–30.
42. Peanne R, de Lonlay P, Foulquier F, Kornak U, Lefeber DJ, Morave E, Perez B, Seta N, Thiel C, van Schaftingen E, Mathijs G, Jaeken J. Congenital disorders of glycosylation (CDG): Quo vadis? Eur J Med Genet. 2018;61:643–63.

Chapter 4
Detection of Carbohydrates by Colorimetric Methods

Abstract Classic colorimetric methods are still often used to detect the presence of carbohydrates. Several of these methods can distinguish between carbohydrate types and are also used for quantitative determinations. This chapter provides 18 protocols of different frequently used colorimetric methods, including their chemical reaction schemes. The protocols range from a simple spot test to spectrophotometric quantitation using standard calibration curves.

Keywords Furfural · Hydroxymethylfurfural · Furan · Chromophore · Fluorophore · Cereal · Wheat · Corn bran · Flavonoids · Anthranol · Molisch · Fehling · Benedict · Lobry de Bruyn-Alberda van Ekenstein · Barfoed · Seliwanoff · Bial · Warren · Aminoff · Dische

4.1 Introduction

In general, the first step in glycoscience is to detect the presence of carbohydrates in a particular sample and, if carbohydrate is present, to determine the total content (percentage of sugars).

For a long time, colorimetric methods have been used for the detection of carbohydrates and also for quantification. Many of the colorimetric methods were developed during the 1950/1960s, some even much earlier, and several are still commonly used, although old protocols have been adjusted to modern conditions during the last decades. Some of the most frequently used methods nowadays will be discussed in this chapter.

In principle, for all these methods, a (salt-free) solution containing carbohydrate material is treated with a specific reagent, generating a characteristic-colored reaction product that is proportional to the sugar concentration. Quantification can often be performed spectrophotometrically by measuring the specific absorbance. The produced chromogen molecules strongly absorb light in the visual/UV spectra. With the availability of a spectrophotometer, most quantitative colorimetric methods are simple and rapidly performed with low costs.

G. J. Gerwig, *The Art of Carbohydrate Analysis*, Techniques in Life Science and Biomedicine for the Non-Expert, https://doi.org/10.1007/978-3-030-77791-3_4

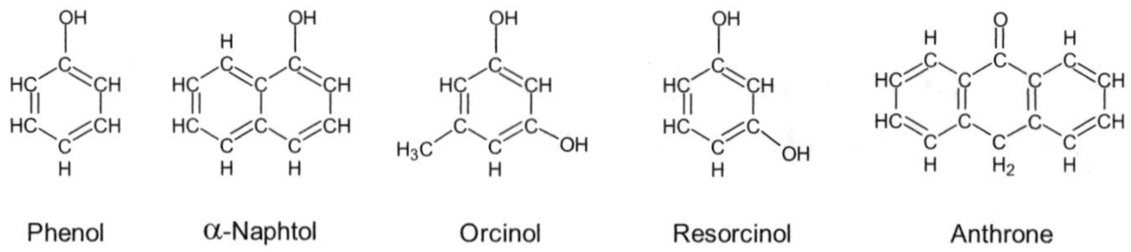

Fig. 4.1 Structures of commonly used chromogens in colorimetric carbohydrate determinations

When performing these colorimetric tests, it is advised to obtain data from three replicates, and care has to be taken that the protocols are optimized to obtain good reproducibility. Always use dry and clean glassware.

Most of the methods rely on the action of a hot strong acid to split the glycosidic linkages in glycans by hydrolysis and subsequent dehydration of the released monosaccharides forming furan derivatives. Pentoses are converted to furfural and hexoses to hydroxymethylfurfural (see Sect. 4.3.2). The dehydrated products react with aromatic compounds, such as phenol, α-naphthol, orcinol, resorcinol, or anthrone (Fig. 4.1), to give a colored product of which UV absorbance can be determined.

The absorbance intensity of the colored solution is proportional to the amount of sugar. Hence, quantification can be obtained by reference to a calibration graph of appropriate standard sugars of known concentration and a blank, treated simultaneously under the same standardized conditions. The blank should contain all the reagents used in the test sample but exclude the sample itself. Values obtained in the blank analysis must be subtracted from those obtained with the sample being analyzed. Furthermore, some colorimetric detection tests make use of the feature that monosaccharides can act as a mild reducing agent. It has to be noted that colorimetric methods for assaying sugar in complex biological samples remain problematic because of the presence of numerous biochemical species that can interfere with sugar quantification.

4.2 Preparing Standard Sugar Solutions for Calibration Curves

Most of the quantitative colorimetric methods presented have sensitivities in the carbohydrate concentration range of 50–100μg/500μL, whereby most of the assays have linear absorbance curves. For each quantitative colorimetric method, it is necessary to construct a calibration graph of standards. It is important to use reference compounds whose core structures most resemble those of the test compounds. For instance, a standard stock solution is prepared by dissolving 50 mg anhydrous glucose (or other monosaccharides appropriate for the chosen test) in 50 mL water (always use Milli-Q or pure distilled water). For the calibration graph, prepare standards by taking 0 (blank), 50µL (50µg), 200µL (200µg), 300µL (300µg), 400µL (400µg), and 500µL (500µg) of stock standard solution and making up the volume for each to the volume used for the target sample (usually 500µL). In the colorimetric method of choice, these standard samples should be treated in the same way as the target sample. The respectively obtained absorbance is plotted against the used sugar concentration (0.1–1.0µg/µL) to draw a calibration curve. The linear part is used for the quantitative determination of carbohydrates in the target sample.

4.3 General Tests

4.3.1 Detection of Carbohydrate by a Spot Test

As a first, rapid, and easy method to test solutions for the presence of carbohydrate (free or bound), the spot test can be used. A clear aqueous solution of the target material is spotted on a small piece of Silicagel TLC plate and dried. Then, the piece of TLC plate, by holding it with a pair of tweezers, is quickly dipped in an orcinol/sulfuric acid solution or an α-naphthol/sulfuric acid solution and directly dried by heating (Protocol 1). The presence of carbohydrates gives purple/orange spot coloration. This method is very practical to detect carbohydrate-containing fractions during chromatographic separations.

Protocol 1. Spot Test to Detect Carbohydrate

Materials

Silicagel 60 plastic TLC plates (no. 5553; Merck, Darmstadt)
10µL syringe (Hamilton)
A pair of tweezers
Handheld hot air blower or vented oven at 120 °C
Concentrated sulfuric acid (H_2SO_4)
Orcinol (5-methylresorcinol)
α-Naphthol
Methanol
Ethanol
Orcinol/H_2SO_4 reagent: Prepare a solution **A**: 0.2 g orcinol in 100 mL methanol and a solution **B**: 20 mL H_2SO_4 in 100 mL methanol. Then, when cooling in an ice bath, carefully and slowly add **B** to **A** and mix (CAUTION: spattering/heat). Store in a dark bottle at room temperature.
α-Naphthol/H_2SO_4 reagent: Prepare a mixture of 35 mL of 15% α-naphthol in ethanol, 20 mL conc. sulfuric acid, 130 mL ethanol, and 15 mL distilled water. Store in a dark bottle at room temperature.

Procedure

1. Cut a piece of plastic TLC plate in the appropriate size.
2. Put a small spot (~1μL) of aqueous target solution on the piece of TLC plate with a Hamilton syringe (eventually, multiple additions of 1μL, with intermediate drying, on the same spot in case of low carbohydrate concentration expected).
3. Dry the spot with a warm air blower and cool to room temperature.
4. Quickly dip the plate in the orcinol/H_2SO_4 reagent or in the α-naphthol/H_2SO_4 reagent.
5. Heat the piece of TLC plate in a vented oven at 100–120 °C for 5–10 min or with a hot air blower (avoid burning) till color appears, which indicates the presence of sugar.

A very antique, but still used, detection method of sugars on a TLC plate is the charring technique. Quickly dip the plate in a solution of 5% sulfuric acid in methanol and then heat it in a vented oven at 120 °C for 5 min. Sugar-containing positions will color black. For real thin-layer chromatography (TLC) and staining of the plates after development, see also Sect. 5.3.1.

4.3.2 Phenol–Sulfuric Acid Assay

The most well-known and widely used colorimetric method for total sugar determination is Dubois's phenol–sulfuric acid assay from 1956. The use of the method is cited in more than 25,000 peer-reviewed articles. It is a simple and rapid method to (quantitatively) determine carbohydrates (<5μg/μL) in a sample. The method is specific for sugars with a hydroxyl group at C2 (Fig. 4.2), that is, hexoses, pentoses, and uronic acids, and therefore not suitable for aminosugars and 2-deoxysugar (Protocol 2). It is mostly used for measuring the sugar content in oligo/polysaccharides, glycoproteins, and glycolipids. The assay can also be carried out at microscale in a multiwall titer plate [1].

The phenol–sulfuric acid method does not distinguish between free and bound monosaccharides. In a hot, strong acidic medium, poly-, oligo-, and disaccharides are hydrolyzed, and the released monosaccharides are dehydrated and converted into different furan derivatives depending on the type of monosaccharide. Aldopentoses give furfural, and hexoses give 5-hydroxymethylfurfural (HMF) (Fig. 4.2). Subsequently, these products condense with an aromatic compound, in this case phenol. The complexes with phenol form orange–yellow–gold (yellow–brown)-colored products (the exact compositions of these products are still indeterminate) showing a maximum light absorbance around 480–490 nm, proportional to the sugar concentration in a linear fashion for low amounts. The typical dehydration reaction shows keto-enol tautomerism:

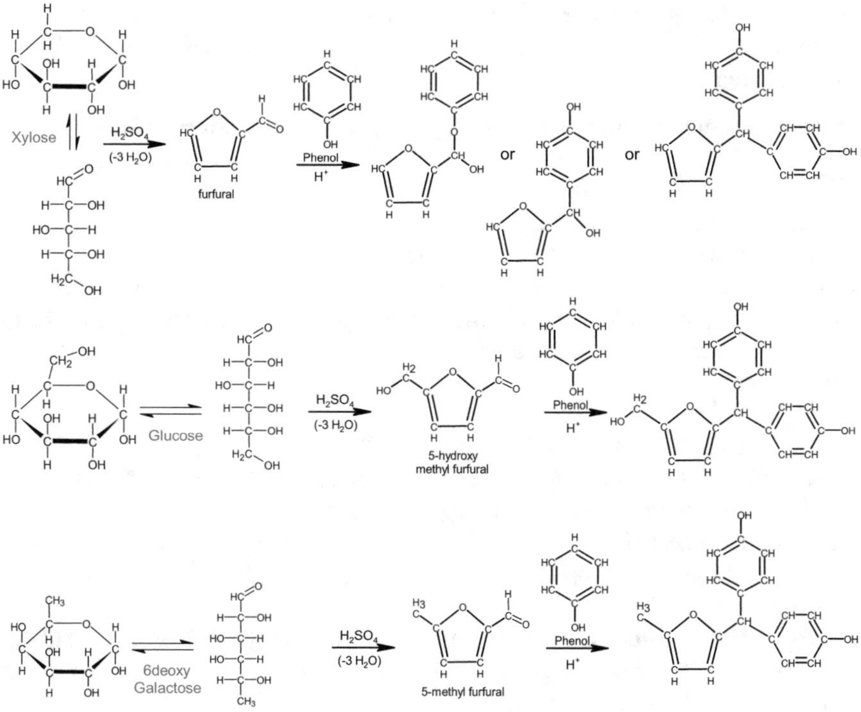

Fig. 4.2 The dehydration reaction of monosaccharides gives furan derivatives. The simplified reaction, during the phenol–sulfuric acid assay, is shown for aldoses. It must be said that the exact composition of the chromophore products is still indeterminate

The phenol–sulfuric acid assay provides an approximate value for the total carbohydrate content, estimated by comparing the absorbance at 480/490 nm with a standard curve, usually prepared with a series of standards of known glucose concentration. Mixing and standing time should be kept the same for all samples to assure reproducible results. For sugar analysis of glycoproteins, usually galactose is used as standard. For material that contains many pentoses (e.g., cereal, wheat, or corn bran), it is better to use xylose to construct the standard curve and measure the absorbance at 480 nm. Usually, ~2 mM of monosaccharides (~35μg/100μL) is required, but scaled-down microplate versions (<5μg/100μL) are used now [1].

Depending on the material to be analyzed, care has to be taken because noncarbohydrate compounds (e.g., flavonoids) and degradation products due to the strong acid conditions may also produce absorbance in the same spectral band with sugars, leading to overestimation of the carbohydrate signal. Molecules with similar functional groups (e.g., –CH=O) as sugars may give positive reactions in colorimetric techniques. Furthermore, results can be influenced by high salt concentration; hence, always check a blank eluate fraction when measuring samples eluted from a chromatographic column.

Protocol 2. Phenol–Sulfuric Acid Assay for Total Sugar Determination

Materials

Visible light spectrophotometer
Glass cuvettes (0.5 or 1 mL, 1-cm path length)
Thick-walled, 125 × 16-mm Pyrex test tubes (Take care that they are very clean!)
Glass Pasteur pipettes (remove a portion of the tip to get a wider opening for rapid
   flow of concentrated sulfuric acid)
Vortex mixer
Water (Milli Q)
Concentrated Sulfuric acid (H_2SO_4 high quality)
Phenol (C_6H_5OH) crystals: stored in a dark bottle at 4 °C to prevent oxidation
   (CAUTION: very corrosive, carcinogenic, neurotoxic)
Standard monosaccharide stock solutions: 1 mg/mL usually glucose but depending
   on the expected composition of the target sample (e.g., galactose for
   glycoproteins)
5% (w/v) phenol solution: Dissolve 5 g of 99% phenol crystals in water and fill up
   to 100 mL with water (use gloves and eye protection). The solution can be stored
   in a dark bottle at room temperature.

Procedure

1. 100–500µL clear aqueous target solution (carbohydrate conc. ~0.05–10µg/mL)
   is placed in a test tube and filled up to 500µL with water.
2. Prepare tubes with (1 mg/mL Glc or Gal or Man) standards [0 (blank), 20, 40,
   60, 80 and 100µL of stock standard] and fill up to 500µL with water.
3. Add 250µL 5% (w/v) phenol solution to each tube [CAUTION: wear gloves
   and eye protection].
4. Mix well (briefly vortexing) and leave for 5 min at room temperature.
5. Rapidly add 500µL concentrated sulfuric acid directly, without touching the
   glass wall, into the solvent with the modified Pasteur pipet [CAUTION: exo-
   thermic reaction].
6. Keep tubes undisturbed for 10 min at room temperature.
7. Mix well (briefly vortexing).
8. Keep tubes 20–30 min in a water bath at 25–30 °C (the solution turns a yellow-
   ish–orange color as a result of the interaction between the carbohydrates and
   the phenol).
9. Transfer the solution to a (1-cm path length) glass cuvette.
10. Read the absorbance at 480 nm (mainly for the presence of pentoses and uronic
    acids) and/or 490 nm (mainly for the presence of hexoses). The absorbance of
    a blank should be subtracted.

11. The amount of sugar is determined by comparing the absorbance to a standard graph of the reference monosaccharide. Mixing and standing times should be kept the same for all samples to assure reproducible results.

For using the colorimetric method in a flat-bottomed multiwall microtiter plate (350-μL well volume), the volumes of sample and reagents can be adjusted as follows:

1. Put 10μL of target sample into the microtiter plate (eventually several aliquots of different concentrations and a blank).
2. Put also five 10-μL aliquots of glucose standard (range 0.5–20μg/μL) into the microtiter plate.
3. Add 15μL of 5% (w/v) phenol solution to each well and mix (aspirate 5x) thoroughly.
4. Leave for 5 min at room temperature.
5. Rapidly add 100μL of concentrated H_2SO_4 to each well and mix (aspirate 5x) [CAUTION: hazardous reaction, do not spill, wear gloves and eye protection].
6. Leave for 20 min at room temperature.
7. Read absorbance on an (ELISA)plate reader at 490/492 nm.
8. The amount of sugar is determined by comparing the absorbance to a standard curve.

This Phenol–Sulfuric Acid assay is very practical during chromatographic separations as a rapid nonspecific method to detect carbohydrate-containing fractions (check blank eluate / remove organic solvent).

To eliminate the use of toxic phenol, recently, a new method was reported [2]. Concentrated sulfuric acid (3 mL) is rapidly added to a 1-mL aliquot of carbohydrate solution and vortexed for 30 s and then cooled to room temperature in an ice bath for 2 min. Finally, UV light absorption at 315 nm is directly read and compared to reference solutions simultaneously measured.

4.3.3 Anthrone Test

This test is typically used for the detection of hexoses (in particular glucose), but aldopentoses and hexuronic acids can also be detected by changing the temperature of heating and the acid concentration. It can be used quantitatively; however, the limit of quantification is ten times lower than for the phenol–sulfuric acid method. The mechanism of the anthrone reaction with carbohydrate basically involves that in a hot acidic medium, carbohydrates are first hydrolyzed into simple monosaccharides, and the hexoses are dehydrated to hydroxymethyl furfural. Subsequently, these furan compounds give rise to the appearance of blue–green color after reacting with anthrone (Fig. 4.3) (Protocol 3). This product has an absorption maximum at 620/630 nm. Pentoses, like xylose and arabinose, also produce furfural products that can be determined at 465 nm instead of 620 nm. Do not use Gal and Man as

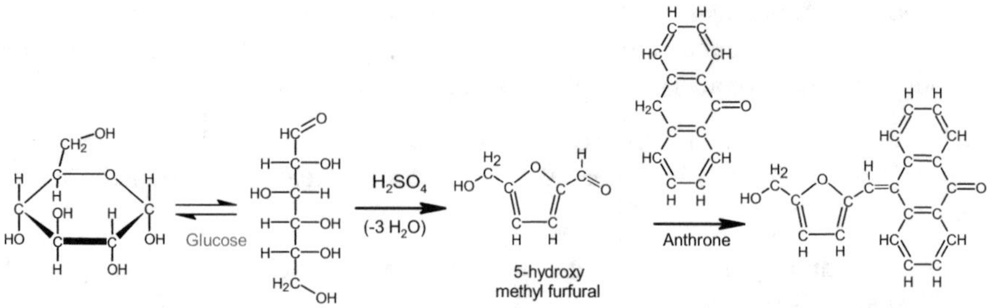

Fig. 4.3 Reaction sequence of the Anthrone assay. It is assumed (not shown) that anthranol (the enol tautomer of anthrone) is involved in the reaction

standards because they are ~60% less respondent to UV spectrophotometric measurement than glucose. A microtiter plate method has been developed to reduce cost, time, and hazards [3].

Protocol 3. Anthrone Assay for Hexose Determination

Materials

UV–Visible light spectrophotometer
Quartz cuvettes (1 mL, 1-cm path length)
Thick-walled 125 × 16-mm Pyrex test tubes (with a srew cap)
Glass Pasteur pipettes
Water bath
Vortex mixer
Water (Milli Q)
Concentrated sulfuric acid (H_2SO_4)
Anthrone
Glucose stock solutions (100 mg/100 mL)
Anthrone reagent: freshly prepared, ~15 min before use.
Dissolve 25 mg anthrone in 25 mL of concentrated sulfuric acid under magnetic stirring and keep stirring for 20 min before use [cool on ice before use].

Procedure

1. Take 500μL of the sample solution in thick-walled 125 × 16-mm Pyrex test tubes (with a screw cap). Also, Glc standards (0, 10, 20, 40, 60, 80μg) in 500μL [before next step, cool samples on ice for 5 min].
2. Add 1.5 mL of freshly prepared Anthrone reagent and close tube tight and vortex.
3. Heat for 15 min at 95 °C in water bath.
4. Cool rapidly at 0°C for 10 min (green to dark green color must appear).
5. Transfer solution in a 1-cm quartz cuvette.
6. Measure the absorbance at 620/630 nm using UV–Vis spectrophotometer.
7. Calculate concentration by comparison to standard graph (calibration curve) prepared with series of glucose standards (and blank).

Standard graph: Plot concentration of the standards on the x-axis versus absorbance on the y-axis. Fructose can also be estimated after incubation for 10 min at 40 °C, cooling, and reading absorbance at 630 nm.

4.3.4 Molisch's Test

The Molisch's Test is a common, sensitive test for determining the presence of most types of carbohydrates and some glycoconjugates in solution (Protocol 4). This is not a quantitative test. During the test, oligo/polysaccharides are hydrolyzed to monosaccharides. Then, pentoses and hexoses are dehydrated by concentrated sulfuric acid to form furfural and 5-hydroxymethylfurfural, respectively (Fig. 4.4). Subsequently, these furfural products react with an aromatic phenolic compound, in this case with two α-naphthol molecules, to form condensation products as a reddish-violet/purple-colored ring at the junction of two liquids.

Protocol 4. The Molisch's Test for Total Sugar Detection

Materials

Thick-walled 125 × 16-mm Pyrex test tubes (take care that they are dry and very clean)
Glass Pasteur pipettes (remove the portion of the tip to get a wider opening for easier flow of concentrated sulfuric acid)
Water (Milli Q)
Ethanol
Concentrated Sulfuric acid (H_2SO_4)
α-Naphthol (CAUTION: very corrosive)
Glucose stock solution (100 mg/100 mL)

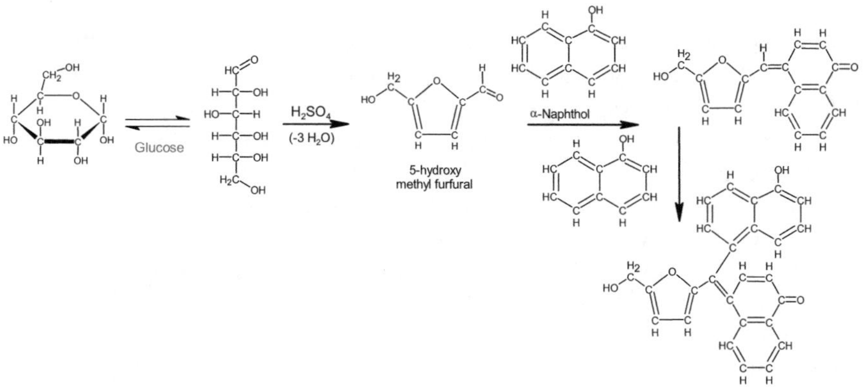

Fig. 4.4 Simplified reaction sequence of the Molisch's test. It must be noted that the exact composition of the chromophore products is still indeterminate

Molisch's reagent: 10% (w/v) α-naphthol in ethanol. Store in a dark bottle at room temperature.

Procedure

1. 500μL clear aqueous target solution (carbohydrate conc. ~0.05–1.0μg/μL) is placed in a clean and dry test tube.
2. Prepare control tube with 500μL standard solution (1 mg/mL Glc).
3. Add 25μL Molisch's reagent to each tube and mix well.
4. Incline the test tube and add slowly/carefully 500μL concentrated sulfuric acid along the side wall of the test tube to form two layers [CAUTION: wear gloves and eye protection].
5. Keep tubes undisturbed for 10 min at room temperature. The appearance of a reddish-violet or purple-colored ring at the junction of the two liquids is observed as a positive result.

4.3.5 Fehling's Test

This is the most commonly used test for the detection of reducing sugars (Protocol 5). Fehling's reagent contains blue alkaline cupric hydroxide solution, which when heated with reducing sugars gets reduced ($Cu^{2+} \rightarrow Cu^{1+}$) to cuprous oxide ($Cu_2O$), giving a brownish-red colored precipitate. Rochelle salt acts as a chelating reagent in the reaction (Fig. 4.5). Aldehydes are easily oxidized and hence are powerful reducing agents. A bi-tartrate Cu^{2+} complex is reduced to Cu^{1+}, leading to a brick-red precipitate (Cu_2O). The net reaction can be written as: $R\text{-}CHO + 2Cu(C_4H_4O_6)_2^{2-} + 5OH^- \rightarrow R\text{-}COO^- + Cu_2O + 4C_4H_4O_6^{2-} + 3H_2O$.

The reaction with Cu^{2+} is not as simple as the equation here implies; in addition to the forming of D-gluconate, a number of shorter chain acids are produced by the fragmentation of glucose.

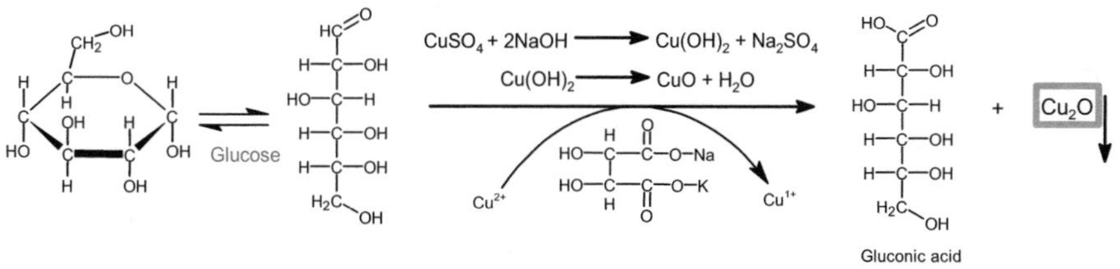

Fig. 4.5 Simplified representation of reactions in the Fehling's test

Protocol 5. Fehling's Test for Reducing Sugars

Materials

Test tubes
Pipettes
Water bath
Copper sulfate ($CuSO_4.5H_2O$)
Sodium potassium tartrate ($KNaC_4H_4O_6.4H_2O$)
Sodium hydroxide (NaOH)

Fehling's reagent:

A: 6.9 g of copper(II)sulfate [$CuSO_4.5H_2O$] in 100 mL water
B: 34.6 g sodium potassium tartrate ($KNaC_4H_4O_6.4H_2O$) (Rochelle salt) plus 16 g sodium hydroxide (NaOH) in 100 mL water (chelating reagent)
Before the experiment, prepare a fresh mix of equal volumes of **A** and **B** (add **B** to **A**) [decant solution from sediment if necessary, before use]

Procedure

1. In a test tube, put 500μL of a diluted carbohydrate solution.
2. Add 1 mL of fresh **Fehling's mix (A + B)**.
3. Mix well and place the tube in a boiling water bath for 10 min. When the content of the test tube comes to boiling, mix again.

The production of yellow/brownish/red precipitate of cuprous oxide indicates the presence of reducing sugars in the given sample. Usually, the resulting solution is colorless.

4.3.6 Benedict's Test

Another test for reducing sugars is known as Benedict's test. As in the Fehling's test, the free aldehyde or ketone group in the reducing sugars will donate an electron to Cu^{2+}, converting it to Cu^{1+}, which will produce cuprous oxide, Cu_2O, as an orange/

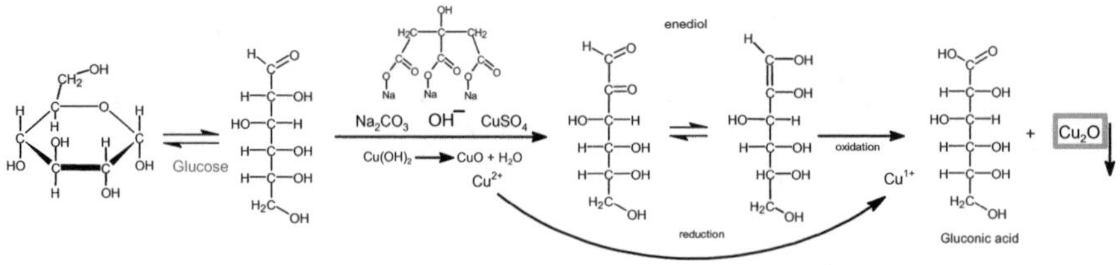

Fig. 4.6 Representation of reactions in Benedict's test. The net reaction can be written as:
$R-CHO + 2Cu(C_6H_5O_7)^- + 5OH^- \rightarrow R-COO^- + Cu_2O + 2C_6H_5O_7^{3-} + 3H_2O$

brick-red–colored precipitate (Fig. 4.6). Since the Cu^{2+} solution ($CuSO_4$) is blue, the change in color provides an easy visual indication of the presence of reducing sugar (Protocol 6).

Depending on the concentration and type of sugars, green to yellow to orange to brick-red color is developed, specific for increased amounts of reducing sugars. However, it is not a real quantitative determination method. Disaccharides having a reducing terminal sugar and are less reactive than monosaccharides. In the first instance, ketones should not give a positive Benedict's test but ketoses do, due to basic conditions. Consequently, fructose (ketose) is converted into glucose and mannose (Lobry de Bruyn-Alberda van Ekenstein rearrangement). In an alkaline medium, reducing sugars tautomerize and form enediols. Enediols are powerful reducing agents. They reduce cupric ions to cuprous form and are themselves converted to sugar acids. The cuprous ions combine directly with OH^- ions to form yellow cuprous hydroxide $[Cu(OH)_2]$ which upon heating is converted to brick-red cuprous oxide (Cu_2O). It has to be noted that other reducing compounds can also give positive results.

Protocol 6. Benedict's Test for Reducing Sugars

Materials

Test tubes
Pipettes
Water bath
Sodium citrate ($Na_3C_6H_5O_7$)
Sodium carbonate (Na_2CO_3)
Copper sulfate ($CuSO_4.5H_2O$)
Benedict's reagent:
A: Dissolve 17.5 g of sodium citrate and 10 g of anhydrous sodium carbonate (Na_2CO_3) in 80 mL water. **B**: Dissolve 1.75 g $CuSO_4.5H_2O$ in 15 mL (hot) water. Mix **A** + **B**, under stirring, and add up to 100 mL with water (pH ~10).

Procedure

1. Take 500µL of the test sample in a dry test tube (take also standard and blank).
2. Add 2 mL of Benedict's reagent and mix well.
3. Heat slowly in boiling water bath for 5 min [CAUTION for spitting].
4. Remove the tubes from the water bath and allow them to cool.
5. Observe the color change from blue to green, yellow, orange to brick-red precipitate depending upon the concentration of reducing sugar in the test sample.

4.3.7   Barfoed's Test

Barfoed's test is used to detect the presence of reducing monosaccharides in solution and is not suitable for oligo/polysaccharides. Barfoed's reagent, an acidic solution containing copper(II)acetate, is combined with the test solution and boiled
(Protocol 7). Reducing monosaccharides are oxidized by the copper ion in solution
to form a carboxylic acid and a reddish precipitate of copper(I)oxide (Cu_2O) within
3 min (Fig. 4.7). Barfoed's reagent is slightly acidic and meant for monosaccharides. Aldoses and ketoses reduce cupric ions. By controlling the pH and time of
heating, the test can distinguish reducing monosaccharides (fast reaction) from
disaccharides (very slow reaction).

Fig. 4.7   Reaction sequence of the Barfoed's test. The net reaction can be written as: R-CHO + $2Cu^{2+}$
+ $2H_2O \rightarrow$ R-COOH + Cu_2O + $4H^+$

Protocol 7. Barfoed's Test for Reducing Monosaccharides

Materials

Test tubes
Pipettes
Acetic acid (CH_3COOH)
Cupric acetate [$Cu(OAc)_2$]
Barfoed's reagent: Prepare freshly, 5 g $Cu(OAc)_2$ in 99 mL water plus 1 mL acetic acid (pH ~4.5)

Procedure

1. Take 500µL sample solution in a small test tube.
2. Add 1 mL of freshly prepared Barfoed's reagent.
3. Mix well and heat for 5 min in a boiling water bath.
4. Allow cooling under running tap water for a few minutes.
5. Fast formation of a deep blue color and a brick-red precipitate of cuprous oxide at the bottom and along the side of the test tube within 3 min indicates the presence of reducing monosaccharides.
6. If the precipitate formation takes more time, then dealing with a reducing disaccharide (same reaction but at a slower rate).

4.4   Specific Tests

The colorimetric methods, discussed earlier, are suitable as tests for determining the presence of most types of carbohydrates but particularly for monosaccharides as aldohexoses. Although, in some cases, differences can be observed between

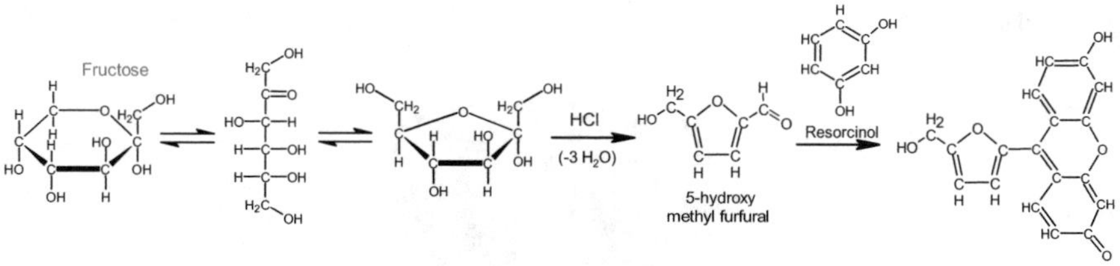

Fig. 4.8 Reaction sequence of the Seliwanoff's test for fructose [4]

aldohexoses, aldopentoses, and disaccharides by changing reaction conditions, some tests for specific monosaccharides have been developed. The most frequently used methods nowadays will be discussed.

4.4.1   Seliwanoff's Test for Ketoses

A rapid test to distinguish ketoses from aldoses is performed with the Seliwanoff's test (Protocol 8). This is a color reaction specific for ketoses (containing a ketone group, C=O). It is not used as a quantitative determination. When concentrated HCl is added, ketoses undergo dehydration to yield furfural derivatives more rapidly than aldoses (containing an aldehyde group) (Fig. 4.8). The test reagent causes the dehydration of ketohexoses to form 5-hydroxymethylfurfural, which reacts further with resorcinol (1,3-dihydroxybenzene), in the presence of HCl, to rapidly produce a product with a deep cherry red color within 0.5–2 min. It has to be noted that aldohexoses can also react to form the same product, but more slowly and giving a faint red/pink color. In this way, the Seliwanoff's test distinguishes between aldoses and ketosugars. The method is extremely sensitive for fructose, sucrose, and other fructose-containing carbohydrates.

Protocol 8. Seliwanoff's Test for Ketoses

Materials

Water bath
Pipettes
Test tubes
Resorcinol
Concentrated hydrochloric acid (HCl, 11.6 M)
Seliwanoff's reagent: Prepare freshly 15 mg resorcinol in 20 mL water [7.5% (w/v) solution] plus 10 mL conc. HCl.

Procedure

1. 250μL clear aqueous target solution (carbohydrate conc. >0.5μg/μL) is placed in a clean and dry test tube.
2. Add 750μL Seliwanoff's reagent and mix well.
3. Place in boiling water bath for 0.5–2 min (do not heat longer).
4. Cool the solution in a water bath at 20 °C.
5. A deep cherry red color within 0.5–2 min indicates the presence of ketoses.
6. Aldoses also react but slowly to produce a faint red/pink color upon continuous heating.

4.4.2   Bial's Test for Pentoses

The Bial's test is used for the detection of pentoses (Protocol 9). Bial's reagent dehydrates pentoses (e.g., ribose or xylose) to form furfural (Fig. 4.9). Furfural reacts with orcinol in the presence of ferric ions (Fe^{3+}) to form a blue/green-colored product in the solution. It has to be noted that other colors, like the formation of yellow/green or muddy brown to gray precipitate, points to the abundant presence of hexoses.

Fig. 4.9 Reaction sequence of the Bial's test

Protocol 9. Bial's Test for the Detection of Pentoses

Materials

Water bath
Tabletop centrifuge
Pyrex glass tubes 125 × 16 mm
Pipettes
Concentrated Hydrochloric acid (HCl, 11.6 M)
Ferric chloride (FW 270.32) ($FeCl_3$)
Orcinol (5-methylresorcinol, MW 124.16)
Bial's reagent: Dissolve 250 mg of orcinol in 95 mL cooled and concentrated HCl.
Add 5 mL of 10% (w/v) aqueous ferric chloride solution. Store at 4 °C.

Procedure

1. Place 500μL of target sugar (~10 mM carbohydrate) solution in a test tube.
2. Add 1.5 mL of Bial's reagent.
3. Heat the sample gently in a boiling water bath for about 1 min.
4. Cool to room temperature.

The formation of a blue/green product is indicative of pentoses. Hexoses generally react to form muddy brown to gray products. Appropriate dilution with water facilitates color detection.

4.4.3 Colorimetric Analysis of (N-Acetyl) Aminosugars

N-acetyl glucosamine (GlcNAc) and *N*-acetyl galactosamine (GalNAc) are two important aminosugars often present in the glycans of animal and human glycoproteins and glycolipids. A regularly used colorimetric method for the determination of (*N*-acetyl) hexosamines is the *p*-dimethylaminobenzaldehyde (**DMAB**) assay (Protocol 10). For conjugated (*N*-acetyl) hexosamines, hydrolysis (4–6 M HCl, 4–8 h at 100–110 °C) has to be performed to obtain free aminosugars. Usually, optimal conditions for complete release have to be determined for a specific target sample. Prior to the colorimetric assay, a re-*N*-acetylation is necessary. This is established by the addition of acetic anhydride in acetone. Further, the reaction involves a condensation with acetylacetone under alkaline conditions to a pyrrole derivative, which reacts with *p*-dimethylaminobenzaldehyde to give a red/pink-colored product. The absorbance of the solution is determined at 585 nm. The sensitivity is about 2μg/100μL.

Protocol 10. DMAB Assay for (N-acetyl) Hexosamines

Materials

UV–Visible spectrophotometer, diode array/variable wavelength
High-purity quartz cuvettes (1 cm)
100 × 13 mm glass test tubes

Glass marbles
Water bath
Vortex mixer
Dipotassium tetraborate.4H$_2$O (K$_2$B$_4$O$_7$.4H$_2$O, MW 305)
Hydrochloric acid (conc. HCl)

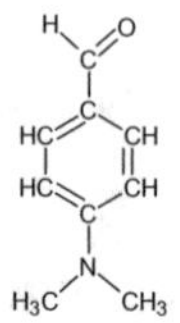

4-(N,N-dimethylamino)benzaldehyde (**DMAB**, MW 149.2)
Glacial acetic acid (CH$_3$COOH)
GalNAc and GlcNAc standard solutions
0.8 M potassium tetraborate solution: Dissolve 24 g K$_2$B$_4$O$_7$.4H$_2$O in 100 mL
   water (takes a while, heat a bit), adjust to pH 9
DMAB solution (Ehrlich's reagent): Dissolve 10 g p-dimethylaminobenzaldehyde
   in 100 mL of glacial acetic acid, containing 12.5% (v/v) HCl. Dilute this solution
   with 9 volumes of glacial acetic acid at least 15 min before use.

Procedure

  1. Put 250μL of test solution in a glass stoppered tube.
  2. Add 50μL of 1.5% (v/v) acetic anhydride in acetone (freshly prepared).
  3. Mix well, keep 5 min at room temperature.
  4. Add 250μL of 0.8 M potassium tetraborate solution.
  5. Close the tube with a glass marble.
  6. Heat for exactly 3 min in a boiling water bath.
  7. Cool in tap water to room temperature.
  8. Add 2.5 mL of DMAB solution.
  9. Close the tube with a glass marble.
 10. Incubate for 20 min at 37 °C.
 11. Color intensity is read at 585 nm after cooling. It has to be noted that GalNAc
      gives one-third response of GlcNAc.

4.4.4   Colorimetric Analysis of Uronic Acids

Hexuronic acids mostly occur in glycosaminoglycans of proteoglycans, abundantly
present in human tissues and cells. Bacterial capsular polysaccharides also contain
uronic acids. After isolation of the uronic acid-containing polymer, it must be totally

hydrolyzed, usually with 0.5 M sulfuric acid, 2–4 h at 100 °C, to free uronic acids before colorimetric estimation. Although uronic acids often give response in the colorimetric methods for neutral sugars (e.g., phenol–sulfuric acid assay and anthrone test), there are two more specific assays for uronic acids, the *meta*-hydroxy diphenyl (**MHDP**) assay and the carbazole assay. The anthrone test, typically used for detection of hexoses (in particular glucose), can easily be modified for uronic acids by altering the conditions (Protocol 11) [5]. The MHDP assay involves acidic hydrolysis of the uronic acid-containing polymer and dehydration of the free hex-uronic acids, followed by conjugation to a chromogen (Protocol 12). The carbazole assay also involves acidic treatment of the sample (often proteoglycans) to hydro-lyze and dehydrate the released hexuronic acids (stabilized with tetraborate), fol-lowed by treatment with carbazole to form a chromogen (Protocol 13). The detection limit for these methods is around 5μg/100μL. Different glycosyluronic acids can give slightly different responses, so appropriate standards must be used for accu-racy. High salt concentrations (>1 M) must be avoided.

Protocol 11. The Modified Anthrone Test for Uronic Acids

Materials

UV–Visible spectrophotometer, diode array/variable wavelength
High-purity quartz cuvettes (1 cm)
100×13 mm glass test tubes
Glass marbles
Water bath
Vortex mixer
Anthrone
Glucose stock solution 1 g/L
Glucuronic acid stock solution 0.5 g/L
0.3 M Tris HCl, pH 8.2 containing 2% EDTA
Anthrone reagent: 0.5 g anthrone in 10 mL concentrated sulfuric acid (96%)

Procedure

1. Take 200μL of the sample solution in a glass test tube.
2. Add 400μL of Anthrone reagent, mix close with a glass marble.
3. Incubate for 30 min at 60 °C.
4. Cool to RT for 10 min.
5. Read absorbance at 620 nm for total sugar content and at 560 nm for HexA.
6. Compare to standard curves, correct for blank.

Protocol 12. The Meta-Hydroxy Diphenyl (MHDP) Assay for Uronic Acids

Materials

UV/VIS spectrophotometer
Disposable glass or plastic cuvettes (0.5–1.0 mL, 1 cm)
125×16-mm Pyrex tubes with a screw cap

Pasteur pipettes
Heating block 100 °C
Water bath
Ice bath
Sulfuric acid (H_2SO_4, high quality)

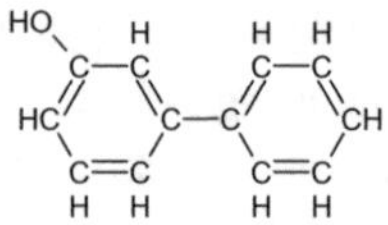

meta-hydroxy diphenyl (**MHDP**, MW 170.2)
4 M ammonium sulfamate [0.1 M reduces color from hexoses]
Uronic acid standard: 1 mM glucuronolactone in water

Reagent A: Dissolve 0.95 g sodium tetraborate ($Na_2B_4O_7.10H_2O$) in 10 mL of water and add 90 mL of ice-cold 98% sulfuric acid carefully to form a layer. Leave undisturbed overnight for slow mixing (~25 mM) but mix thoroughly at RT before use. Store under N_2.

Reagent B: Dissolve 150 mg of *meta*-hydroxy diphenyl in 100 mL of 5% (w/v) NaOH solution

Procedure

1. Put 250μL aqueous target sample (0.5–20μg of total uronic acid) in a tube [also take a series of standards containing 10, 30, 50, and 100 nmol glucuronolactone in 250μL water and a blank of 250μL water].
2. Cool on an ice bath.
3. Add 40μL of cooled 4 M ammonium sulfamate, vortex, and cool the samples in an ice bath.
4. Carefully add 1.5 mL of ice-cold reagent **A** with mixing and cooling in ice bath (avoid spattering).
5. Screwcap the tubes and heat the mixture for 10 min at 100 °C.
6. Cool rapidly in an ice bath to room temperature.
7. Add 50μL of reagent **B**.
8. Close the tubes and keep for 15 min at room temperature while color develops.
9. Transfer the solution to 1.0-mL disposable cuvette using a Pasteur pipette.
10. Read absorbance at 520 nm.
11. Prepare a standard calibration curve of A_{520} versus nmol glucuronic acid and determine the amount of uronic acid in the target sample by reference to the standard curve.

Protocol 13. The Carbazole Assay for Uronic Acids

Materials

UV/VIS spectrophotometer
Disposable glass or plastic cuvettes (0.5–1.0 mL, 1 cm)
125 × 16-mm Pyrex tubes with screwcap
Pasteur pipettes
Heating block 100 °C
Water bath
Ice bath
Sulfuric acid (H_2SO_4 high quality)
Sodium tetraborate (Sigma, $Na_2B_4O_7.10H_2O$)

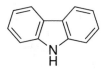

Carbazole (Kodak)
Ammonium sulfamate [reduces the color yield from neutral sugars to <5%]
Glycosyluronic acid standards
Reagent A: Dissolve 0.95 g sodium tetraborate ($Na_2B_4O_7.10H_2O$) in 10 mL of hot
   water and add 90 mL of ice-cold 98% sulfuric acid carefully to form a layer.
   Leave undisturbed overnight for slow mixing (~25 mM) but mix thoroughly at
   room temperature before use. Store under N_2.
Reagent B: Dissolve 100 mg of carbazole (recrystallized from ethanol) in 100 mL
   of absolute ethanol (0.1% w/v solution)

Procedure

1. Put the target sample (250µL) in the test tube and cool in an ice bath.
2. Put standards (0, 10, 30, 50, and 100 nmol glucuronolactone in 250µL water) in
   test tubes.
3. Add 40µL of cooled 4 M ammonium sulfamate, vortex, and cool the samples in
   an ice bath.
4. Carefully add 1250µL of ice-cold reagent **A** with mixing and cooling in ice bath
   (avoid spattering).
5. Screwcap the tubes and heat mixture for 5–10 min at 100 °C.
6. Cool rapidly in ice bath (5 min).
7. Add 50µL of reagent **B** and mix well.
8. Close the tubes and heat for 10 min at 100 °C.
9. Cool rapidly to RT while color develops.
10. Transfer the solution to cuvette using a Pasteur pipette.

11. Determine absorbance at 525 nm.
12. Prepare a standard calibration curve of A_{525} versus concentration of glucuronic acid standards and determine the amount of uronic acid in the target sample by reference to the standard curve.

4.4.5 Colorimetric Analysis of Sialic Acids

Sialic acids are a very important family of acidic, negatively charged sugars (see Chap. 9). In general, they are present at terminal positions in N- and O-glycans of glycoproteins and in carbohydrate chains of glycolipids. In the first instance, free sialic acid can be quickly detected by a spot test with resorcinol on a TLC plate (Protocol 14). Free sialic acids are usually obtained by mild acid hydrolysis (0.01–0.1 N H_2SO_4, 1 h, 80–100 °C or 2 M acetic acid, 3 h, 80 °C) of the glycoconjugate. It must be noted that strong hydrolysis conditions remove N-acetyl or N-glycolyl groups and O-acyl groups. Different spectrophotometric assays are available to determine sialic acid.

Bound and free sialic acids can be colorimetrically detected by a slightly modified Bial's test (Protocol 15, Fig. 4.10). Here, sialic acids are oxidized with acid to form a blue–purple chromophore in the presence of orcinol/Fe^{3+} (comparable with the spot test/Cu^{2+}). The chromophoric compound is extracted into isoamyl alcohol to determine the absorbance at 570 nm. The minimum detection limit is about

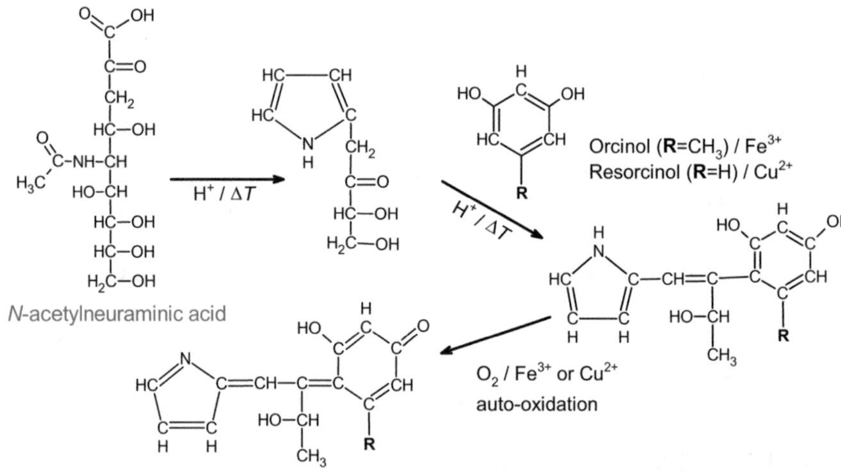

Fig. 4.10 Supposed reaction sequence of the spot test and Bial assay for the sialic acid Neu5Ac. Strong acidic conditions lead to de-N-acetylation, yielding an internal Schiff base formed by the condensation of the free amino group at C5 and the C2 carbonyl group, followed by decarboxylation and dehydration, to yield a pyrrole derivative, which then gives a colored condensation product with (re)orcinol

5 nmol, but care has to be taken because other sugars can interfere, limiting the quantification. *N*-acetyl neuraminic acid (Neu5Ac) is used for a standard graph. The amount of sialic acid in a target solution is determined by reference to a standard curve. Samples containing a known exact amount of sialic acid (e.g., containing 5, 10, 20, 30, 40, and 50 nmol Neu5Ac) are analyzed simultaneously, and a curve is plotted of the absorbance versus nanomole using the average values of the standard solutions. The linear part of the curve is used to calculate the concentration in the target sample. The presence of pentoses, hexoses, and uronic acids can strongly disturb the determination.

The thiobarbituric acid (**TBA**) assay (Protocol 16) (also called Warren or Aminoff method) is a sensitive and more reliable method for the determination of the presence of free sialic acid and for quantification by absorbance measurements. Free sialic acid is oxidized with sodium periodate (periodate oxidation, see Sect. 5.2.7) in concentrated phosphoric acid to form β-formylpyruvic acid, which then reacts with 2′-thiobarbituric acid, resulting in a chromophoric compound (giving a salmon-pink color). After extraction into cyclohexanone, the absorbance at 549 nm is measured. The minimum detection limit is about 1.5 nmol. Modified sialic acids can give reduced color response due to substituents (e.g., *O*-acetyl groups) at the C7–C8–C9 side chain that prevents periodate oxidation at this part.

Note that sialic acid has to be free for the TBA assay. There is no reaction with bound sialic acid, so a prior hydrolysis step is necessary for the analysis of glyco-conjugates. However, this colorimetric assay is still one of the most used methods to determine the amount of sialic acid in a given sample [6].

Free and conjugated sialic acids can also be determined based on the reaction of chromogens obtained in an acidic medium with reagents such as *p*-dimethylaminobenzaldehyde (**DMAB**) (not discussed here) and diphenylamine (**DPA**) (Protocol 17). Nowadays, different Sialic Acid Quantitation Kits are commercially available (e.g., Sigma-Aldrich) for rapid and accurate determination. In that case, sialic acid is released from glycoconjugates using neuraminidase (sialidase) which cleaves all sialic acid linkages, including (α2→3,6,8,9). The detailed protocols and chemicals for labeling are usually supplied with the kits.

Protocol 14. Spot Test for Sialic Acids and Sialic Acid-Containing Glycoconjugates

Materials

Silicagel 60 TLC plates (Kiesel gel 60; Merck)
Clean glass plate
Hamilton 10µL syringe
Warm/hot air blower (hair dryer)
Copper sulfate ($CuSO_4$)
Hydrochloric acid (HCl)
Resorcinol
Resorcinol/HCl/CuSO$_4$ reagent: Mix 5 mL 2% resorcinol in water with 45 mL 5 M HCl and add 125µL of 0.1 M Cu(II)SO$_4$ (freshly prepared 4 h before use)

Procedure

1. Put a spot of aqueous target solution on the TLC plate (eventually, multiple additions of small volumes in case of low concentration expected).
2. Dry the spot with a warm air blower.
3. Dip the plate quickly in resorcinol/HCl/CuSO$_4$ reagent.
4. Put the TLC plate horizontally and cover the TLC plate with a clean glass plate (to prevent fast evaporation).
5. Heat in a ventilated oven at 120 °C for 5 min.
6. Presence of sialic acid gives purple coloration.

Protocol 15. Bial's Test for Sialic Acids (Ferric-Orcinol Assay)

Materials

Visible light spectrophotometer
Cuvettes (0.5 or 1.0 mL, 1-cm path length)
Glass thick-walled Pyrex test tubes (125 × 16 mm)
Glass marbles
Hot water bath (96 °C)
Ice/water bath (0 °C)
Tabletop centrifuge
Hydrochloric acid (concentrated, 11.6 M)
Ferric chloride (FeCl$_3$)
Orcinol (5-methylresorcinol)
Isoamyl alcohol [(CH$_3$)$_2$CH-CH$_2$-CH$_2$-OH]
N-acetylneuraminic acid (Neu5Ac)
Prepare Sialic acid standard stock: 1 mM (3.09 mg/10 mL) solution of Neu5Ac in water (Store frozen).
Bial's reagent: Dissolve 200 mg of orcinol in 82 mL cooled, concentrated HCl. Add 2 mL of 1% (w/v) aqueous FeCl$_3$ solution, and make up to 100 mL with water. Stable for 1 week when stored at 4 °C.

Procedure

1. Take 200µL of target sample in glass Pyrex test tube.
2. Prepare standard samples: 5–10–20–40 nmol Neu5Ac in 200µL water and blank 200µL water.
3. Add 200µL of Bial's reagent to each tube, vortex, close tubes with glass marbles.
4. Heat for 15 min at 96 °C and then cool to RT in a cold water bath.
5. Add 1 mL of isoamyl alcohol, vortex, and keep for 5 min at 0 °C (ice bath).
6. Centrifuge tubes for 3 min at 500 ×g to separate the phases.
7. Transfer upper phase solution to the cuvette, read absorbance at 572 nm, and subtract blank.
8. Calculate concentration compared to the standard curve of Neu5Ac (blue–purple/red–violet color).

Protocol 16. Thiobarbituric Acid (TBA) Assay for Sialic Acid

Materials

Spectrophotometer
Cuvettes (1.0 mL, 1 cm)
75 × 12-mm glass tubes/glass marbles
Tabletop centrifuge
Heating block 100 °C
Sodium hydroxide (NaOH, MW 40.00, anhydrous pellets)
Concentrated sulfuric acid (H$_2$SO$_4$, MW 98)
Concentrated *ortho*phosphoric acid (H$_3$PO$_4$, MW 97)
Sodium *meta*-periodate (NaIO$_4$, MW 213.9; Sigma)

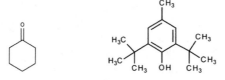

2′-thiobarbituric acid (**TBA**, 4,6-dihydroxypyrimidine-2-thiol, FW 144.15)
Potassium iodide (KI, MW 166)
Sodium arsenite (NaAsO$_2$, MW 129.9; Sigma)
Sodium sulfate (Na$_2$SO$_4$, MW 142, anhydrous; Fisher)
N-acetylneuraminic acid (Neu5Ac, MW 309.3)

Cyclohexanone   Butylated hydroxytoluene
(C$_6$H$_{10}$O, MW 98, Sigma)   (BHT, MW 220.4)

Prepare Sialic acid standard stock: 1 mM (3.09 mg/10 mL) solution of Neu5Ac in
 water (Store frozen)

Reagents
A (Periodate reagent): Dissolve 4.3 g of sodium metaperiodate in 4.0 mL water.
 Add 58 mL of concentrated *ortho*-phosphoric acid and make up to 100 mL with
 water. (Store in the dark at 4 °C).
B (Arsenite reagent): Dissolve 10 g of sodium arsenite, 7.1 g of Na$_2$SO$_4$, and 10 mg
 potassium iodide in 100 mL of 0.1 M H$_2$SO$_4$ (= 570µL conc. sulfuric acid in
 100 mL water). (Store at room temperature).

C (**TBA reagent**): Dissolve 0.8 g of 2-thiobarbituric acid (TBA) and 7.1 g of Na_2SO_4 in 100 mL water (complete dissolution is obtained by dropwise addition of 1 M NaOH solution to pH ~9). (Store in the dark at room temperature).

D (**Extraction reagent**): Cyclohexanone, containing 0.1% (w/v) Na_2SO_4

Usually, a prior mild hydrolysis is performed in 200μL of 0.1 M H_2SO_4 (containing 100μL 1% BHT in ethanol). Heat at 80 °C for 1 h and, then, cool to room temperature. The presence of BHT (butylated hydroxytoluene) minimizes lipid peroxidation and the formation of interfering malondialdehyde during hydrolysis. Ester groups can be lost.

Procedure

1. Prepare standard samples containing 5–10–20–30–40 nmol Neu5Ac in 200μL water.
2. To 200μL hydrolysate, and to each standard sample, add 100μL reagent **A** (avoid touching the sides of the tube) and mix well (vortex briefly at low speed).
3. Incubate for 20 min at 37 °C (periodate oxidation).
4. Add 800μL of reagent **B** along the side of the tube (to destroy the excess of periodate, brown color develops) [color is due to a reduction of IO_4^- by AsO_2^- to I^- via I_2].
5. Vortex vigorously for 1–2 min to expel the yellow iodine color.
6. Leave for 5 min at RT (stopping oxidation reaction).
7. Add slowly 1.5 mL reagent **C**, along the side of the tube, shake and cap the tubes with marble.
8. Incubate for 15 min at 98 °C.
9. Cool rapidly to RT (cold water bath).
10. Add 1.5 mL of reagent **D**, and vortex vigorously [sodium sulfate facilitates the extraction with cyclohexanone].
11. Centrifuge for 2 min at 1000 ×g to separate the two layers.
12. Take the upper cyclohexanone layer into a cuvette and determine the absorbance at 549 nm (pink/red color) within 2 h.
13. Determine the sialic acid concentration by comparison to a standard curve.

Protocol 17. The DPA Assay for Free and Conjugated Sialic Acid

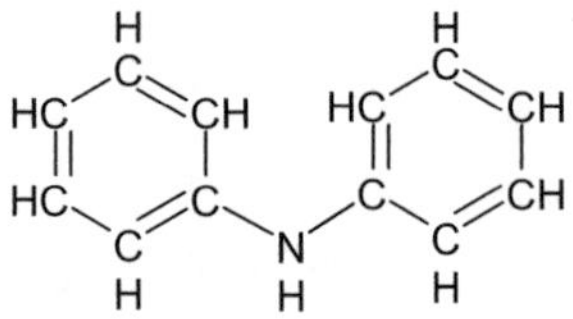

Diphenylamine (**DPA**, MW 169.22)

Reagents

Dische's reagent: Dissolve 1 g of diphenylamine (recrystallized from ethanol) in a mixture of 90 mL of glacial acetic acid and 10 mL of concentrated sulfuric acid [store at 4 °C in dark but warm at 40 °C before use to redissolve crystals]

Solution of 7.5% (w/v) trichloroacetic acid in water
Standards 100µg sialic acid / 1.0 mL water

Procedure

1. Put 100µL target sample (containing 1–20µg sialic acid) in the test tube.
2. Add 200µL of 7.5% (w/v) trichloroacetic acid (TCA) solution and mix.
3. Add 6 mL of Dische's reagent and mix.
4. Heat for 30 min in boiling water bath.
5. Cool the tubes in water to room temperature.
6. Keep tubes for 30 min at room temperature in dark (violet–blue coloration).
7. Determine the absorbance at 530 nm.
8. Calculate sialic acid concentration by comparison to the standard curve.

N.B. High concentrations of hexoses interfere with yielding blue color, glycerol produces green color, but (*N*-acetyl) hexosamines do not interfere.

4.4.6 Iodine Test for Starch

The iodine test is a method to detect polysaccharides, in particular for the detection of starch in an aqueous solution (Protocol 18). Starch is a glucose polymer consisting of α-amylose and amylopectin (see Sect. 2.3). Amylose in starch is responsible for the formation of a deep blue–black color in the presence of iodine. The amylose (linear helix chain) forms a coordinate complex with iodine centrally located within the helix due to adsorption. The color obtained depends upon the length of the unbranched linear chain. Amylopectin (branched) gives a purple color. Dextrins, being intermediates during hydrolysis of starch, like amylo-, erythro-, and achro-dextrins, give violet, red, and no color with iodine, respectively. Glycogen, comparable with amylopectin, gives a reddish-brown/violet color.

Iodine is not very soluble in water; therefore, the iodine reagent is made by dissolving iodine in water in the presence of potassium iodide. This makes a linear tri-iodide ion complex, which is soluble ($I_2 + I^- \rightarrow I_3^-$). The tri-iodide ion slips inside of the amylose coil.

Protocol 18. Iodine Test for Starch

Materials

Iodine (I_2)
Potassium iodide (KI)
Iodine solution: Dissolve 500 mg of KI in 20 mL of distilled water and add few crystals of iodine to obtain a deep orange/yellow solution.

Procedure

1. Add three drops of iodine solution to about 1 mL of the carbohydrate-containing test solution (~0.5–10 mg/mL).
2. Mix gently.
3. Heat for 1 min in a water bath at 95 °C. A blue–black color indicates the presence of starch polysaccharides. Different polysaccharides can give different color results (e.g., glycogen gives a red-brown color).

In this Chap. 4, several colorimetric detection and quantification methods for carbohydrates have been discussed. Some of them are still frequently used, for instance, in the food industry. During the past years, colorimetric assays have been more and more transferred to 96-well format (using adapted amounts of reagents) for the increase of throughput to determine different sugars in various applications. However, in the biotechnological industry, the use of these benchtop wet chemical techniques has become of limited use. Advanced analytical instruments are more easily commercially available now. In many cases, the process of separating, detecting, quantifying, and identifying individual sugars has become simpler by GLC, HPLC, and HPAEC methodologies. Even, mass spectrometers and Nuclear Magnetic Resonance (NMR) spectroscopy instruments have become regular-used equipment, often semiautomatic applied, in carbohydrate research laboratories, as we will see in the next chapters.

References

1. Masuko T, Minami A, Iwasaki N, Majima T, Nishimura S-I, Lee YC. Carbohydrate analysis by a phenol-sulfuric acid method in microplate format. Anal Biochem. 2005;339:69–72.
2. Albalasmeh AA, Berhe AA, Ghezzehei TA. A new method for rapid determination of carbohydrate and total carbon concentrations using UV spectrophotometry. Carbohydr Polym. 2013;97:253–61.
3. Laurentin A, Edwards CA. A microtiter modification of the anthrone-sulfuric acid colorimetric assay for glucose-based carbohydrates. Anal Biochem. 2003;315:143–5.
4. Sánchez-Viesca F, Gómez R. Reactivities involved in the Seliwanoff reaction. Modern Chemistry. 2018;6:1–5.
5. Rondel C, Marcato-Romain C-E, Girbal-Neuhauser E. Development and validation of a colorimetric assay for simultaneous quantification of neutral and uronic sugars. Water Res. 2013;47:2901–8.
6. Schauer R, Kamerling JP. Exploration of the sialic acid world. Adv Carb Chem Biochem. 2018;75:1–213.

Chapter 5
Analytical Techniques to Study Carbohydrates

Abstract The description of a complex carbohydrate structure includes, in the first instance, monosaccharide composition, anomeric configurations, and the type of glycosidic linkages. To obtain these data, several techniques are available. This chapter describes chemical solvolytic cleavage methods to generate monosaccharides and/or oligosaccharide fragments and discusses, in detail, the chromatographic separation methods to isolate them for further study. Several detection techniques are discussed. Electrophoretic separation of carbohydrates is briefly discussed. A typical flow chart for the structural analysis of carbohydrates from biological samples is presented. Furthermore, protocols for fluorescent or chromophore labeling of glycans are provided.

Keywords Glycomics · HPLC · HPAEC-PAD · MALDI-TOF-MS · ESI-MS · GC-MS · NMR · TLC · SEC · HILIC · PGC · CZE · Glycosidic bond · Hydrolysis · Methanolysis · Formolysis · Acetolysis · Periodate oxidation · IgG · Affinity chromatography · Lectins · Reductive amination · Michael addition · PMP labeling · Permethylation

5.1 Structural Parameters

In comparison to genomics and proteomics—where automated synthesis, amplification, expression, and characterization have become routine—automated tools available for the study of glycomics are still scarce. There is an enormous diversity of carbohydrate structures that can be created using a limited number of monosaccharide building blocks. The biosynthetic nontemplate-driven nature, as well as the frequent occurrence at low concentrations, has made system-wide glycan profiling a challenging task [1]. Setting up a scheme for glycan analysis is not straightforward. Since carbohydrates can be present in free form or in chemically bound form, the preparation of an adequate sample and the choice of the carbohydrate analysis methods depend on the nature and available amount of the target material being analyzed. For a mixture of monosaccharides in solution, the method of choice for analysis is not too complicated, but for chemically bound carbohydrates, there are several options. In that case, the first goal is to get an accurate compositional

© Springer Nature Switzerland AG 2021

G. J. Gerwig, *The Art of Carbohydrate Analysis*, Techniques in Life Science and Biomedicine for the Non-Expert, https://doi.org/10.1007/978-3-030-77791-3_5

Table 5.1 Primary structural features to be determined of complex carbohydrate/glycan

General description	Examples
1. Nature and number of constituent monosaccharides	Gal, Man, GlcNAc, Fuc (2:3:4:1)
2. Their absolute configuration (D or L)	D-Gal, D-Man, D-GlcNAc, L-Fuc
3. Their ring size [pyranose (*p*) or furanose (*f*)]	Man*p*, Gal*f*
4. Anomeric configuration of glycosidic linkages	α, β
5. Position and type of glycosidic linkages	$(1{\to}2)$, $(1{\to}3)$, $(1{\to}4)$, $(1{\to}6)$
6. Sequence of monosaccharides, including branching topology	Man($\alpha1{\to}6$)[Man($\alpha1{\to}3$)] Man($\beta1{\to}4$)GlcNAc
7. Type of carbohydrate–peptide linkage	N-linked to Asn, O-linked to Ser/Thr
8. Nature, number, and location of appended noncarbohydrate groups: phosphate, sulfate, acetate	Glc6P, IdoA2S, 9*O*Ac-Neu5Ac

analysis in terms of the molar ratio of the various monosaccharide constituents, for instance, in a glycoconjugate or in a specific polysaccharide. With this information, the approach to a detailed structural characterization can be decided. The ideal analytical method for a complex glycoconjugate would comprise that (1) all types of glycans can be detected and quantified, (2) isomers can be separated, and (3) the components can be assigned a particular structure, including the overall topology of the molecule and all linkages. Several parameters, as listed in Table 5.1, have to be investigated leading to the final primary chemical structure of a complex carbohydrate.

The characterization of the chemical structure of oligosaccharides on glycoproteins, for instance, can be generally undertaken as:

1. analysis of still-attached glycans on intact glycoproteins
2. analysis of glycopeptides (glycans attached to the peptides after protease digestion) or
3. analysis of free glycans (oligosaccharides, chemically or enzymatically cleaved from the proteins/peptides).

The most common procedure is the analysis of fragments obtained by chemical and/or enzymatic degradation. Oligosaccharides are generally removed from their protein or lipid conjugates and then subjected to analytical methods. The bottleneck for biological samples is the small amount, often yielding femto- to picomoles of individual components. A summarized overview of methodologies that are currently in use in glycoconjugate glycan and/or polysaccharide analysis is presented in Table 5.2. It is important to note that the characterization of a complete structure of a glycan using enzymes and chromatography involves meticulous and time-consuming efforts.

Table 5.2 Commonly used techniques/methods for obtaining information about specific carbohydrate features of glycoconjugate glycans and/or polysaccharides

Information	Methods
Carbohydrate content, type, and quantity of constituent monosaccharides, D/L-configuration	Colorimetric assays, monosaccharide analysis by GLC, HPAEC, or HPLC, capillary electrophoresis (CE), GLC-absolute configuration determination, NMR spectroscopy
Nature of glycan–peptide linkage (N or O)	Amino acid analysis, proteolytic digestion, enzymatic glycan release (N), alkaline glycan release (O), hydrazinolysis (N/O)
Type of glycans (high mannose, complex, hybrid)	GLC/HPAEC monosaccharide analysis, size/charge profile analysis by HPLC, CE, LC-MS profiling
Number/proportions of glycans present	Size/charge profile analysis, mapping by HPLC (UV/ fluorescent derivatives), HPAEC-PAD, MALDI-MS
Sequence of monosaccharide residues	Exoglycosidase digestions, partial hydrolysis, periodate oxidation, LC-MS (tandem MS), NMR
Positions of glycosidic linkages	Methylation analysis, GLC-EIMS, HPLC-MS, NMR
Anomeric configurations	Exoglycosidase digestions, NMR
Certain structural determinants	Lectin microarrays, endo/exoglycosidase digestion, lectin affinity chromatography, antibody responses
Type of (charged) substituents	Size/charge profile analysis by HPLC, HPAEC, NMR
Spatial structure of the glycan	X-ray analysis, 2D/3D NMR, molecular dynamics (MD), mechanics and modeling

All methods in Table 5.2 may provide limited partial structural information. There is no general method available that can characterize the complete glycan structure with their oligosaccharide sequence and linkage information at once. It is therefore essential to apply several methods to measure individual parameters, such as glycosylation site analysis, oligosaccharide sequence, and monosaccharide content. A combination of the various analytical methods allows the complete glycan architecture. Even classic techniques, such as refractometry, density measurement, and polarimetry, are in use to obtain confirmatory data.

Modern separation techniques (see Sect. 5.3) are centered around high-performance liquid chromatography (HPLC) of UV- or fluorescently labeled and native glycans or glycopeptides. The chromatography is often conjugated online with mass spectrometry (LC-MS/MS) The analytical techniques include chromatographic separations (e.g., SEC, HPLC, and CE), mass spectrometry (e.g., MALDI-TOF MS, ESI-MS), NMR spectroscopy, and lectin arrays [2]. However, high-throughput methods are still scarce [3]. Sequential degradation with (linkage-specific) exoglycosidases affords information about the monosaccharide sequences and linkage types (see Sect. 8.2.3). Lectin affinity studies (see Sect. 5.3.7) and immunological reactivity studies can contribute to the structural characterization of a glycan. A typical flow chart, illustrating the strategy for the compositional and structural analysis of a carbohydrate-containing compound, is shown in Fig. 5.1.

Due to the complexity of a glycan structure, structural elucidation is usually performed after the release of the glycans. The strategies for the analysis of glycans

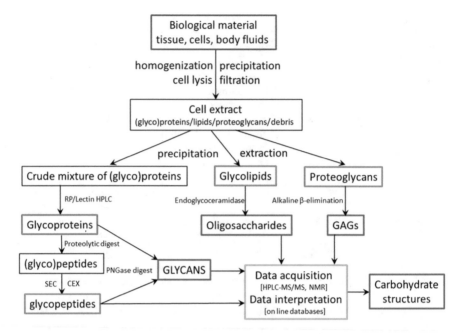

Fig. 5.1 A typical flowchart for the structural analysis of carbohydrates from biological samples

have their own strong and weak points, and appropriate methods should be selected for each glycan sample. Advanced physical techniques, like mass spectrometry and NMR spectroscopy, are generally used for data acquisition leading to the definitive identification and detailed structural elucidation of the glycans.

For studying the glycans, a possible workflow can be summarized as follows:

1. release glycans from the protein/lipid,
2. labeling of the free glycans for detection,
3. separation of the complex pool of glycans/oligosaccharides, and
4. assignment of structures to the peaks.

Ultimately, the primary structure of a glycan is defined by the nature and order of constituent monosaccharide residues, by the configuration and position of glycosidic linkages, and by the nature and location of non-glycan entities to which they are attached (aglycones). Clearly, a comprehensive elucidation of intact glycoform structures and their biological significance requires the involvement of the application of several complementary technologies. In view of the concept of this book and space limitations, it is not possible to discuss all existing analytical methods in detail. This means that in the following chapters and sections, selections will be made according to the relevance and application frequency of the methods. An effort is made to describe the methods in a comprehensible way for the nonexpert in carbohydrate chemistry. References are made to recent scientific papers and books on carbohydrate research to encourage further study.

When the presence of carbohydrates, for instance in a protein or lipid, is confirmed, the next step is to find out the composition of the glycans by a so-called monosaccharide analysis. In principle, such a sugar analysis consists of three steps: (1) release of the constituent monosaccharides, (2) separation of the mixture of monosaccharides, and (3) identification and quantification of the individual monosaccharides.

5.2   Glycosidic-Bond Cleavage Techniques

After isolation and purification of the desired target material (e.g., glycoprotein/ peptide, glycolipid or polysaccharide, etc.), to determine the monosaccharide constituents, a solvolytic cleavage of all glycosidic linkages must be performed, yielding a pool of soluble monosaccharides [4]. The choice of method to release the monosaccharides is directed by the sample species and the method of subsequent separation/analysis (e.g., HPLC, CE or GLC) and the technique of identification (e.g., MS or NMR).

When using acidic hydrolysis for the liberation of the constituent monosaccharides, a delicate balance between the risk of incomplete hydrolysis and partial destruction of the product must be maintained. The hydrolysis conditions depend on the structural properties of the carbohydrates analyzed. All glycosidic bonds must be cleaved while avoiding the destruction of the liberated monosaccharides. It is advised to keep the concentration of carbohydrates relatively low during acid hydrolysis (<1 mg/mL). The susceptibility to acid hydrolysis of the glycosidic bonds, as well as the stability of the released unit, is different for each monosaccharide. Note that the most fragile sugars are pentoses and deoxy sugars. Incomplete hydrolysis often occurs for aminosugars and aldonic and uronic acids. Furthermore, pyranosides and furanosides hydrolyze at different rates (f more easily than p), and also, α and β anomers have different rates of hydrolysis (α more easily than β). It seems that (1→3)-linkages are more resistant to acid than (1→2)- and (1→4)-linkages. Moreover, the presence of substituents affects the rate of hydrolysis, and substituents may be lost during hydrolysis (e.g., de-N-acetylation). In some cases, more than one set of hydrolytic conditions, experimentally determined, may be required to achieve a complete analysis. This is of particular concern to get complete hydrolysis of polysaccharides containing uronic acids. Frequently, reduction of the carboxylic groups of the uronic acids giving their corresponding hexoses is carried out before hydrolysis to allow the complete liberation of monosaccharides. It has to be noted that the acidic conditions, used for cleavage, yield a mixture of α,β-pyranose and α,β-furanose ring forms for each monosaccharide.

Several types of acid, such as sulfuric acid (H_2SO_4), hydrochloric acid (HCl), and trifluoroacetic acid (TFA) are used under different conditions (concentration, temperature, time). Furthermore, two special methods, formolysis and acetolysis, are occasionally used to cleave glycosidic linkages, during structural analysis of polysaccharides, in particular.

5.2.1   Sulfuric Acid Hydrolysis

Sulfuric acid (H_2SO_4) hydrolysis (0.1–2 N, 2–12 h at 100 °C) is not that often used for hydrolysis of glycoconjugates. For neutral polysaccharides, using 1 N H_2SO_4 for maximum 4 h at 100 °C is a good option. The drawback is the laborious neutralization of the acid with barium carbonate after hydrolysis and removal of the precipitate, which may absorb some carbohydrates.

5.2.2   Hydrochloric Acid Hydrolysis

Hydrochloric acid (HCl) is used more often for hydrolysis of glycoconjugates, and the conditions range from 0.5 M to 6 M, 1–12 h at 100 °C; however, concentrations >2 M with reaction times >3 h significantly increase the decomposition of sugars. The hydrolysis must be performed under nitrogen because oxygen stimulates the decomposition of sugars. Neutralization of HCl after hydrolysis before lyophilization will reduce decomposition. Markedly, milder conditions [0.01–0.05 M HCl, 0.5–1 h, 90 °C] are used to release terminal fucose or sialic acids from glycoconjugates. Hydrochloric acid hydrolysis is often used to prepare the glucose α-oligomer standard mixture (DP 1–30), used in glycan structure determination (GU values) by HPLC (see Sect. 8.2.2).

5.2.3   Trifluoroacetic Acid Hydrolysis

An elegant method seems to be hydrolysis with TFA. But also, different conditions must be tested, ranging from 4 M TFA for 1–4 h at 100–125 °C to 2 M TFA for 1–2 h at 120 °C, depending on the type of glycoconjugates and polysaccharides. For example, fructofuranosyl linkages of fructans (polysaccharides consisting of fructose residues) are acid labile and are hydrolyzed already with 1 M TFA within 30 min at 70 °C. For sugar analysis of an unknown glycoprotein, the following starting hydrolysis conditions could be used: 2–4 M TFA, 3–6 h at 100 °C, under nitrogen, in glass tubes fitted with Teflon-lined screw caps.

It has to be mentioned that hydrolysis with TFA can result in the epimerization of mannose to glucose, and high TFA concentrations will destroy sialic acids. Furthermore, incomplete release of 2-acetamide sugars and glycosyluronic acids can occur (e.g., for glycosaminoglycans). Release of only sialic acid from glycans can be obtained by hydrolysis with 0.05 M or 0.1 M TFA at 80°C for 1–2 h. TFA has the advantage of being easily eliminated by evaporation with a stream of dry nitrogen. De-N-acetylation during acidic hydrolysis makes a re-N-acetylation of aminosugars necessary. The next conditions [0.15 N TFA, 1 h, 100 °C or 2 M TFA,

30 min, 60 °C] are required for hydrolysis of ketoses, e.g., fructose polysaccharides (inulins, levans).

For monosaccharide analysis of polysaccharides, the following hydrolysis conditions are mostly used: 2–4 M TFA, 100–121 °C, 1–18 h. For all solvolysis, re-*N*-acetylation of aminosugars has to be performed.

5.2.4   Methanolysis

Instead of cleaving the glycosidic linkages with an acid catalyst in an aqueous solution, acidified methanol can be used, leading to methanolysis. Methanolysis is less destructive than aqueous hydrolysis but also needs a re-*N*-acetylation step. Sialic acids are not destroyed. Methanolysis is usually performed with anhydrous methanolic hydrochloric acid (1 N methanolic HCl, 16 h, 85 °C), under nitrogen atmosphere, giving methyl glycosides (and methyl ester methyl glycosides in the case of uronic acids), having α,β-pyranose and α,β-furanose ring forms for each monosaccharide in a fixed ratio (see Sect. 6.2).

5.2.5   Formolysis

This hydrolysis technique is regularly applied for polysaccharides, which are not soluble in hot water. They can be treated for a short period with formic acid (90% formic acid, 2 h at 100 °C) as a first step. Cleavage of some glycosidic linkages will reduce the size of the polysaccharide, which increases solubility. The formic acid can be removed by evaporation.

5.2.6   Acetolysis

Acetolysis is a useful supplementary method for structural studies of polysaccharides to obtain specific oligosaccharides. Acetolysis involves complete acetylation of the free hydroxyl groups of the polysaccharide, followed by selective cleavage of glycosidic bonds. Acetolysis can selectively cleave (1→6) glycosidic linkages, leaving mostly intact other linkages (1→2, 1→3, 1→4). Glycosidic linkages in an α-configuration are more susceptible to acetolysis than those in a β-configuration. As such, the reaction conditions determine the fragments formed. After de-O-acetylation, the resulting oligosaccharide fragments are usually isolated by SEC or HPLC and used for further analysis.

The acetolysis method is performed as follows:

1. Put 100μg dried polysaccharide in a conical, Teflon-lined, screw-capped Pyrex test tube.
2. Add 200μL acetic anhydride/glacial acetic acid/conc. H_2SO_4 (10:10:1, v/v), close tube.
3. Incubate 3–12 h at 35–40 °C (obtain optimal conditions experimentally).
4. Add 50μL pyridine, mix well (neutralization).
5. Evaporate under stream of nitrogen at 40 °C.
6. Repeat evaporation 2× with 100μL of toluene till dryness.
7. Add 500μL methanol, containing 0.1% sodium methoxide, mix well.
8. Incubate 20 min at room temperature (de-O-acetylation).
9. Adjust the pH to 8 by dropwise addition of 1 M acetic acid.
10. Evaporate 5× with 250μL methanol till dryness.
11. Separate/isolate acetolysis fragments by SEC or HPLC.

5.2.7   Specific Degradations

In special cases, more selective chemical degradation procedures are used to pre-pare fragments from oligosaccharides and, in particular, from polysaccharides. When analyzing structurally complex polysaccharides, next to partial acid hydroly-sis, sometimes additional techniques to partially degrade the polysaccharide into more accessible fragments for analysis are needed. These methods can include par-tial alkaline hydrolysis, uronic acid degradation, periodate oxidation (Smith) degra-dation, as well as enzymatic degradations [5]. Since these specialized techniques require some expertise and proficiency from the analyst, the methods will only be discussed briefly here.

Partial degradation of polysaccharides by acid hydrolysis is based on the fact that some glycosidic linkages are more labile to acid than others. As mentioned earlier, furanose and deoxy sugars are easily hydrolyzed by acid. The glycosidic linkages of uronic acids and 2-amino-2 deoxyhexoses are highly resistant to acid hydrolysis.

A Smith-degradation involves periodate oxidation with a sodium periodate ($NaIO_4$) reagent (at 4 °C), followed by reduction with borohydride ($NaBH_4$), and mild acid hydrolysis of the oligo/polysaccharide. The principle comes to that sugar residues containing carbons with free hydroxyl groups in vicinal position are cleaved quantitatively between the involved carbons to give dialdehydes, as shown below. So, the bonds between C2 and C3 and C4 are broken in an unsubstituted monosaccharide. The aldehydes are subsequently reduced to the corresponding alcohols. However, sugars that are, for instance, 3-O-substituted or 2,4-di-O-substituted do not contain free vicinal hydroxyl groups and are not subject to peri-odate oxidation.

The determination of the composition of oligosaccharide fragments created by the Smith degradation can give important information about an oligo/polysaccharide structure. The experiment must be carried out in the dark to avoid unspecific oxidation. Oligosaccharides are isolated by SEC.

Oligosaccharides can also be created by enzymatic cleavages. Commercially available endo-enzymes, specific for certain linkage and sugar types, are useful to release oligosaccharides from polysaccharides. For instance, lichenase for β-glucans or alginase for the study of alginates.

5.3   Common Separation Techniques for Mono/Oligosaccharides

Carbohydrate characterization exploits various methods of separation techniques. High resolution is desired to separate underivatized or derivatized monosaccharides and oligosaccharides either as analytical profiling or as a preparatory tool. Chromatographic methods have evolved considerably over the past years, and several techniques are available (Table 5.3).

Table 5.3 A selection of chromatographic separation methods used in carbohydrate research

Method	Abbreviation
(High-performance) thin-layer chromatography	(HP)TLC
Size-exclusion chromatography	SEC
Gel-permeation chromatography	GPC
High-performance liquid chromatography	HPLC
Ultrahigh-performance liquid chromatography	UHPLC
High-pH anion-exchange chromatography	HPAEC
Strong or weak anion-exchange chromatography	SAX or WAX
Reversed-phase chromatography (HPLC)	RPC
Straight-phase chromatography (HPLC)	SPC
Reversed-phase ion-pairing chromatography	RP-IPC
Porous graphitized-carbon chromatography	PGC
Hydrophilic-interaction liquid chromatography	HILIC
Enhanced fluidity liquid chromatography	EFLC
Lectin affinity chromatography	LAC
Gas–liquid chromatography	GLC

A combination of separation techniques will probably be necessary to isolate pure oligosaccharide isomers from an oligosaccharide library. Through the use of sometimes different chromatographic methods, it is possible to separate mixtures into their component sugars, identify each component by retention time (usually compared to standards), and provide a determination of the relative quantities of each component (using peak areas). Chromatographically separated and isolated peak fractions can be further analyzed by chemical (e.g., monosaccharide/methylation analysis) and physical methods (e.g., mass spectrometry and NMR spectroscopy).

5.3.1   Thin-Layer Chromatography

Traditional thin-layer chromatography (TLC), nowadays as high-performance thin-layer chromatography (HPTLC), is a simple, rapid technique with low cost, easy maintenance, and good selectivity of detection to analyze mono-, di-, and oligosaccharide mixtures although essentially qualitative [6, 7]. Silica-Gel 60 TLC plates (Kieselgel G) allow the separation by using different solvent systems. Different precoated TLC plates are commercially available. TLC is an easy method to monitor the progress of any reaction with carbohydrates, for instance, during hydrolysis or enzymatic activity experiments or extractions. Several carbohydrate samples can be analyzed in parallel in one run. Samples are applied to the TLC plate as short streaks (1 cm) using a drawn-out capillary glass tube or syringe. Depending on the concentration, streaks can be piled up with intermediate drying.

The choice of a suitable solvent system is often trial and error, but the following are a good start. For instance, n-butanol/acetic acid/water (2:1:1 v/v) or n-butanol/ethanol/acetic acid/pyridine/water (1:10:0.3:1:3 v/v) or butanol/ethanol/water (5:3:2 v/v) can be used for separation of oligosaccharides (DP 1–10) on silica-gel TLC plates (Merck SiO_2/G plates).

For fructo-oligosaccharides (DP 2–6), chloroform/acetic acid/water (6:7:1 v/v) can be used. Cellulose-coated TLC plates are generally used for the analysis of monosaccharides. Here, a typically used solvent system for monosaccharides is butanol/pyridine/0.1 M HCl (5:3:2 v/v). Sometimes, two or three successive developments (ascents) with intermediate drying are performed in order to improve resolution. The final positions of the bands, compared to pure standard sugars, usually permit identification.

For visualization of the sugar positions after development, a staining method is necessary. Before staining, make sure that the TLC plate is completely dry. The simplest staining is a nonselective charring technique by heating (2–5 min, 120 °C) of the TLC plate after wetting it with a reagent consisting of ethanol/conc. H_2SO_4 (9:1 v/v). Nowadays, spraying with staining reagents is replaced by quickly dipping the developed and well-dried TLC plate in the staining solution (see Sect. 4.3.1), followed by heating in a horizontal position.

Another, often used, staining reagent for carbohydrates on silica TLC plates is the naphthoresorcinol/ethanol/sulfuric acid solution. The procedure is as follows:

1. Prepare solution **A** by dissolving 0.2 g of naphthoresorcinol (naphthalene-1,3-diol) and 0.4 g diphenylamine in 100 mL of 95% ethanol
2. Prepare reagent **B**, by carefully adding 4 mL of concentrated sulfuric acid to 96 mL of solution **A,** just prior to use
3. Dip the TLC plate in reagent **B** and then heat the plate for 5 min at 100–150 °C. In this case, different sugars can give different colors, e.g., aldoses produce blue or violet bands, ketoses produce pink or red bands, and uronic acids produce blue bands.

Nowadays, there is growing interest in the coupling of HPTLC with mass spectrometry [8, 9].

5.3.2 Size-Exclusion Chromatography

The traditional low-pressure liquid chromatography in the form of size-exclusion chromatography (SEC) or gel-permeation chromatography (GPC) is still used for preparative purposes to separate oligosaccharide mixtures. It is also a popular approach for the determination of molecular weight (M_w) and distribution of oligo/polysaccharides. SEC coupled with multi-angle laser-light scattering (MALLS) detection is a powerful technique to investigate macromolecules.

SEC separates molecules based on their hydrodynamic volume (effective size). Aqueous eluents containing a low salt concentration (e.g., 0.05 M NaCl) are used to prevent adsorption phenomena. The chromatography column contains a tridimensional cross-linked gel (molecular sieve gel) based on dextran polysaccharide (Superdex, Sephadex) or polyacrylamide (Bio-Gel). Small-molecular-size (low-molecular weight) carbohydrate molecules are retarded as they are traveling through the pores of the gel, while high-molecular size molecules are excluded from penetrating the gel pores and elute with the void volume of the column. Molecules of intermediate size can penetrate some of the pores of the gel matrix. By choosing the degree of cross-linking pores (e.g., Biogel P-2, P-4, P6 and P-10 [e.g., column size 100 × 1 cm]), separations of oligosaccharides ranging DP2-DP20 can be achieved. SEC on BioGel P-4 columns is very suitable to separate neutral oligosaccharides and cleaved N- and O-glycans from glycoproteins. Furthermore, small Biogel P-2 (200–400 mesh) columns [15 × 1 cm] can be used for desalting carbohydrate samples (not monosaccharides). The elution can be monitored by refractive index (RI) detection. Although native sugars do not contain a chromophore, detection by UV absorbance (around 195 nm) is possible but insensitive. Consequently, detection is frequently performed manually by colorimetric/spot tests (see Chap. 4) on eluate fractions.

5.3.3 High-Performance Liquid Chromatography

High-performance liquid chromatography (HPLC), now well developed to ultrahigh-performance liquid chromatography (UHPLC), is the most used technique to separate carbohydrates, including monosaccharides and oligosaccharides (glycans), but also glycopeptides. The separation is based on physicochemical parameters such as hydrophilicity, hydrophobicity, or charge. Nowadays, HPLC equipped with a UV detector is a common analytical tool in laboratories worldwide. By reducing the stationary-phase particles from 3.5–5.0μm to <2μm, UHPLC significantly improves the resolution in relatively shorter analysis times (however, needing higher pressure, >600 bar). As such, it is used as a quantitative and qualitative analysis method [10]. Over 50 different commercial HPLC columns are ready-packed available, with instruction protocols, from many different companies, divided in straight-phase, normal-phase (NP/HILIC), reversed-phase (RP), porous graphitized carbon (PGC), size-exclusion (SEC), and ion-exchange types (IC), all having several typical names. To find a suitable column, a search in recent literature is advised and to consult the advertisements from column suppliers for appropriate application [11, 12].

Reversed-phase columns (RP-HPLC), containing alkylated (amino-bonded) silica-based packing materials as nonpolar, hydrophobic stationary phase, are regularly used for bottom-up peptide mapping and are still widely used to separate derivatized oligosaccharides. The columns are relatively cheap and available from various suppliers. Glycopeptide studies are frequently carried out by RP-HPLC. However, for glycans being highly polar, often total derivatization (methylation or acetylation) must be used to improve separation ability and detection sensitivity [13–15]. C18 (octadecyl-silyl silica) or C8 (octa-silyl silica) reversed-phase columns separate glycans based on hydrophobicity in the order of polar to nonpolar and can result in isomer separation. The principle involves the application of derivatized glycans in water to the column and elution with increasing concentrations of organic solvent [e.g., 0–50% acetonitrile (v/v), containing 50 mM formate, adjusted to pH 5 with triethylamine]. To minimize unwanted sample adsorption, in the case of large molecule separation, higher than ambient column temperatures (up to 80 °C) can be employed. However, acidic conditions with increased temperature dramatically influence the risk of sialic acid loss.

Alternatively, carbohydrates, usually derivatized with an aromatic tag rendering them hydrophobic, interact with the hydrophobic alkyl groups on the stationary phase of the column and are gradient eluted as the polarity of the mobile phase (H_2O) decreases by increasing concentrations of organic solvent (acetonitrile). The separation is based on hydrophobicity, the more hydrophobic the sugar is the longer it remains in interaction on the column, increasing its elution time. An increase in the number of polar glycan units reduces retention in RP-LC, thus, polar glycans are eluted earlier than less polar glycans. Elution typically consists of a gradient of increasing acetonitrile concentration in water. The use of relatively weak organic elution enhances the compatibility of this separation mode with MS. Permethylation of glycans increases the hydrophobicity and thus are more readily separated by RP-LC. Since HPLC is often used in conjunction with mass spectrometry to give more detailed structural information, this is an added advantage [16–18]. Recently, new modifications for rapid permethylation were introduced, in particular, aimed at the derivatization of glycans for HPLC-MS/MS studies [19–21]. Principle and permethylation methods will be discussed later.

RP-HPLC of fluorescent-labeled glycans can be combined with ESI-MS detection. Often used labels are pyridylamino (2-PA, 2-aminopyridine), 2-aminobenzamide (2-AB), or 2-anthralic acid (2-AA, 2-aminobenzoic acid), introducing enhanced hydrophobicity necessary for RP retention (see Sect. 5.4.2). However, due to the relatively early elution of acidic glycans, RP-HPLC is not very suitable for the quantification of sialic acid-containing glycans. On the other hand, released monosaccharides, derivatized with 2-AB or 2-AA, are frequently analyzed by RP-HPLC.

A different strategy for RP-HPLC analysis of charged carbohydrates is *ion-pairing* [22]. Reversed-phase ion-pairing chromatography (RP-IPC) is a technique in which charged additives [e.g., diethylamine (DEA), tetra-*n*-butyl ammonium chloride, or TFA] are used as ion-pairing agents to enhance volatility and to increase the retention of oppositely charged analytes to the reversed phase. This technique is used for separating carbohydrates with negatively charged groups, such as sialic acid, phosphate, and sulfate, in particular with glycosaminoglycans (GAGs) [22, 23]. However, ion-pairing agents like amines cause ion suppression leading to decreased MS sensitivity [24].

Normal-phase columns (NP-HPLC) contain a polar stationary phase having primary, secondary, or quaternary amines (amino/amide-bonded C18/C8 silica). NP-HPLC is a useful technique for profiling both N- and O-glycan pools. Operating at room temperature, they separate carbohydrates based on polar interactions (adsorption and hydrogen-bond interactions) and are often used for the separation of monosaccharides and glycoprotein-derived glycans, eluting in the order of nonpolar to polar and with increasing retention time for bi-, tri-, and tetra-antennary N-glycans. Glycans are usually prederivatized. Samples are applied in an organic solvent solution to the column. Elution is mostly performed with increasing water gradient in acetonitrile, containing, for instance, 15 mM KH_2PO_4. In general, always filter and degas all HPLC solvents and buffers on the day of use.

Hydrophilic interaction liquid chromatography (HILIC) is also a form of normal-phase chromatography, and it is easily coupled with MS. The separation potential for (acidic) oligosaccharides (often released glycans) observed with HILIC is better than with other techniques, such as RP and porous graphitized carbon (PGC) chromatography. Nowadays, HILIC as ultra-performance liquid chromatography (HILIC-UPLC) is a favored separation method (for very polar compounds). Most glycans are highly hydrophilic and polar substances. The principle of HILIC involves the interaction of the charged silanol groups on the amide column with the hydroxyl groups on the carbohydrates. Injected in a highly organic phase (e.g., ACN), the neutral glycans interact with the stationary phase via hydrogen bonding and are retained; subsequently, the sugars elute from a HILIC column in order of increasing polarity (increasing water concentration). Thus, the separation is based on the number of polar groups and their configuration as well as the size of the gly-can, bringing about elution in the order of hydrophilicity and increased size. In other words, retention derives from a combination of partitioning processes and electro-static interactions (ion exchange and hydrogen bonding) between the analytes and the surface of the hydrophilic stationary phase [25]. This separation technique (pos-sible in femtomole range) is popular in the analysis of fluorescent-labeled (e.g., 2-AB) glycans, glycopeptides from nonglycopeptides, and glycoproteins due to its high throughput and automation. HILIC separates neutral and sialylated glycans in one chromatographic run. At present, HILIC with fluorescence detection (FLD) has been well recognized as a standard method for the separation and quantification of released N-glycans. As an example, the HILIC separation (TSK-Amide-80 column) of N-glycans released from human IgG and labeled with 2-aminobenzamide (2-AB) is shown in Fig. 5.2 and peak assignment in Table 5.4. Glycans with ($\alpha2{\rightarrow}3$) sialylation have shorter HILIC retention times than isomeric N-glycans with ($\alpha2{\rightarrow}6$) sialic acid linkages. The stoichiometric attachment of one label per glycan allows the relative quantification of the different glycan species.

Charged and uncharged carbohydrates can be separated by comparable, highly sensitive, reproducible, and robust methods [15, 24–31]. The retention times of the glycans are compared to an external 2-AB-labeled dextran ladder standard (glucose units, GUs) using well-established databases for structural assignments (see Sect. 8.2.2 and Chap. 13) [32, 33]. Chip-based HILIC techniques have been developed [34]. For a highly informative review on HILIC, see [35]. A disadvantage is, of course, the often use of environmental polluting solvents (e.g., acetonitrile).

Recently, in an attempt to fulfill the demand for using greener solvents in carbo-hydrate chemistry, a new kind of chromatography based on HILIC, called Enhanced Fluidity Liquid Chromatography (EFLC-HILIC), was used for oligosaccharide

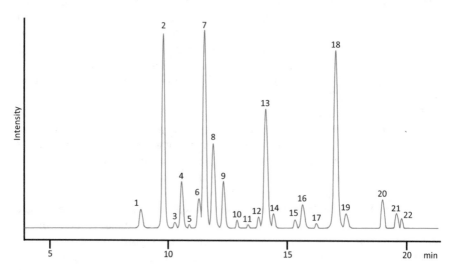

Fig. 5.2 The HILIC-FLD profile of 2-AB-labeled N-glycans released from human IgG. Fluorescence detection at 330 nm excitation and 420 nm emission

Table 5.4 Assignment of IgG glycan structures to peaks in the HILIC profile of Fig. 5.2

Peak number	Structure	Peak number	Structure
1	(glycan structure, 2AB)	12	(glycan structure, 2AB)
2	(glycan structure, 2AB)	13	(glycan structure, 2AB)
3	N.D.	14	(glycan structure, 2AB)
4	(glycan structure, 2AB)	15	N.D.
5	N.D.	16	(glycan structure, 2AB)
6	(glycan structure, 2AB)	17	(glycan structure, 2AB)
7	(glycan structure, 2AB)	18	(glycan structure, 2AB)
8	(glycan structure, 2AB)	19	(glycan structure, 2AB)
9	(glycan structure, 2AB)	20	(glycan structure, 2AB)
10	(glycan structure, 2AB)	21	(glycan structure, 2AB)
11	N.D.	22	(glycan structure, 2AB)

Symbol assignment: ○ Man, ○ Gal, ■ GlcNAc, ▼ Fuc, ◆ Neu5Ac
N.D. not determined

separation [36]. A significant proportion (30–40%) of a liquefied gas, most commonly carbon dioxide (CO_2), is dissolved in the mobile phase (usually MeOH/H_2O), allowing a wider solubility range for larger polar carbohydrates (more hydroxyl groups). The addition of liquid CO_2 enhances diffusivity and decreases viscosity while maintaining mixture polarity, which typically results in a reduced time of analysis and higher efficiency. The quality of separations of small oligosaccharides (DP 1–7) using MeOH/H_2O/CO_2 was comparable with separations using the more environmentally polluting ACN/H_2O solvents. Fructo-oligosaccharides (DP 10–40) could also be separated.

5.3.4 Porous Graphitized Carbon Chromatography

Porous graphitized carbon chromatography (PGC) can be used for fractionation of (underivatized) N- and O-glycans, oligosaccharide alditols, glycopeptides, and glycoproteins. As a solid-phase extraction (SPE) application, carbon columns are used for desalting carbohydrate samples, but it has to be noted that PGC-SPE columns have almost no retaining capacity for monosaccharides and most disaccharides. However, oligosaccharides can easily be desalted on a nonporous graphitized carbon cartridge, washed with 80% acetonitrile and 0.1% (v/v) TFA in water. The oligosaccharide sample is loaded and washed with five column volumes of water. The oligosaccharides are eluted with 40% acetonitrile with 0.05% (v/v) TFA.

The most interesting feature of the PGC technique for oligosaccharide analysis is its efficacy in separating structural isomers (isobaric structures). A PGC-HPLC instrument online connected with a mass spectrometer makes it a powerful carbohydrate analytical system, especially with the MS^n technique [37–41]. Retention times in combination with glycan mass analyses are often sufficient for structural assignment of glycans, aided by comparison to a hydrolyzed dextran ladder as a standard [42].

Nowadays, a commonly used separation method for native glycan analysis at the isomeric level is graphitic carbon chromatography. The combination of hydrophobic interaction, ionic interaction, and polarization interaction means that retention of glycans is greatly dependent on size, monosaccharide composition, and linkages. That makes carbon separation unique and powerful. Porous graphitic carbon (PGC)-based HPLC is used to study N-glycan profiling [43]. It seems that PGC is the most potent stationary phase for isomeric separation of derivatized glycans [44]. However, it has to be noted that PGC columns suffer from relatively low reproducibility due to the fouling on the stationary surface, but this issue can be addressed by regular regeneration of columns.

PGC can separate native and reductively aminated glycans. Separation occurs by an adsorption mechanism in which planar molecules exhibit more retention than nonplanar ones [45]. Oligosaccharides undergo both hydrophobic interactions with graphitized carbon, although anionic interactions also contribute to retention. To date, the exact interactions are still not well understood. Also, the retention/resolution of glycans on PGC increases at higher temperatures. For an excellent review on PGC chromatography, see [46]. The method is robust, using increasing organic solvent concentrations (water/acetonitrile) without salts, allowing for use over a broad pH range. Sialylated species are retained more at acidic pH, and, furthermore, ($\alpha2\rightarrow3$)-linked isomers of sialylated N-glycans elute later compared to ($\alpha2\rightarrow6$)-linked isomers. Free and reduced oligosaccharides (mostly used) are strongly retained by graphitized carbon stationary phases, but the monosaccharide composition has a significant influence on retention times. For instance, it is important to note that 3-fucosyllactose (3′-FL) compared to 2-fucosyllactose (2′-FL) is not retained during PGC-SPE, which could explain the absence of 3′-FL in some milk oligosaccharide (MOS) studies [47]. Furthermore, N-glycans carrying a core fucose unit show generally a significantly increased retention time on porous graphitized carbon.

Different PGC columns are commercially available. For instance, HPLC Hypercarb® 3–5µm particles graphitized carbon is mostly used with column sizes of 0.2×75 mm, 0.18×100 mm, or 0.32×150 mm (Thermo Fisher Scientific) and often directly connected to ESI-MS. Mobile phases used for separation are composed of 5–10 mM ammonium bicarbonate with increasing acetonitrile concentration. Nowadays, microsystems have been developed. HPLC chips are microfluidics-based technologies for LC-MS, especially suited to samples, which are only available in extremely small amounts [48, 49]. Agilent Technologies (Santa Clara, CA) and SGE (Ringwood, Australia) both have introduced HPLC chips consisting of a separation column in graphitized carbon, designed specifically for oligosaccharide application. A nanoLC(PGC)-chip system was used for serum N-glycan separation [50]. PGC LC-MS was successfully used to analyze heparin and heparan sulfate oligosaccharides [41] and human milk oligosaccharides (HMOs) [51, 52]. The broad application of PGC is clearly demonstrated for the separation of permethylated O-glycans, free oligosaccharides, and glycosphingolipid glycans [53]. An inherent limitation of a PGC column is its relatively high cost and the requirement of a long equilibration time and relatively low retention reproducibility due to the column's high susceptibility to contamination. So, samples may not be too dirty and must contain an internal standard in the case of quantifications.

5.3.5   Anion/Cation-Exchange Chromatography

Low-pressure anion- and cation-exchange columns have, nowadays, been replaced by HPLC methods, offering faster separation modes and advanced automatization. Cation-exchange chromatography, sometimes used for desalting, is rarely used for analytical separation of oligosaccharides due to several drawbacks such as fast efficiency loss, long analysis times, operation at high column temperature, and need for specialized column regeneration.

In contrast, weak anion-exchange (WAX) chromatography is often used for the analysis of glycans labeled with PA or 2-AB [54]. At neutral or slightly alkaline pH, the negative charge of most glycans is determined by the presence of sialic acids, carboxylic acids, phosphate, and sulfate groups. Furthermore, liberated sialyl-oligosaccharide-alditols from glycoproteins can easily be fractionated and isolated using anion-exchange HPLC, usually into mono-, di-, tri-, and tetra-sialylated glycan fractions. However, elutions are often performed with high salt gradients, which hinder the combination with MS detection. Anion exchange-hydrophilic interaction chromatography (AEX-HILIC), of which the working principle is based on the combination of negative charge and polarity, is frequently used for sialylated glycan separations [55].

5.3.6   High-pH Anion-Exchange Chromatography

A special type of ion-exchange HPLC, called high-pH (or high-performance) anion-exchange chromatography (HPAEC) with pulsed amperometric detection (PAD), has become a commonly used technique in glycoscience due to its exceptional resolving power and sensitive detection [56–59]. HPAEC-PAD can analyze mono- and oligosaccharides at around the 1–10 nmol range. The technique is capable of resolving glycans, such as released N- and O-glycans, based on charge, size, composition, isomers, and linkages, providing a profile of the overall glycosylation. It has great advantages, not only in terms of high resolution and short analysis times, but, in particular, it can be used for underivatized carbohydrates dissolved in water/0.1 M NaOH (not containing salt).

HPAEC is performed on a polymer-based pellicular anion-exchange matrix resin (the active component of which is a quaternary ammonium ion), equilibrated by an alkaline mobile phase. Separation is achieved by sodium hydroxide/sodium acetate gradient elution at a high pH of 11–14. Carbohydrates, as polyhydric compounds, obtain weakly acidic properties under these alkaline conditions forming sugar into negatively charged oxyanions, which can bind to the amino groups of the stationary phase of the column (e.g., CarboPac columns offered by ThermoFisher/Dionex Corp.). The selectivity of the chromatographic behavior may be determined by the particular hydroxyl groups of the oligosaccharides that become deprotonated and undergo interaction with the quaternary ammonium groups of the column resin. Thus, the separation depends on the charge, but molecular size, sugar composition, branching, and linkage of the monosaccharides also play a role. Consequently, the prediction of the elution order is largely empirical. By using a strong base (high pH) such as NaOH (range 12–25 mM) as an eluent, sugars are either partially or completely ionized (deprotonation of hydroxyl groups). The slight difference in relative pK_a values (12–14) of hydroxyl groups in sugars aids the chromatographic separation of individual mono/oligosaccharides by making use of a salt gradient (e.g., 0–300 mM sodium acetate in100 mM sodium hydroxide) for elution. The pK_a difference is sufficient to separate structural N-glycan isomers with only a single linkage difference. Elution times can be reduced by using a steeper acetate gradient.

Pulsed amperometric detection (PAD) is an electrochemical method for high-sensitive detection of carbohydrates (sensitivities in the low-picomole range) without pre- or post-column derivatization. Eluting carbohydrates are detected by measuring the electrical current generated by their oxidation at the surface of a gold electrode. It applies a triple sequence of potential to a gold electrode, catalyzing the electro-oxidation of –C–OH containing compounds in a high-pH solution (even carbohydrates without reducing groups). The resulting gold-oxide layer is converted back (cleaned) to native gold on reversing the voltage on the electrode, and the potential is cycled back to the analytical voltage. It has to be noted that nonoxidizable analytes (could be present as contaminants) will not be detected by PAD. In case of low-concentration NaOH gradients, the sensitivity can decrease, which can be retrieved by post-column addition of NaOH. Peak areas afford quantification; however, there are significant differences in response factors linked to the same properties that allow separation of isomers. This requires real standard curve determination for each glycan of interest for quantification.

Thorough cleaning and reconditioning of the column with high-molar (0.5–1 M) NaOH after each analytical run is essential for reproducible results. The eluents must be prepared from helium-sparged water and maintained in a helium atmosphere at all times to minimize the accumulation of carbonate. The carbonate will cause changes in the retention times of oligosaccharides.

Since <2% of the sample is oxidized for detection, eluent fractions can be collected for further characterization (after desalting) with MS and/or NMR. Desalting is performed using a carbon column.

As the mutarotation (α/β) of the reducing sugars is very rapid at high pH, the oxyanions elute as a single peak for each compound [60]. The elution order of the carbohydrates correlates with their pK_a values and capacity factors. Capacity factors for monosaccharides increase with the increasing number of carbon atoms and are lowest for sugar alcohols (alditols) and higher for analogous aldoses and ketoses. Before the separation of oligosaccharides, regularly, at first, the reducing end residues (the aldehyde in open-ring conformation) are reduced with sodium borohydride to form oligosaccharide alditols. The capacity factors of oligosaccharides increase with their chain length.

HPAEC-PAD has several advantages: (1) separation of monosaccharides to large oligosaccharides, (2) simultaneous separation of both neutral and charged glycans, (3) both isocratic and gradient elution possible, and (4) both amino acids and carbohydrates can be detected in a single run. Typically, glycans with a reduced end (alditols) or tagged (e.g., 2-AB) exhibit less retention than glycans with a reducing end. Furthermore, retention times increase with bi-, tri-, and tetra-antennary N-glycans. Extra-linked fucose residues have a decreasing influence on retention time. Sialylated oligosaccharides with ($\alpha2\rightarrow3$)-linked Neu5Ac residues are more retained than those with ($\alpha2\rightarrow6$)-linked Neu5Ac. Peak fractions can be isolated for further analysis, but these samples then have to be desalted. Automated desalting systems make direct coupling to mass spectrometers possible. Oligosaccharide alditols can be separated on a CarboPac PA-100 column, eluted by a linear gradient from 0 to 250 mM Na acetate in 0.2 M NaOH. Extended equilibration periods between runs are advised to obtain reproducible results. But, in general, HPAEC-PAD is a very convenient technique for monosaccharide analysis (see Sect. 6.5). The analysis technique is frequently used in the food industry [61]. Recently, ThermoFisher Scientific introduced a novel advanced HPAEC-PAD system, called ThermoScientific™ Dionex™ ICS-6000 HPIC™ using a Dual Eluent Generation Cartridges (EGC) mode, for analyzing complex carbohydrates (consult: thermofisher.com/IC).

5.3.7   Affinity Chromatography

Affinity chromatography (AFC) has to be mentioned as a special chromatographic technique for the isolation and fractionation of oligosaccharides, glycopeptides, and glycoproteins. The separation, usually performed in a small column, is based on the specific interaction of carbohydrates with an immobilized ligand on a solid silica matrix (e.g., agarose). These ligands can be antibodies or carbohydrate-binding proteins (CBPs) called lectins. Lectins possess a specificity directed toward certain sugars, but a complicating factor is that lectins have at least two sugar-binding sites and not only recognize one specific monosaccharide but can also recognize oligosaccharide structures with specific monosaccharide sequences [62]. Lectins can recognize carbohydrates (glycan motif) linked to proteins and lipids or free monosaccharides or polysaccharides. Lectins are isolated from plants, where they play vital roles in physiology, development, and stress response.

Regarding lectin affinity chromatography (LAC), the use of a series of columns containing immobilized lectins, whose sugar-binding specificities have been precisely elucidated, enables to fractionate oligosaccharides or glycopeptides into structurally distinct groups or can be used for the enrichment of glycoprotein/peptides [63]. The process is called, multi-lectin affinity chromatography (MLAC) [64]. An overview of lectins with their carbohydrate specificities can be found in Chap. 10, Table 10.1.

Although many lectins, isolated from plant sources, are commercially available, as well as several lectin-immobilized gels, the lectin-affinity method is a difficult technique for the nonexpert because the preparation of columns with immobilized lectins demands specific knowledge, expertise, and skills. Moreover, columns of different immobilized lectins must be well standardized to ensure activity, specificity, and reproducibility. There are more parameters than simple oligosaccharide recognition that influence the binding (or no binding) of glycoproteins to lectins [65]. Lectins are also known to exhibit nonspecific protein binding. For information, concerning lectins and their use in affinity chromatography, the following literature is recommended [66]. A special kind of affinity chromatography is immunoaffinity chromatography using immobilized antibodies, in particular, monoclonal antibodies. Lectin affinity chromatography, followed by a specific antibody immunoaffinity chromatography step, is a very useful strategy for effective isolation and purification of a specific glycoprotein [67, 68].

5.4 Detection Techniques for Liquid Chromatography

5.4.1 Detectors

When using advanced separation techniques, the detection of sugar compounds in the eluent can be a challenging feature. Appropriate detectors are required for the identification and quantification of the individual carbohydrates. The most common detection methods in liquid chromatography are ultraviolet light (UV), diode-array (DAD), fluorescence (FLD), and mass spectrometry (MS). Since free carbohydrates do not contain a chromophore that can be used for UV/Vis detection, they may only be detected with a refractive index (RI) or evaporating light scattering (ELS) detector.

Pulsed amperometric detection (PAD) used with HPAEC is the only sensitive detector for working with nonderivatized sugars (see Sects. 5.3.6 and 6.5.1). However, this detection method is not very selective, as other compounds that can be oxidized or reduced also give rise to signals. Furthermore, disadvantages are that a high pH is necessary and that the differential deprotonation of oligosaccharides induces a vast range of response factors on the detector and a relatively limited linearity in detector response. Low-wavelength UV detection (<200 nm) can also be used for underivatized carbohydrates, but the sensitivity is poor and depending on elution solvents. Detection of underivatized sugars after chromatographic separation is, of course, possible with mass spectrometry, but financially more expensive.

Evaporative Light Scattering detection (ELSD) is regularly used during SEC, but rarely with (RP)HPLC. It is a destructive detection technique that requires the nebulization and evaporation of the mobile phase using nitrogen gas, leaving behind sugar that is not evaporated. The sugar microparticles scatter light onto a photomultiplier tube which measures the intensity as voltage. However, although sensitive, the calibration curve is not linear over large concentrations and the detector is also susceptible to mobile phase contaminants, influencing the signal–noise ratio. ELS detection exhibits maximum sensitivity under aqueous conditions.

Refractive index detection (RID) is often used as a universal detector during the preparative SEC of carbohydrates. RI detectors measure changes in the index of refraction of the eluent induced by the solute. The sensitivity can vary depending on the eluent composition and temperature changes. Furthermore, RID precludes gradient elution.

Carbohydrates in column eluates can also be detected by direct coupling of the HPLC to a mass spectrometer, for example, equipped with a thermo-spray interface (LC-ESI-MS) (see Chap. 11). Liquid chromatography coupled with mass spectrometry (e.g., UHPLC-MS) has revolutionized carbohydrate analysis and has become the dominant tool due to its high sensitivity and fast separation.

5.4.2   Labeling of Glycans for Detection

After being released from a glycoprotein, the separation of the glycan mixture into homogeneous oligosaccharide fractions or for direct analysis is usually performed by HPLC. The complex heterogeneity of N-glycans demands refined systems for the separation of the oligosaccharide mixtures. To detect sugars in the HPLC column eluates, the introduction of an easily detectable label in the carbohydrate to be analyzed is often a prerequisite. Free carbohydrates are difficult to monitor because they have an inherent low UV adsorption in the solution. Moreover, often only limited amounts of samples are available. The coupling of a chromophore or fluorophore, usually introduced as an aromatic or heterocyclic substituent to the glycan,

will facilitate the detection by UV or fluorescence [69]. Moreover, the stoichiometric attachment of one label per glycan allows a direct quantification based on fluorescence or UV-absorbance intensity. The benefits of labeling glycans are not merely
from a detection perspective; the labeled glycans are often more easily retained on
the column, which improves separation during chromatographic processes.
Furthermore, labels can enhance the ionization yield during MS analysis (in
positive-ion mode).

A variety of labels (most commercially available) can be used (Fig. 5.3). The
choice mostly depends on the objective of the analysis. For an excellent comprehensive review on glycan labeling strategies and their use in identification and quantification, see [70]. There are some issues to consider when choosing suitable tagging
reagents:

1. rapid and robust production requiring minimum specialized equipment
2. nonselective labeling necessary for the correct molar proportion to be detected
3. efficiency of purification
4. stability of the label and
5. suitability for subsequent processes such as MS analysis.

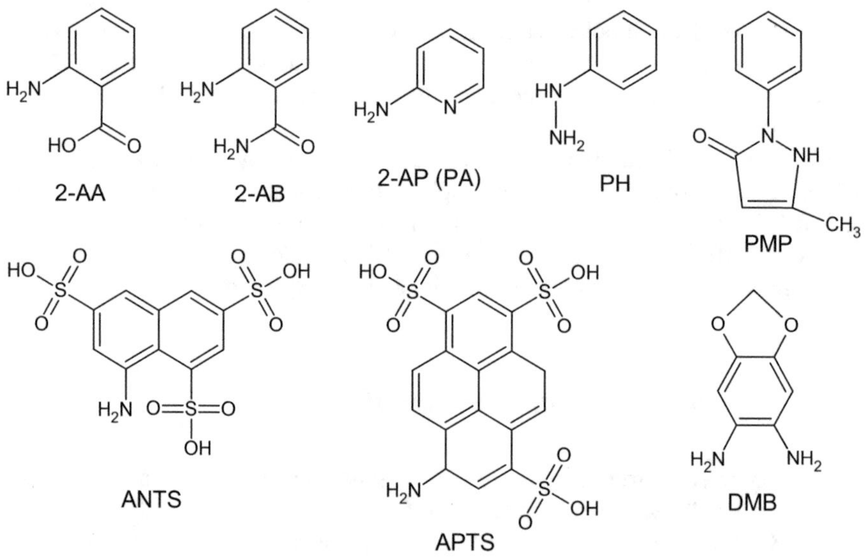

Fig. 5.3 Structures of fluorescent labels frequently used for HPLC and CE of carbohydrates.
2-AA anthranilic acid or 2-aminobenzoic acid, *2-AB* 2-aminobenzamide, *2-AP* 2-aminopyridine,
PH phenylhydrazine, *PMP* 1-phenyl-3 methyl-5-pyrazolone, *ANTS* 8-aminonaphtalene-
1,3,6-trisulfonic acid, *APTS* 8-aminopyrene-1,3,6-trisulfonate, *DMB* 1,2-diamino-4,5-methy-
lenedioxybezene

5.4.2.1   Labeling by Reductive Amination

In general, the labeling of oligosaccharides or released glycans is performed by a reductive amination reaction. The reducing-end residue of an oligosaccharide is always in equilibrium between its closed ring (hemi-acetal) form and its open ring (aldehyde) form (Sect. 2.1.1, Fig. 2.1). The glycans are usually dissolved in DMSO containing acetic acid to promote ring opening. Then, the reducing-end aldehyde (carbonyl group) of the oligosaccharide is treated with a labeling compound containing a primary amino group. Acidic conditions create a condensation reaction to form an imine, otherwise known as a Schiff base. The Schiff base is unstable and is usually reduced with sodium cyanoborohydride to form a stable (secondary) alkyl amine in which the label is permanently attached to the reducing end of the oligosaccharide (Fig. 5.4, e.g., 2-AB labeling). Since reduction with sodium

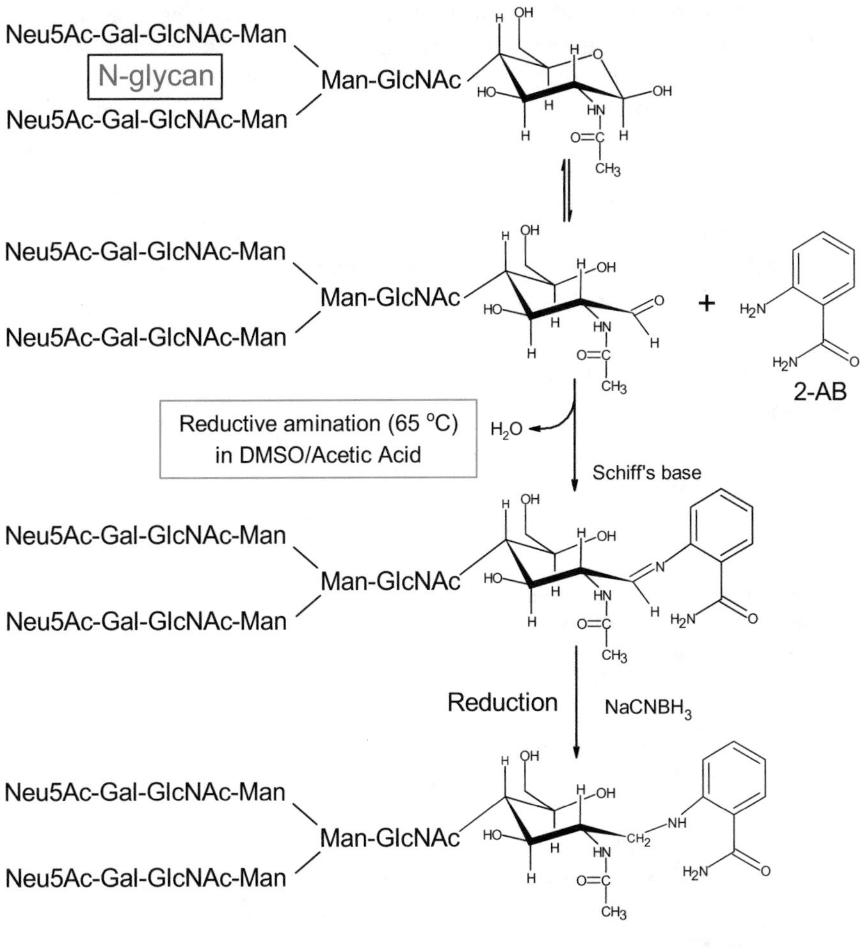

Fig. 5.4 Example of labeling of N-glycan by the reductive amination reaction with 2-aminobenzamide (2-AB). The reaction is initiated by the attack of the lone pair of the amino group of the label on the carbon of the aldehyde of the reducing end of the glycan. Note that in the final product, the reducing-end ring structure is broken

cyanoborohydride results in the release of highly toxic hydrogen cyanide, other reducing agents, like borane-diethyl amine and 2-picoline borane, have been suggested as alternatives [37, 71].

The search for faster sample preparation with new labeling techniques still continues. For instance, Agilent company recently released Gly-X, an N-glycan workflow for LC/FLD/MS, including an automated 5-min PNGase F digestion combined with instant dye/labeling, clean-up, HILIC-UHPLC on a 1290 Infinity II LC system coupled to a 6545XT AdvanceBio LC/Q-TOF mass spectrometer and data processing with MassHunter BioConfirm software.

N-glycans, enzymatically liberated from glycoproteins, are often labeled with a fluorophore, e.g., 2-aminobenzamide (2-AB) or 2-aminobenzoic acid (AA), (Protocol 19), which allows their detection at femtomole levels [14]. Most labeling procedures require cleanup to remove excess labels, reagents, salts, and other contaminants prior to chromatographic analysis because the excess of the free label gives problems with overload, column contamination, or quenching of signals during mass spectrometry. The salts and labeling reagent must be removed, which is usually done by cartridge technologies, using hydrophilic SPE (RP-C_{18}, PGC or HILIC), liquid–liquid extraction (water/chloroform), or gel filtration chromatography (Sephadex G-10, G-15, LH-20). Protocols for effective cleaning of the labeled glycans are usually provided with the nowadays commercially available labeling kits. Care has to be taken for sample losses during cleanup procedures using tips and cartridge technologies. The application of cellulose for the purification of 2-AA-labeled glycans on a high-throughput platform is often used. The tags on glycans, introduced by reductive amination, can be removed by treatment with N-bromosuccinimide (NBS) to generate the free, reducing glycans again [72].

Protocol 19. 2-AB/2-AA Labeling of Reducing Oligosaccharides and N-Glycans

Materials
Eppendorf cup with a screw cap
Water bath or heating block or Thermomixer (Eppendorf)
SpeedVac
Graphitized carbon SPE cartridges
Dimethyl sulfoxide (DMSO)
Glacial acetic acid (CH_3COOH)
2-aminobenzamide (2-AB, Aldrich)
2-anthranilic acid (2-AA, Aldrich)
Sodium cyanoborohydride ($NaCNBH_3$, toxic, corrosive, hygroscopic)
2-picoline borane
acetonitrile
Freshly prepared **labeling solution**:
   1. Mix 420μL DMSO and 180μL acetic acid in a 2-mL screw cap Eppendorf cup.
   2. Add 24 mg of 2-aminobenzamide (**2-AB**) or add 24 mg of 2-anthranilic acid (**2-AA**).

3. Add 32 mg of NaCNBH$_3$ [or 2-picoline borane to 1 M] (reducing reagent), close the cup.
4. Warm with intermittent shaking at 37 °C to obtain complete dissolving (keep in dark).

Labeling

1. Dissolve dry N-glycans (~50–100µg) in 50µL of water in 0.5-mL Eppendorf cup.
2. Add 50µL of freshly prepared labeling solution (see above).
3. Incubate for 2 h at 65 °C with intermittent shaking in the dark.
4. Cool slowly to room temperature.
5. Purification using graphitized carbon SPE (PGC-SPE)

   preconditioned with 5 mL of ACN and 5 mL of water/ACN (50/50, containing 0.1% TFA) and equilibrated with 10 mL of water.

6. Apply sample to the graphitized carbon SPE cartridge.
7. Wash with 8 mL of water to elute labeling reagents.
8. Then, the labeled glycans are eluted with 5 mL of water/ACN (50:50 v/v, containing 0.1% TFA).
9. Dry by SpeedVac/lyophilization [keep sample in dark].

2-AB-labeled glycans are well suited to a variety of glyco-analytical methods, including hydrophilic interaction liquid chromatography (HILIC) (Sect. 5.3.3, Fig. 5.2), weak anion-exchange (WAX) HPLC, reversed-phase HPLC (RP-HPLC), MALDI-MS, and ESI-MS as it is predictive. Negatively charged labels, such as 2-AA (specific absorbance at 310 nm and giving negative incremental value for sialic acids), are preferred for analytical electrophoretic (CE) analysis. Furthermore, for HPLC-MS/MS, glycan labeling by reductive amination with 2-aminopyridine (2-AP/PA) is regularly used (Fig. 5.5) (Protocol 20 for N-glycans).

Protocol 20. 2-AP (PA) Labeling of N-Glycans

Materials
Reacti vials (Pierce) 1 mL with Teflon seal or glass tubes (100 × 10 mm)
Water bath
Heating block
Centrifuge
Glacial acetic acid (CH$_3$COOH)
2-aminopyridine [toxic, irritant, keep in the freezer under inert N$_2$]
Borane-dimethylamine complex [(CH$_3$)$_2$NH·BH$_3$, Aldrich]
Bio-Gel P-2/Sephadex G-15 column (45 × 1 cm), elute 10 mM NH$_4$HCO$_3$
Coupling reagent: dissolve 550 mg 2-aminopyridine (2-AP) in 200µL of acetic acid. Dilute with 1.8 mL of water, pH must be 6.8. (store at −20 °C)

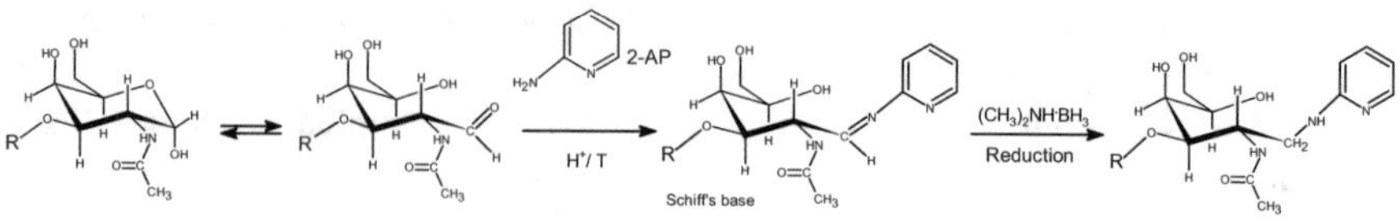

Fig. 5.5 Labeling of carbohydrate chain with 2-AP

Reducing reagent: freshly prepare 200 mg of $(CH_3)_2NH \cdot BH_3$ in a mixture of 50μL of water plus 80μL of acetic acid. [toxic, corrosive, wear skin and eye protection]

Procedure

1. Lyophilize (10–100 nmol) oligosaccharide (having a reducing end) in Reactivial or glass tube with a screw cap.
2. Add 20μL of coupling reagent, mix well, and close the tube.
3. Heat for 1 h at 90°C under stirring (to form Schiff base).
4. Cool to room temperature.
5. Add 70μL of reducing reagent, close tube, and mix well.
6. Heat for 40 min at 80 °C.
7. Cool to room temperature.
8. Dilute with 500μL of water.
9. Remove excess reagent and isolate the PA-oligosaccharides by chromatography on Bio-Gel P-2 or Sephadex G-10/15 column (1.5 × 30 cm), eluted with 10 mM ammonium bicarbonate. UV detection at 214 nm (A 254 nm).
10. Collect void vol fraction and lyophilize.

5.4.2.2   Labeling by Michael Addition with PMP

Besides reductive amination, another way to label the reducing end of glycans and monosaccharides is known as the Michael addition. The reaction is a nucleophilic addition of a carbanion to an α,β-unsaturated carbonyl compound, resulting in the formation of a C–C bond. Labeling is performed with 1-phenyl-3-methyl-5--pyrazolone (PMP) under mild alkaline conditions instead of acidic conditions (Fig. 5.6) (Protocol 21) [73]. The derivatization reaction itself is a base-catalyzed Michael-type addition involving a two-step labeling process in which donor molecules of the labeling reagent are formed and consecutively added to the reducing end of the glycan with a stoichiometry of two label molecules per glycan. After the addition of the first label molecule, a water molecule is lost, which results in the formation of an α,β-unsaturated carbonyl compound. The following step yields a Michael 1,4-addition product by conjugation of the second label molecule.

After derivatization, the reducing-end monosaccharide is labeled with two PMP molecules. It is important to note that the incorporation of the two hydrophobic labels per glycan strongly influences the chromatographic properties and restricts isomer separation. PMP labeling was also performed on released O-glycans [74].

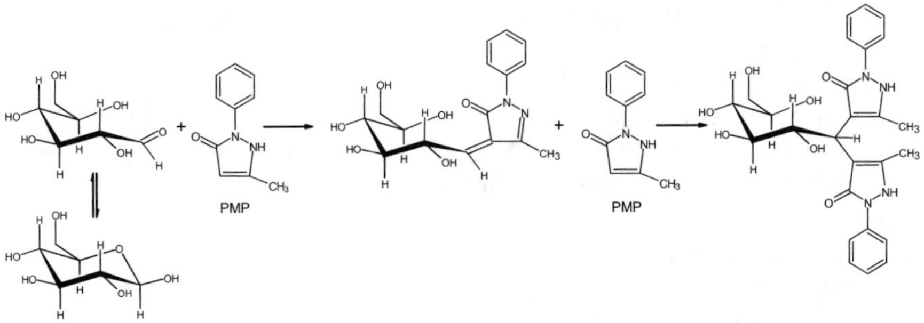

Fig. 5.6 Labeling of monosaccharide with PMP via a Michael-type addition

The labeling facilitates the separation by reversed-phase liquid chromatography employing C18 as the stationary phase (RP-LC). Isomeric disaccharides could also be separated by LC-ESI-MS/MS when PMP-labeled [75]. PMP derivatives can be up to 100 times more sensitive during MS analysis.

PMP derivatization is also used for a rapid monosaccharide analysis by CZE (Sect. 5.5, Fig. 5.7). Usually, for oligo/polysaccharides, the monosaccharides are obtained by acid hydrolysis, followed by derivatization. The PMP derivatives of pentoses, deoxyhexoses, hexoses, hexuronic acids, and N-acetylhexosamines can also be analyzed by ultrahigh-performance liquid chromatography triple quadrupole mass spectrometry (UHPLC/QqQ-MS) (Sect. 6.5.2). Additionally, also for linkage analysis, PMP derivatization can be used. After permethylation and hydrolysis, the resulting partially methylated monosaccharides are labeled with PMP and separated/analyzed by UHPLC/dMRM-MS. In multiple reaction monitoring (MRM), the concentration of an unknown sample is determined by comparing its MS response to that of a known standard (see Chap. 11) [77–79].

Protocol 21. PMP Labeling for Reducing Monosaccharides

Materials
1-mL Reacti-vials (Pierce) with Teflon seal
Heating block
Centrifuge
Lyophilizer
1-phenyl-3-methyl-5-pyrazolone (PMP) (Sigma-Aldrich, St. Louis, MO)
Methanol
Chloroform
Sodium hydroxide (NaOH)
Hydrochloric acid (HCl)

Procedure

1. Take 10μL of aqueous monosaccharide solution (~10μg monosaccharide).
2. Add 250μL of 300 mM aqueous NaOH, mix well [or 28% (v/v) NH₄OH].

3. Add 250µL of 500 mM PMP-methanol solution (freshly prepared).
4. Vortex for 10 s.
5. Incubate for 30 min at 70 °C.
6. Cool to room temperature.
7. Neutralize by adding 250µL of 0.3 M HCl (in case of NH$_4$OH solution, dry by SpeedVac and reconstituted in 500µL water).
8. Add 250µL water.
9. Add 1 mL of chloroform, vortex well.
10. Centrifuge slowly.
11. Discard the organic (lower) phase.
12. Repeat 2× steps 9–11.
13. Collect the aqueous (upper) layer (containing labeled glycans) and filtrate through a 0.22µm membrane.
14. Lyophilize [redissolve in water prior to HPLC or CZE analysis (Sect. 5.5, Fig. 5.7)].

5.4.2.3   Permethylation of Glycans for Mass Spectrometry

Compared to unlabeled glycans, mass spectrometric detection is usually enhanced for glycans tagged with a label at the reducing end. Moreover, complete derivatization by permethylation benefits mass spectrometric fragmentation. The hydrogens on the highly polar hydroxyl groups, amine groups, and carboxyl groups are converted to nonpolar methyl groups. Permethylation protocols will be discussed in Sect. 6.6, but for mass spectrometric analysis of small quantities of glycan, a quick and efficient solid-phase permethylation method has been developed [80, 81]. A fast permethylation procedure for released N-glycans is presented with Protocol 22. In a solid-phase permethylation, the reaction is processed in micro-spin columns. These are packed with sodium hydroxide beads, glycans, and reaction reagents in a minimized reaction volume. After the reaction, liquid–liquid extraction is utilized to isolate the pure methylated products.

Protocol 22. Fast Permethylation and Purification of Released N-Glycans for MS

Materials
DMSO (Sigma-Aldrich)
Iodomethane (Sigma)
Acetonitrile
Sodium hydroxide beads, 20–40 mesh (Sigma-Aldrich)
Eppendorf centrifuge
Pierce C18 microspin column (ThermoFisher)
Empty micro spin columns, 5µm frit
Eppendorf cups with column holder

Procedure: [Work in Low-Humidity Environment]

1. Soak sodium hydroxide beads in DMSO.
2. Transfer sodium hydroxide/DMSO mixture into an empty spin column (2 cm height).
3. Place the spin column, using column holder, in a clean Eppendorf tube.
4. Centrifuge the spin column until no visible DMSO is left in the column.
5. Add 200µL of fresh DMSO.
6. Place the spin column, using column holder, in a clean Eppendorf tube.
7. Centrifuge the spin column until no visible DMSO is left in the column.
8. Add sample: ~10 ug glycan suspended in 30µL DMSO, 1.2µL water, 20µL iodomethane.
9. Incubate at room temperature for 25 min.
10. Add again 20µL of iodomethane.
11. Incubate at room temperature for 10 min.
12. Place the spin column, using column holder, in a clean Eppendorf tube.
13. Centrifuge the spin column to elute the mixture.
14. Dry the eluent in SpeedVac (takes a long time).
15. Desalt by SPE, using a Pierce C18 microspin column (ThermoFisher) or C18 Sep-Pak cartridge (Alltech), equilibrated with 85% acetonitrile.
16. Apply sample.
17. Wash with water.
18. Elute permethylated glycans with 80% acetonitrile.
19. Dry with SpeedVac/lyophilization (Samples ready e.g., MALDI-TOF-MS, see Chap. 11).

5.5   Electrophoretic Separation of Carbohydrates

In addition to the wide variety of chromatographic methods to separate carbohydrates, some electrophoretic methods are in use for the analysis of both derivatized and underivatized carbohydrates [82, 83]. The application of these techniques in glycoscience started in the early 1990s. Electrophoretic separation relies on the migration of charged molecules in a conducting solution under the influence of an applied electrical field. The instrument consists of a capillary connecting two buffer solutions, a high-voltage power supply and a detector.

Charged (ionic) carbohydrates are (1) acidic polysaccharides, (2) glycosaminoglycans (GAGs), and (3) oligosaccharides, containing sialyl, sulfate, or phosphate groups. This means that these compounds may be separated directly using electrophoresis due to their negative charge. Neutral sugars (e.g., monosaccharides) can be given a charge by high pH, complexation (e.g., with borate ions), or, very common for oligosaccharides, conjugation to a charged chromophore or fluorophore (commonly APTS or ANTS) via reductive amination, which also aids in optical/UV detection (e.g., 200 nm).

Capillary electrophoresis (CE), usually applied as Capillary Zone Electrophoresis (CZE), is a technique that enables the analysis of carbohydrate monomers, oligomers, and polymers, and it offers a very rapid, simple method for exhaustive carbohydrate screening, including glycopeptides and even intact glycoproteins [84–87]. CE has the distinction of high speed and the requirement of very low sample volumes (nanoliter range). However, it must be noted that CE is not currently routinely used in analytical research laboratories, but the use of CE for the analysis of glycoprotein pharmaceuticals is growing [88, 89]. Highly sensitive detection of labeled saccharides is achieved by connecting the CE system with a fluorescence detector or mass spectrometer, but improvements with respect to CE-MS interfacing are required because of the low flow rate and highly ionic effluent [90]. The advantage of CE is that detailed glycosylation analysis can be performed with minimal amounts (~5 pmol) of the sample.

CE is performed using a narrow-bore fused silica capillary (i.d. 25–75μm, 20–100 cm length) filled with an electrolyte (phosphate or borate) buffer solution. The two ends of the capillary rest in two separated electrolyte reservoirs, containing high-voltage platinum electrodes. The inner surface of the bare fused silica (BFS) capillary has a negative charge due to silanol (Si–OH) groups, which are ionized (SiO$^-$) above pH 3. Applying a high voltage (10–30 kV), a positive ion buffer layer (H$^+$) adjacent to the inner surface migrates to the cathode and produces an electroosmotic flow (EOF). This flow acts as the driving force for migrating ionized sugars separated based on their charge-to-mass ratios, in combination with the viscosity of the electrolyte buffer (background electrolyte, BGE), to the detection side (usually a photo-diode array detector). Positive sugars elute first, followed by neutral, and finally negative sugars. For sugars with the same ion charge, the one with the smallest size migrates faster. It has to be noted that CE can suffer from poor reproducibility due to inconsistent flow rates.

Nowadays, advanced CE instruments are commercially available and are equipped with UV or laser-induced fluorescence detectors or even with a mass spectrometer. For CE analysis of sugars, derivatization of their reducing ends with a fluorescent dye by reductive amination is the most general and useful technique for highly sensitive detection. Derivatization is generally performed with fluorescent tags containing negative charges for very sensitive CE with laser-induced fluorescence detection (CE-LIF). A variety of reagents, including 8-aminopyrene-1,3,6-trisulfonate (APTS), 3- and 8-aminonaphthalene-1,3,6-trisulfonic acid (ANTS), are in use, imparting a triple-negative charge on the mono/oligosaccharide. The three sulfonic groups provide a nearly pH-independent high anionic charge.

Sugar samples are stirred with an amine solution in water at 90 °C for 10 min and then the resulting solution of Schiff base is treated with sodium cyanoborohydride solution at 90 °C for 1 h to form a stable amine. After the addition of water, the reaction mixture is stored at −20 °C. Diluted samples are analyzed by CE at pH 10.3 in a 50 mM borate buffer using acetonitrile as an organic modifier and detected by UV detector at 200 nm.

CZE separations of APTS- or ANTS-labeled glycans, combined with LIF detection, are frequently attained using nonvolatile phosphate, citrate, and borate buffers.

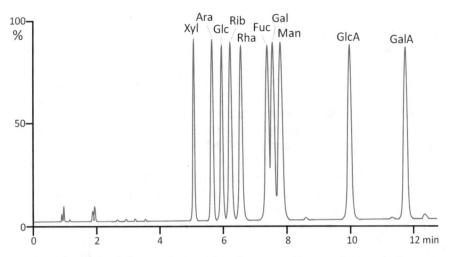

Fig. 5.7 Example of separation of PMP-labeled monosaccharides by CZE with UV detection (conditions conform [76]. In MS analysis, characteristic fragments are: [monosaccharide mass + mass of two PMP residues + H] = [M + 330 + H]$^+$ and m/z 175 = [PMP + H]$^+$

Particularly with borate buffers, linkage-specific separation of glycans could be achieved due to the complexation by borate. A rapid analytical CZE method was established for monosaccharide analysis [76]. The monosaccharides were labeled with 1-phenyl-3-methyl-5-pyrazolone (PMP) (Sect. 5.4.2.2, Protocol 21, Fig. 5.6). Separation was performed using an uncoated capillary (50μm id × 58.5 cm) with detection by UV at 245 nm (Fig. 5.7).

CE can be coupled with MS (CZE-MS) [91, 92]. For the connection of CZE to ESI-MS, a special interface is needed, consisting of a coaxial triple-tube sprayer, the innermost tube being the separation capillary, the middle tube the sheath-flow, and the outermost the spray gas tube. Also, combinations of CE with MALDI-MS are in use, making use of automated spotting [93]. CZE-MS is also used for both glycoprotein and glycopeptide analysis. In general, these analyses are carried out at 10 kV using 50 cm fused silica capillaries and buffers compatible with MS detection. Using a mass spectrometer as an online detector not only assures the analysis of every component eluting (mass mapping) but also at the same time diagnostic carbohydrate ions can be generated by collisional activation that permits the selective and specific detection of glycopeptides. CZE-ESI-MS(MS) offers a fast analysis of glycoprotein microheterogeneity and characterization of the oligosaccharides. However, the interpretation of CZE-MS data tends to be time-consuming and requires considerable expertise [91, 94].

New CE techniques for carbohydrate separations are frequently developed, such as capillary gel electrophoresis (CGE), in which the capillary inner surface is modified to suppress EOF. The commonly used gel solutions are linear neutral polymers such as polyacrylamide (PAA) and polyethylene glycol (PEG) for separation of

(2-AA/APTS-labeled) oligosaccharides. Unfortunately, CGE-based methods are incompatible with MS coupling.

However, capillary gel electrophoresis with laser-induced fluorescence detection (CGE-LIF) can be used to obtain structure and linkage information from heterogeneous mixtures of glycoforms which are, for instance, APTS-labeled [95]. Nowadays, microchip electrophoresis (ME) methods are used. This means that electrophoresis is performed in microchannels fabricated on chips.

Specific CZE methods to separate underivatized sugars have also been developed. The carbohydrate hydroxyl groups are ionized in strongly alkaline conditions, then the anionic carbohydrate enediolate structures are suitable for electrophoretic analysis and identification through indirect UV 195/270 nm detection (usually in complex formation with borate) or electrochemical detection [96]. Additionally, capillary affinity electrophoresis (CAE) using a set of lectins with different specificities exists. CAE allows the characterization of the structures of carbohydrates based on their affinities for an appropriate set of carbohydrate-binding proteins. In this case, the carbohydrate that shows affinity to the used lectin will be retarded during the run [97].

A quite old electrophoretic technique is Fluorophore-Assisted Carbohydrate Electrophoresis (FACE®) to separate both mono- and oligosaccharides. The technique lends itself to the analysis of multiple samples in one run. In principle, it is polyacrylamide gel electrophoresis (PAGE) of fluorescent-labeled carbohydrates. Typically, the relevant oligosaccharides (N-glycans) are enzymatically cleaved from the glycoprotein/glycopeptides using PNGase F and then, utilizing the reductive amination reaction, labeled with low-molecular-weight 8-aminonaphtalene-1,3,6-trisulfonic acid (ANTS), giving a net negative charge and fluorophore to the glycan (Fig. 5.8).

Usually, the labeled oligosaccharides are separated by electrophoresis on precast 30–60% acrylamide slab gels, followed by isolation of the individual bands visible under UV light. Subsequently, the isolated band(s) are treated with a series of exoglycosidases, followed by re-electrophoresis on a polyacrylamide gel to obtain

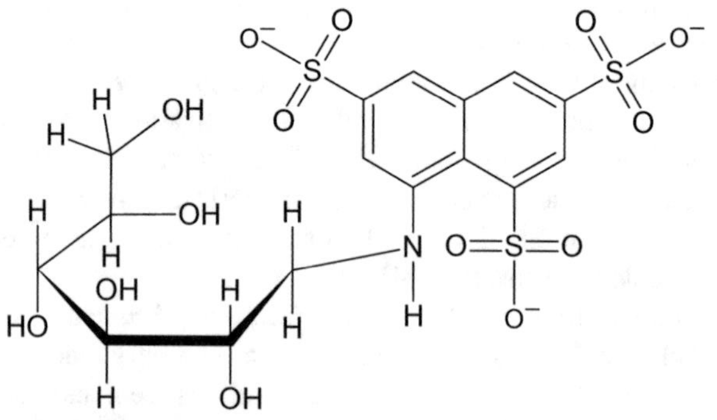

Fig. 5.8 ANTS-labeled monosaccharide

structural data. The band patterns are detected in a fluorescent imager and compared to a simultaneously run ANTS-labeled glucose ladder (DP 2–15). A FACE imager (SE 2000 FACE workstation) with accessory software is commercially available (Glyko, Inc., Novato, CA).

Since all the abovementioned electrophoretic separation/MS methods need special equipment (e.g., Prince CE system, Prince Technologies, The Netherlands), and the techniques require substantial expertise for handling and for the interpretation of the data, it is beyond the scope of this book to discuss protocols in detail here. The interested reader is directed for additional information to other sources, for example, "Capillary Electrophoresis of Carbohydrates" [98] and an excellent review [99]. It is also advised to consult the protocols provided with the commercially available instruments.

5.6   Gas-Liquid Chromatography

Gas-liquid chromatography (GLC or GC) is still one of the most reliable methods to separate, analyze, and quantify monosaccharides. It presents several favorable characteristics, such as easy instrument handling and high resolution, which makes it suitable for the analysis of complex mixtures. Furthermore, highly sensitive detection is reached by the easy coupling of different detectors. A drawback of GLC analysis in carbohydrate research could be that it is limited to derivatized sugars of low molecular mass, usually mono-, di-, and maximal trisaccharides.

The technique relies upon an inert carrier gas (helium or hydrogen) to carry monosaccharide derivatives through a glass/fused silica capillary tube that is coated with a thin layer of a stationary (polar or nonpolar) liquid phase [long capillary wall-coated open tubular (WCOT) columns]. A number of different detectors may be coupled to a GLC system, for example, a flame ionization detector (FID) (capable of detecting sub-nanomolar amounts) or a mass spectrometer (MS). GC-MS is a vital method for the identification of constituent monosaccharides and linkage determination in poly- and oligomeric carbohydrates.

However, due to the polar hydroxyl, amino, and carboxylic groups, free monosaccharides/alditols are not volatile enough for gas chromatographic separation. They have to be derivatized to increase volatility, primarily by avoiding hydrogen bond forming.

Classic derivatization methods consist of the substitution of the polar groups (i.e., the acidic hydrogens of hydroxyl groups and carboxylic groups) of the sugars by trimethylsilylation, acetylation, trifluoroacetylation, or methylation in order to increase the volatility at practical temperatures. For comprehensive reviews of derivatization of carbohydrates, the reader is referred to [100] and [101]. The search for new derivatization for GC(MS) analysis of (mono)saccharides continues. Recently, a GCMS procedure based on trimethylsilyl-dithioacetal (TMSD) derivatization was established [102]; however, a drawback is the use of thiol as a reagent and the long GC running time.

The two most common derivatization methods for monosaccharides, being trimethylsilylation and acetylation, are often used for gas chromatographic analysis of monosaccharides. These derivatives can easily be separated on different capillary fused silica GLC columns, such as EC-1 or AT-1 (Alltech), DB-1, DB-5 or HP-5 (Agilent), Rtx-1 or Rtx-5 (Restek).

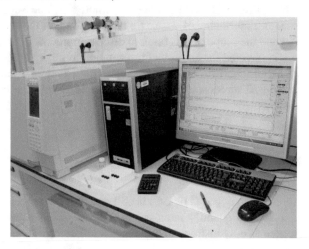

References

1. Prescher JA, Bertozzi CR. Chemical technologies for profiling glycans. Cell. 2006;126:851–4.
2. Pabst M, Altmann F. Glycan analysis by modern instrumental methods. Proteomics. 2011;11:631–43.
3. Stöckmann H, Adamczyk B, Hayes J, Rudd PM. Automated, high-throughput IgG-antibody glycoprofiling platform. Anal Chem. 2013;85:8841–9.
4. Sanz ML, Martínez-Castro I. Recent developments in sample preparation for chromatographic analysis of carbohydrates. J Chromatogr A. 2007;1153:74–89.
5. Kamerling JP, Gerwig GJ. Strategies for the structural analysis of carbohydrates. In: Kamerling JP, editor. Comprehensive glycoscience-from chemistry to systems biology. Amsterdam: Elsevier; 2007. p. 1–68.
6. Zhang Z, Xiao ZP, Linhardt RJ. Thin layer chromatography for the separation analysis of acidic carbohydrates. J Liq Chromatogr Relat Technol. 2009;32:1711–32.
7. Islam MK, Sostaric T, Lim LY, Hammer K, Locher C. Sugar Profiling of honeys for authentication and detection of adulterants using high-performance thin layer chromatography. Molecules. 2020;25:5289.
8. Fuchs B, Süss R, Nimptsch A, Schiller J. MALDI-TOF-MS directly combined with TLC: a review of the current state. Chromatographia. 2009;69:S95–S105.
9. Mernie EG, Tolesa LD, Lee M-J, Tseng M-C, Chen Y-J. Direct oligosaccharide profiling using thin-layer chromatography coupled with ionic liquid-stabilized nanomatrix-assisted laser desorption-ionization mass spectrometry. Anal Chem. 2019;91:11544–52.
10. Rowe L, Burkhart G. Analyzing protein glycosylation using UHPLC: a review. Bioanalysis. 2018;10:1691–703.
11. Melmer M, Stangler T, Premstaller A, Lindner W. Comparison of hydrophilic-interaction, reversed-phase and porous graphitic carbon chromatography for glycan analysis. J Chromatogr A. 2011;1218:118–23.

12. Nagy G, Peng T, Pohl NL. Recent liquid chromatographic approaches and developments for the separation and purification of carbohydrates. Anal Methods. 2017;9:3579–93.
13. Lowenthal MS, Kilpatrick EL, Phinney KW. Separation of monosaccharides hydrolyzed from glycoproteins without the need for derivatization. Anal Bioanal Chem. 2015;407:5453–62.
14. Yan H, Yalagala RS, Yan F. Fluorescently labelled glycans and their applications. Glycoconj J. 2015;32:559–74.
15. Vreeker GCM, Wuhrer M. Reversed-phase separation methods for glycan analysis. Anal Bioanal Chem. 2017;409:359–78.
16. Raessler M. Sample preparation and current applications of liquid chromatography for the determination of non-structural carbohydrates in plants. Trends Anal Chem. 2011;30:1833–43.
17. Lin Z, Lubman DM. Permethylated N-glycan analysis with mass spectrometry. Methods Mol Biol. 2013;1007:289–300.
18. Zhou S, Dong X, Veillon L, Huang Y, Mechref Y. LC-MS/MS analysis of permethylated N-glycans facilitating isomeric characterization. Anal Bioanal Chem. 2017;409:453–66.
19. Gao X, Zhang L, Zhang W, Zhao I. Design and application of an open tubular capillary reactor for solid-phase permethylation of glycans in glycoproteins. Analyst. 2015;140:1566–71.
20. Hu Y, Borges CR. A spin column-free approach to sodium hydroxide-based glycan permethylation. Analyst. 2017;142:2748–59.
21. Shajahan A, Supekar NT, Heiss C, Azadi PJ. High-throughput automated micropermethylation for glycan structure analysis. Anal Chem. 2019;91:1237–40.
22. Zaia J. On-line separations combined with MS for analysis of glycosaminoglycans. Mass Spectrom Rev. 2009;28:254–72.
23. Du JY, Chen LR, Liu S, Lin JH, Liang QT, Lyon M, Wei Z. Ion-pairing liquid chromatography with on-line electro-spray ion trap mass spectrometry for the structural analysis of N-unsubstituted heparin/heparan sulfate. J Chromatogr B. 2016;1028:71–6.
24. Melmer M, Stangler T, Schiefermeier M, Brunner W, Toll H, Rupprechter A, Lindner W, Premstaller A. HILIC analysis of fluorescence-labeled N-glycans from recombinant biopharmaceuticals. Anal Bioanal Chem. 2010;398:905–14.
25. Zauner G, Deelder AM, Wuhrer M. Recent advances in hydrophilic interaction liquid chromatography (HILIC) for structural glycomics. Electrophoresis. 2011;32:3456–66.
26. Cao L, Zhang Y, Chen L, Shen A, Zhang X, Ren S, et al. Sample preparation for mass spectrometric analysis of human serum N-glycans using hydrophilic interaction chromatography-based solid phase extraction. Analyst. 2014;139:4538–46.
27. Sandra K, Vandenheede I, Sandra P. Modern chromatographic and mass spectrometric techniques for protein biopharmaceutical characterization. J Chromatogr A. 2014;1335:81–103.
28. Saldova R, Kilcoyne M, Stöckman H, Martin SM, Lewis AM, Tuite CME, Gerlach JQ, Le Berre M, Borys MC, Li ZJ, Abu-Absi NR, Leister K, Joshi L, Rudd PM. Advances in analytical methodologies to guide bioprocess engineering for bio-therapeutics. Methods. 2017;116:63–83.
29. Kozlic P, Goldman R, Sanda M. Hydrophilic interaction liquid chromatography in the separation of glycopeptides and their isomers. Anal Bioanal Chem. 2018;410:5001–8.
30. Kim W, Kim J, You S, Do J, Jang Y, Kim D, Lee J, Ha J, Kim HH. Qualitative and quantitative characterization of sialylated N-glycans using three fluorophores, two columns, and two instrumentions. Anal. Biochem. 2019;571:40–8.
31. Van Schaick G, Pirok BWJ, Haselberg R, Somsen GW, Gargano AFG. Computer-aided gradient optimization of hydrophilic interaction liquid chromatographic separations of intact proteins and protein glycoforms. J Chromatogr A. 2019;1598:67–76.
32. Campbell MP, Royle L, Radcliffe CM, Dwek RA, Rudd PM. GlycoBase and autoGU: tools for HPLC-based glycan analysis. Bioinformatics. 2008;24:1214–6.
33. Royle L, Campbell MP, Radcliffe CM, White DW, Harvey DJ, Abrahams JL, Kim YG, Henry GW, Shadick NA, Weinblatt ME, Lee DM, Rudd PM, Dwek RA. HPLC-based analysis of serum N-glycans on a 96-well plate platform with dedicated database software. Anal Biochem. 2008;376:1–12.
34. Staples GO, Bowman MJ, Costello CE, Hitchcock AM, et al. A chip-based amide-HILIC LC/MS platform for glycosaminoglycan glycomics profiling. Proteomics. 2009;9:686–95.

35. Jandera P. Stationary and mobile phases in hydrophilic interaction chromatography: a review. Anal Chim Acta. 2011;692:1–25.

36. Bennett R, Olesik SV. Gradient separation of oligosaccharides and suppressing anomeric mutarotation with enhanced-fluidity liquid hydrophilic interaction chromatography. Anal Chim Acta. 2017;960:151–9.

37. Ruhaak LR, Deelder AM, Wuhrer M. Oligosaccharide analysis by graphitized carbon liquid chromatography-mass spectrometry. Anal Bioanal Chem. 2009;394:163–74.

38. Kolarich D, Windwarder M, Alagesan K, Altmann F. Isomer-specific analysis of released N-glycans by LC-ESI MS/MS with porous graphitized carbon. Methods Mol Biol. 2015;1321:427–35.

39. Stavenhagen K, Kolarich D, Wuhrer M. Clinical glycomics employing graphitized carbon liquid chromatography-mass spectrometry. Chromatographia. 2015;78:307–20.

40. Stavenhagen K, Plomp R, Wuhrer M. Site-specific protein N- and O-glycosylation analysis by a C18-porous graphitized carbon-liquid chromatography-electrospray ionization mass spectrometry approach using pronase treated glycopeptides. Anal Chem. 2015;87:11691–9.

41. Miller RL, Guimond SE, Prescott M, Turnbull JE, Karlsson N. Versatile separation and analysis of heparan sulfate oligosaccharides using graphitized carbon liquid chromatography and electrospray mass spectrometry. Anal Chem. 2017;89:8942–50.

42. Ashwood C, Pratt B, MacLean BX, Gundry RL, Packer NH. Standardization of PGC-LC-MS-based glycomics for sample specific glycotyping. Analyst. 2019;144:3601–12.

43. Abrahams JL, Campbell MP, Packer NH. Building a PGC-LC-MS N-glycan retention library and elution mapping resource. Glycoconj J. 2018;35:15–29.

44. Bapiro TE, Richards FM, Jodrell DI. Understanding the complexity of porous graphitic carbon (PGC) chromatography: modulation of mobile-stationary phase interactions overcomes loss of retention and reduces variability. Anal Chem. 2016;88:6190–4.

45. Pereira L. Porous graphitic carbon as a stationary phase in HPLC: theory and applications. J Liq Chromatogr Relat Technol. 2008;31:1687–731.

46. West C, Elfakir C, Lafosse M. Porous graphitic carbon: a versatile stationary phase for liquid chromatography. J Chromatogr A. 2010;1217:3201–16.

47. Van Leeuwen SS. Challenges and pitfalls in human milk oligosaccharide analysis. Nutrients 11 (2019) 2684

48. Wu S, Grimm R, German JB, Lebrilla CB. Annotation and structural analysis of sialylated human milk oligosaccharides. J Proteome Res. 2011;10:856–68.

49. Oedit A, Vulto P, Ramautar R, Lindenberg PW, Hankemeier T. Lab-on-a-chip hyphenation with mass spectrometry: strategies for bioanalytical applications. Curr Opin Biotechnol. 2015;31:79–85.

50. Aldredge D, An HJ, Tang N, Waddell K, Lebrilla CB. Annotation of a serum N-glycan library for rapid identification of structures. J Proteome Res. 2012;11:1958–68.

51. Tonon KM, Miranda A, Abrao ACFV, de Morais MB, Morais TB. Validation and application of a method for the simultaneous absolute quantification of 16 neutral and acidic human milk oligosaccharides by graphitized carbon liquid chromatography—electrospray ionization—mass spectrometry. Food Chem. 2019;274:691–7.

52. Porfirio S, Archer-Hartmann S, Moreau GB, Ramakrishnan G, Haque R, Kirkpatrick BD, Petri WA, Azadi P. New strategies for profiling and characterization of human milk oligosaccharides. Glycobiology. 2020;30:774–86.

53. Cho BG, Peng W, Mechref Y. Separation of permethylated O-glycans, free oligosaccharides, and glycosphingolipid-glycans using porous graphitized carbon (PGC) column. Metabolites. 2020;10:433–44.

54. Song K, Moon DB, Kim NY, Shin YK. Glycosylation heterogeneity of hyperglycosylated recombinant human interferon-β (rhIFN-β). ACS Omega. 2020;5:6619–27.

55. Neville DCA, Dwek RA, Butters TD. Development of a single column method for the separation of lipid- and protein-derived oligosaccharides. J Proteome Res. 2009;8:681–7.

56. Hardy MR, Rohrer JS. High-pH anion-exchange chromatography (HPAEC) and pulsed amperometric detection (PAD) for carbohydrate analysis. In: Kamerling JP, editor. Comprehensive glycoscience. Amsterdam: Elsevier; 2007. p. 303–27.

57. Behan JL, Smith KD. The analysis of glycosylation: a continued need for high pH anion exchange chromatography. Biomed Chromatogr. 2011;25:39–46.
58. Rohrer JS, Basumallick L, Hurum D. High-performance anion-exchange chromatography with pulsed amperometric detection for carbohydrate analysis of glycoproteins. Biochemistry. 2013;78:697–709.
59. Carabetta S, Di Sanzo R, Campone L, Fuds S, Rastrelli L, Russo M. High-Performance anion exchange chromatography with pulsed amperometric detection (HPAEC–PAD) and chemometrics for geographical and floral authentication of honeys from southern Italy (*Calabria region*). Foods. 2020;9:1625.
60. Zhang ZQ, Khan NM, Nunez KM, Chess EK, Szabo CM. Complete monosaccharide analysis by high-performance anion-exchange chromatography with pulsed amperometric detection. Anal Chem. 2012;84:4104–10.
61. Corradini C, Cavazza A, Bignardi C. High-performance anion-exchange chromatography coupled with pulsed electrochemical detection as a powerful tool to evaluate carbohydrates of food interest: principles and applications. Int J Carb Chem. (1012):1–13
62. André S, Kaltner H, Manning J, Murphy P, Gabius H-J. Lectins: getting familiar with translators of the sugar code. Molecules. 2015;20:1788–823.
63. O'Connor BF, Monaghan D, Cawley J. Lectin affinity chromatography (LAC). Methods Mol Biol. 2017;1485:411–20.
64. Kulloli M, Hancock WS, Hincapie M. Automated platform for fractionation of human plasma glycoproteome in clinical proteomics. Anal Chem. 2010;82:115–20.
65. Lee A, Nakano M, Hincapie M, Kolarich D, Baker MS, Hancock WS, Packer NH. The lectin riddle: glycoproteins fractionated from complex mixtures have similar glycomic profiles. OMICS. 2010;14:487–99.
66. Sharon N. Lectins: past, present and future. Biochem Soc Trans. 2008;36:1457–60.
67. Simpson RJ, editor. Purifying proteins for proteomics: a laboratory manual. New York: Cold Spring Harbor Laboratory Press; 2004.
68. Ayyar BV, Arora S, Murphy C, O'Kennedy R. Affinity chromatography as a tool for antibody purification. Methods. 2012;56:116–29.
69. Domann PJ, Pardos-Pardos AC, Fernandes DL, Spencer DI, Radcliffe CM, et al. Separation-based glycoprofiling approaches using fluorescent labels. Proteomics. 2007;7(suppl. 1):70–6.
70. Ruhaak LR, Zauner G, Huhn C, Bruggink C, Deelder AM, Wuhrer M. Glycan labeling strategies and their use in identification and quantification. Anal Bioanal Chem. 2010;397:3457–81.
71. Ruhaak LR, Steenvoorden E, Koeleman CAM, Deelder AM, Wuhrer M. 2-Picoline-borane: a non-toxic reducing agent for oligosaccharide labeling by reductive amination. Proteomics. 2010;10:2330–6.
72. Song X, Johns BA, Ju H, Lasanajak Y, Zhao C, Smith DF, et al. Novel cleavage of reductively aminated glycan-tags by N-bromosuccinimide to regenerate free, reducing glycans. ACS Chem Biol. 2013;8:2478–83.
73. Wang CJ, Fan WC, Zhang P, Wang ZF, Huang LJ. One pot nonreductive O-glycan release and labeling with 1-phenyl-3-methyl-5-pyrazolone followed by ESI-MS analysis. Proteomics. 2011;11:4229–42.
74. Zauner G, Koeleman CA, Deelder AM, Wuhrer M. Mass spectrometric O-glycan analysis after combined O-glycan release by β-elimination and 1-phenyl-3-methyl-5-pyrazolone labeling. Biochim Biophys Acta. 2012;1820:1420–8.
75. Wan D, Yang H, Song F, Liu Z, Liu S. Identification of isomeric disaccharides in mixture by the 1-phenyl-3-methyl-5-pyrazolone labeling technique in conjunction with the electrospray ionization tandem mass spectrometry. Anal Chim Acta. 2013;780:36–45.
76. Hu Y, Wang T, Yang X, Zhao Y. Analysis of compositional monosaccharides in fungus polysaccharide by capillary zone electrophoresis. Carbohydr Polym. 2014;102:481–8.
77. Song E, Pyreddy S, Mechref Y. Quantification of glycopeptides by multiple reaction monitoring liquid chromatography/tandem mass spectrometry. Rapid Commun Mass Spectrom. 2012;26:1941–54.
78. Hong Q, Lebrilla CB, Miyamoto S, Ruhaak LR. Absolute quantitation of immunoglobulin G and its glycoforms using multiple reaction monitoring. Anal Chem. 2013;85:8585–93.

79. Xu G, Amicucci MJ, Cheng Z, Galermo AG, Lebrilla CB. Revisiting monosaccharide analysis—Quantitation of a comprehensive set of monosaccharides using dynamic multiple reaction monitoring. Analyst. 2017;143:200–7.
80. Kang P, Mechref Y, Novotny MV. High-throughput solid-phase permethylation of glycans prior to mass spectrometry. Rapid Commun Mass Spectrom. 2008;22:721–34.
81. Zhou S, Wooding K, Mechref Y. Analysis of permethylated glycan by liquid chromatography (LC) and mass spectrometry (MS). Methods Mol Biol. 2017;1503:83–96.
82. Lu G, Crihfield CI, Gattu S, Veltri LM, Holland LA. Capillary electrophoresis separations of glycans. Chem Rev. 2018;118:7867–85.
83. Voeten RLC, Ventouri IK, Haselberg R, Somsen GW. Capillary electrophoresis: trends and recent advances. Anal Chem. 2018;90:1464–81.
84. Rovio S, Yli-Kauhaluoma J, Siren H. Determination of neutral carbohydrates by CZE with direct UV detection. Electrophoresis. 2007;28:3129–35.
85. Campa C, Rossi M. Capillary electrophoresis of neutral carbohydrates: mono-, oligosaccharides, glycosides. Methods Mol Biol. 2008;384:247–305.
86. Suzuki S. Recent developments in liquid chromatography and capillary electrophoresis for the analysis of glycoprotein glycans. Anal Sci. 2013;29:1117–28.
87. Zhao L, Chanon AM, Chattopadhyay N, Dami IE, Blakeslee JJ. Quantification of carbohydrates in grape tissues using capillary zone electrophoresis. Front Plant Sci. 2016;7:818.
88. Kamoda S, Kakehi K. Capillary electrophoresis for the analysis of glycoprotein pharmaceuticals. Electrophoresis. 2006;27:2495–504.
89. Mantovani V, Galeotti F, Maccari F, Volpi N. Recent advances in capillary electrophoresis separation of monosaccharides, oligosaccharides, and polysaccharides. Electrophoresis. 2018;39:179–89.
90. Zamfir AD. Applications of capillary electrophoresis electrospray ionization mass spectrometry in glycosaminoglycan analysis. Electrophoresis. 2016;37:973–86.
91. Amon S, Zamfir AD, Rizzi A. Glycosylation analysis of glycoproteins and proteoglycans using capillary electrophoresis-mass spectrometry strategies. Electrophoresis. 2008;29:2485–507.
92. Pioch M, Bunz SC, Neususs C. Capillary electrophoresis/mass spectrometry relevant to pharmaceutical and biotechnological applications. Electrophoresis. 2012;33:1517–30.
93. Mechref Y, Novotny MV. Glycomic analysis by capillary electrophoresis-mass spectrometry. Mass Spectrom Rev. 2009;28:207–22.
94. Qu Y, Sun L, Zhang Z, Dovichi NJ. Site-specific glycan heterogeneity characterization by HILIC solid-phase extraction, RPLC fractionation, and capillary zone electrophoresis-electrospray ionization-tandem mass spectrometry. Anal Chem. 2018;90:1223–33.
95. Olajos M, Hayos P, Bonn GK, Guttman A. Sample preparation for the analysis of complex carbohydrates by multicapillary gel electrophoresis with light-emitting diode induced fluorescence detection. Anal Chem. 2008;80:4241–6.
96. Sarazin C, Delaunay N, Costanza C, Eudes V, Gareil P. Application of a new capillary electrophoretic method for the determination of carbohydrates in forensic, pharmaceutical, and beverage samples. Talanta. 2012;99:202–6.
97. Kinoshita M, Kakehi K. Capillary-based lectin affinity electrophoresis for interaction analysis between lectins and glycans. Methods Mol Biol. 2014;1200:131–46.
98. Volpi N, editor. Capillary electrophoresis of carbohydrates. Totowa: Humana Press; 2011.
99. Toraño JS, Ramautar R, De Jong G. Advances in capillary electrophoresis for the life sciences. J. Chromatogr B. 2019;1118–1119:116–36.
100. Ruiz-Matute AI, Hernández-Hernández O, Rodríguez-Sánchez S, Sanz ML, Martínez-Castro I. Derivatization of carbohydrates for GC and GC-MS analyses. J Chromatogr B. 2011;879:1226–40.
101. Harvey DJ. Derivatization of carbohydrates for analysis by chromatography; electrophoresis and mass spectrometry. J Chromatogr B. 2011;879:1196–225.
102. Xia Y-G, Sun H-M, Wang T-L, Liang J, Yang B-Y, Kuang H-X. A modified GC-MS analytical procedure for separation and detection of multiple classes of carbohydrates. Molecules. 2018;23:1284.

Chapter 6
Monosaccharide Composition Analysis

Abstract A monosaccharide composition determination is the first step in the process of characterizing complex carbohydrate. Two frequently used methods, gas chromatography and high-performance liquid chromatography, will be discussed. Usually, the second step in carbohydrate structure characterization is the determination of the linkage types of the constituting monosaccharides. These assignments are obtained by gas–liquid chromatography-mass spectrometry. This chapter provides detailed protocols for monosaccharide and linkage analysis, as well as the absolute configuration (D/L) determination.

Keywords HPLC · HPAEC · GC-MS · CE · TLC · EI-MS · Hydrolysis · Methanolysis · Trimethylsilylation · TMS methyl glycosides · Permethylation · Alditol acetates · PMAA

6.1   Introduction

The colorimetric methods, as discussed in Chap. 4, provide an estimate of the total carbohydrates present in a sample according to the various sugar classes (aldohexoses, aldopentoses, aminosugars, glycosyluronic acids, etc.), but the real identity of the individual monosaccharides is not provided. For an exact carbohydrate analysis (i.e., identification and quantification of the constituting monosaccharides), it is often necessary to isolate the carbohydrate-containing biological sample in highly pure form. It reaches too far for this book to discuss the protocols of the specific methods for isolation and purification of target glycoconjugates, such as glycopeptides, glycoproteins, glycolipids, proteoglycans, or polysaccharides from biological sources. Many methods and protocols can be found in the books, dedicated to these subjects, like *"Protein Purification Protocols"* [1], *"Purifying Proteins for Proteomics: A Laboratory Manual"* [2], or *"Guide to protein purification"* [3].

In the case of the availability of a pure carbohydrate-containing compound, there are two ways to analyze the carbohydrate part. On the one hand, direct determination of the monosaccharide composition without the prior release of glycans. On the other hand, first cleave and isolate the complete glycan part(s), which is then partly used for the determination of the molar monomer composition and partly used for

© Springer Nature Switzerland AG 2021   127
G. J. Gerwig, *The Art of Carbohydrate Analysis*, Techniques in Life Science and
Biomedicine for the Non-Expert, https://doi.org/10.1007/978-3-030-77791-3_6

further analyses. The method of isolation of the carbohydrate part depends on the aglycone, the glycan types, and the purpose of analysis.

Often, when the presence of carbohydrate, for instance in a protein or lipid, is confirmed colorimetrically, the next step is to find out directly the composition of the glycans by a so-called monosaccharide analysis. Such a sugar analysis consists of three steps:

1. solvolysis of the constituent monosaccharides using aqueous acidic hydrolysis or methanolysis
2. separation of the mixture of monosaccharides using chromatography and
3. identification of the individual monosaccharides using retention times and mass spectrometry, and quantification using peak areas.

A known amount of internal standard (IS) similar to the components of interest, but not present as a constituent in the analytical sample, is usually added. The purpose of the internal standard is to correct for any physical or chemical losses that might occur during sample processing and subsequent chromatographic analysis and to standardize the results across separate analysis runs.

High-performance liquid chromatography (HPLC), high-pH anion-exchange chromatography (HPAEC), and gas-liquid chromatography (GLC) are the most widely used techniques for the qualitative and quantitative analysis of mixtures of monosaccharides [4]. But capillary electrophoresis (CE) is also applied (see Sect. 5.5, Fig. 5.7). In several cases, TLC is still used for the separation and identification of mono-/oligosaccharides by reference to known standards. Most frequently, gas-liquid chromatography (GLC) is used for monosaccharide analysis. GLC excellently separates monosaccharides after converting them into volatile derivatives. GLC has the disposal of sensitive detection techniques (e.g., flame-ionization detector (FID) or mass spectrometer (MS)) to quantify and identify the monosaccharides. GLC(-MS) is indispensable for absolute configuration (D/L) determination and linkage analysis. Next to GLC, simple monosaccharide composition analysis is increasingly performed by using the HPAEC-PAD method [5].

6.2 Gas-Chromatographic Monosaccharide Analysis Using TMS Methyl Glycosides

The monosaccharide analysis method performed by GLC consists of (1) quantitative methanolysis of carbohydrates into monosaccharide components (methyl glycosides); (2) volatilization of the methyl glycosides; and (3) separation and quantitation of the derivatives on a GC column. A flame ionization detector (FID) is the most commonly used detector. GC-FID is a robust and easy-to-use technique that shows high sensitivity, a large linear response range and low noise, and is very efficient for quantitative work. Quantification of the individual monosaccharide by peak area is usually calculated compared to an internal standard using predetermined response factors (RF) by standards for each sugar, analyzed under identical conditions. Instead of flame ionization detection, a gas chromatograph can easily be coupled to a mass spectrometer (MS) which affords rather complete important structural information. GLC-MS is a necessity in the standard technique for linkage analysis of carbohydrates.

Since the cleavage of the glycosidic linkages by methanolysis is very effective and causes less destruction of the monosaccharides than aqueous acid hydrolysis (see Sect. 5.2.4), this solvolytic technique was introduced as an elegant method for sugar analysis by gas chromatography about 45 years ago. At this moment, it is still one of the most used methods to analyze the monosaccharide composition of carbohydrate-containing samples (Protocol 23).

Methanolysis is carried out in anhydrous methanol-containing hydrogen chloride at elevated temperatures. During methanolysis, the liberated monosaccharides are converted into methyl glycosides. Sialic acids and uronic acids are stable toward methanolysis and are converted to their methyl esters methyl glycosides. To make the methyl glycosides volatile for gas chromatographic determination, the polar hydroxyl (–OH) groups are converted to trimethylsilyl ethers [–OSi(CH$_3$)$_3$]. The identification of the TMS methyl glycosides (methyl esters) is achieved according to their GC retention times and mass spectra by using GLC with flame ionization detector and/or mass spectrometry (MS) with electron impact ionization (EIMS) [6].

Remember that free monosaccharides in solution exist in five different forms: one acyclic form and two anomers (α/β) of the furanose form and two anomers (α/β) of the pyranose form (see Sect. 2.1.1, Fig. 2.1). After methanolysis, the different cyclic tautomeric forms are fixated as methyl glycosides (shown for galactose in Fig. 6.1), resulting in multiple peaks in the gas chromatogram. Many investigators consider this as an advantage because monosaccharides can be identified according to their fixed ratio of the α/β furanose and α/β pyranose forms (a characteristic pattern of the peaks of which the areas are constant). However, samples containing several different monosaccharides can give complicated gas chromatograms with some overlapping peaks, which may interfere with quantitative measurements. Nevertheless, the methanolysis/TMS/GLC method allows the simultaneous identification of neutral, amino, and acidic sugars (e.g., uronic acids and sialic acids).

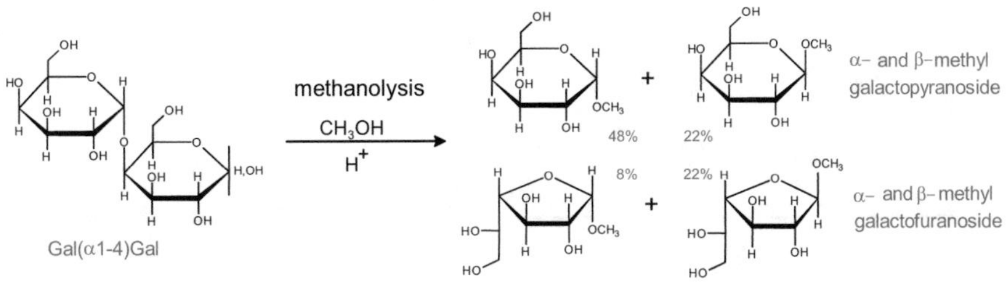

Fig. 6.1 After methanolysis, the different tautomeric forms give multiple methyl glycosides in a fixed ratio, as demonstrated for galactose

The method is generally applied for the qualitative and quantitative sugar analysis of glycoconjugates (glycoproteins and glycolipids) and oligo-/polysaccharides. A minimum amount of sample, containing ~100–250 ng of each monosaccharide, is usually required for reliable results. For quantitative determinations, a known amount of an internal standard (IS), such as mannitol or (*myo*/*meso*-)inositol, is added to the target sample at the beginning of the analysis procedure.

Quantification is achieved by comparison of peak areas with the peak area of the IS. Usually, a set of standards, containing a mix of appropriate monosaccharides (50 nmol), including IS (10 nmol), is methanolized and analyzed in parallel for comparison with the target samples. The detector response for the individual monosaccharides relative to the IS in standard run is determined and used as a correction factor for accurate quantitative determination.

Applying the methanolysis methodology, several features should be taken into account:

1. Samples should be essentially salt- and detergent free
2. Carboxyl groups of acidic monosaccharides will be converted to methyl esters
3. *O*-acetyl and *N*-acyl groups are eliminated; sulfate and phosphate esters are lost
4. Acetyl groups of *N*-acetyl amino sugars are cleaved nearly completely, and therefore a re-*N*-acetylation step has to be incorporated in the procedure
5. Cleavage of the uronic acid glycosidic linkage may be incomplete in some cases
6. Several glycosyluronic acids produce a certain percentage of 3,6-lactones
7. In glycoprotein N-glycan analysis, the linkage between GlcNAc and Asn is split only to a very limited extent, and mainly this GlcNAc residue is liberated as free monosaccharide instead of its methyl glycoside

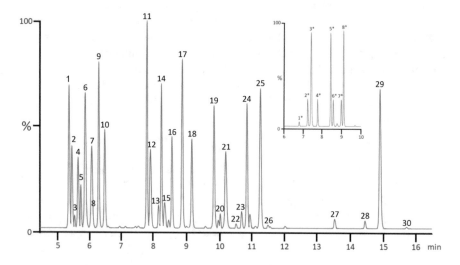

Fig. 6.2 Gas chromatogram of TMS (methyl ester) methyl glycosides of commonly occurring monosaccharides in glycoproteins and glycolipids on an EC-1 column (30 m). Oven temperature program: 140 °C for 3 min, then 140–250 °C, 8°/min. Insert: Glycosyluronic acids

8. In oligosaccharide-alditol analysis, the alditol residues give rise to anhydro derivatives to a certain extent (dehydration)
9. In glycolipid glycan analysis, the ceramide residues are degraded to sphingosine and fatty acid methyl esters, which may interfere with the sugar analysis (methanolysis/TMS/GLC)
10. 2-ketoses (e.g., fructose) are completely degraded
11. Samples to be trimethylsilylated must be very dry because water reacts with the trimethylsilylation reagents and hydrolyzes the trimethylsilylated products.

Typically, the procedure is as follows. After methanolysis, samples are neutralized with silver carbonate, re-N-acetylated, dried, trimethylsilylated, and analyzed by GLC-(FID or MS). A typical gas chromatogram of a standard mixture of monosaccharides as TMS (methyl ester) methyl glycosides is depicted in Fig. 6.2.

Peak assignment

Nr.	Peak	Nr.	Peak	Nr.	Peak	Nr.	Peak
1	Ara (α-p)	11	Man (α-p)	21	GalNAc (α,β-f)	1*	GlcA (lactone-f)
2	Ara (β-p)	12	Gal (β-f)	22	Man-ol (1-OAcetyl)	2*	GlcA (lactone-f)
3	Xyl (α,β-f)	13	Man (β-p)	23	GlcNAc (β-p)	3*	GalA (β-f)
4	Fuc (β-f)	14	Gal (α-p)	24	GalNAc (α,β-p)	4*	GalA (α-f)
5	Ara (f)	15	Gal (α-f)	25	GlcNAc (α-p)	5*	GalA (α-p)
6	Fuc (α-p)	16	Gal(β-p)	26	GlcNAc (no methyl)	6*	GalA (β-p)
7	Fuc (β-p)	17	Glc (α-p)	27	2,7-anhydroNeu5Ac	7*	GlcA (β-p)
8	Fuc (α-f)	18	Glc (β-p)	28	Neu5Ac (α-p)	8*	GlcA (α-p)
9	Xyl (α-p)	19	IS = Mannitol	29	Neu5Ac (β-p)		
10	Xyl (β-p)	20	GlcNAc (α-f)	30	Neu5,9Ac2 (β-p)		

p pyranoside, f furanoside

Protocol 23. Monosaccharide Composition Analysis Using Methanolysis and GLC (TMS Derivatives)

Materials
Teflon-faced, screw-capped glass tubes (100 × 13 mm; Reacti-Vials, Duran Group Mainz)
Pasteur pipettes (long size 230 mm, WU Mainz)
Lyophilizer
Vacuum desiccator with phosphorus pentoxide (P_2O_5)
Heating block (Pierce Reacti-Therm, Rockford, IL)
Evaporation system with Nitrogen (N_2) flow
Bench-top centrifuge (suitable for 13 × 100 mm tubes)
10-μL syringe (sharp needle)

Gas chromatograph coupled with a flame-ionization detector (FID) or electron-impact mass spectrometer (GLC-EIMS)

GC column EC-1 (30 m × 0.32 mm, Alltech Associates Inc., IL, USA) or DB-1 (30 m × 0.25 mm, J&W Scientific) or ZB-1HT (30 m × 0.25 mm, Phenomenex, Torrance, CA)

Chemicals and Solvents

Silver carbonate (Ag_2CO_3)

Anhydrous methanol (CH_3OH, MeOH)

Anhydrous pyridine (C_5H_5N)

Acetic anhydride [Ac_2O; $(CH_3CO)_2O$]

Acetyl chloride (p.a.; 78.5; 98%; Aldrich)

Pyridine, anhydrous (Sigma Aldrich)

Anhydrous sodium sulfate (Merck, Darmstadt, Germany)

Phosphorus pentoxide (P_2O_5)

Monosaccharide standards: Usually *Std1*, a mixture of Ara, Fuc, Xyl, Man, Gal, Glc, GalNAc, GlcNAc, and Neu5Ac, each 1 mg/mL water (or 1 nmol/μL), and *Std2*, a mixture of Rha, Xyl, GlcA, GalA, and ManNAc, each 1 mg/mL water (or 1 nmol/μL), but standard mixtures can be prepared relevant to sugars that are expected in the target sample.

Internal standard (IS): Mannitol or (*myo/meso*)inositol, 1 mg/mL water (or 1 nmol/μL).

1.0 M methanolic HCl [prepared by dilution of commercial ampulla containing 3 M methanolic HCl (Supelco, USA)] or self-made: Carefully add 3.5 mL of acetyl chloride under stirring to 50 mL of anhydrous methanol (= ~1 M) [take care that no water is introduced].

Trimethylsilylation (TMS) reagent is a freshly prepared mixture of anhydrous pyridine-hexamethyldisilazane-trimethylchlorosilane, 5:1:1 (v/v/v). [CAUTION: corrosive].

Procedure

1. Prepare the exact amount of target sample (containing 5–50μg of total sugars) in a screw-capped tube with Teflon lining (screw-top septum vial).
2. Take 100μL of *Std1* in another tube and 100μL *Std2* in another tube.
3. Add an exact amount of Internal Standard solution (mannitol or inositol, 10–100 nmol or 1–10μg) to the target sample and each standard sample.
4. Lyophilize and then dry, overnight, in a vacuum desiccator (over P_2O_5) [samples must be completely dry!].
5. Add 500μL of 1.0 M HCl–methanol under nitrogen atmosphere, firmly close the tubes, and heat overnight (16–18 h) at 85 °C (vortex once after first 15 min).
6. Cool to room temperature.
7. Neutralize by adding ~15–20 mg of solid silver carbonate (Ag_2CO_3), check with a droplet on pH paper.
8. Re-*N*-acetylation: Add 20μL of acetic anhydride to the sample, mix well and leave at room temperature for at least 4 h in the dark. (If samples do not contain *N*-acetylamino sugars, this step can be skipped). [avoid using too much acetic anhydride, otherwise *O*-acetylation of primary hydroxyl groups occurs].

9. Centrifuge (1 min, ~300 ×g) and collect supernatant in a clean glass tube.
10. Wash silver salts pellet with 500µL methanol, centrifuge, and combine super-
    natant with the previous supernatant.
11. Dry up the supernatant pool with a stream of N_2 and dry further, overnight, in a
    vacuum desiccator (over P_2O_5).
12. Derivatization: Add 200µL of trimethylsilylation (TMS) reagent, close tube,
    vortex, and keep the mixture for 30 min at room temperature (or 10 min, 70 °C).
    [Chemical reaction: (sugar)-OH + Me_3Si-NH-$SiMe_3$ + Me_3Si-Cl→(sugar)-
    $OSiMe_3$ + $NH_4Cl↓$] (then, eventually, centrifuge for 5 sec to sediment the white
    precipitate).
13. Analyze about 1–3µL clear sample solvent by GLC-FID/EIMS on an EC-1 (or
    DB-1or ZB-1HT) column (30 m × 0.32 mm i.d.) with a temperature program:
    140 °C for 3 min, then 140–250 °C, 8°/min. The injector temperature is 230 °C
    and FID 280 °C; in case of EI-MS, the source temperature is 200 °C, and the
    electron voltage is 70 eV. Helium (He) flow rate: 3 mL/min. The total run time
    is 17 min (Fig. 6.2).
14. Identification of the TMS methyl glycosides (methyl esters) is based upon the
    comparison of relative retention times with those of standards run under the
    same conditions. The relative retention time is the quotient of the retention time
    of the relevant peak and the IS peak. Quantification can be calculated by the
    total peak areas of each monosaccharide compared to the internal standard. The
    EI-mass spectrum of the peaks assists their identification. As an example, the
    mass spectra of some TMS methyl glycosides are depicted in Fig. 6.3.

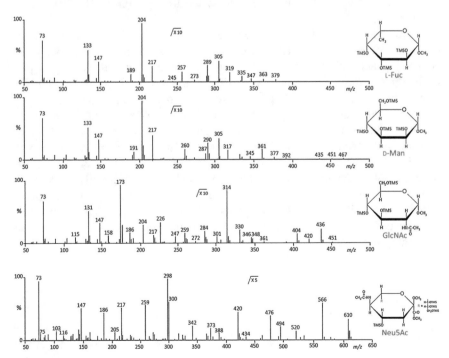

Fig. 6.3 Examples of (GC-)EI-mass spectra of some TMS methyl glycosides. For assignment of
m/z fragmentation peaks, see Chap. 11, Table 11.1

6.3   Gas-Chromatographic Monosaccharide Analysis Using Alditol Acetates

The alditol acetate method (Protocol 24) is another gas chromatographic method for the determination of the monosaccharide composition of complex carbohydrates, including polysaccharides, glycoproteins, and glycolipids. Cleavage of the glycosidic linkages is obtained by aqueous acid hydrolysis, which produces monosaccharides consisting of α/β pyranose/furanose ring forms. Direct derivatization of these monosaccharide forms (e.g., by trimethylsilylation) followed by GLC analysis gives rise to multiple overlapping peak patterns in the chromatogram, even more than observed with TMS methyl glycosides (see Sect. 6.2, Fig. 6.2). To eliminate the multiple peak patterns of the α/β pyranose/furanose forms, the monosaccharides are reduced directly after the cleavage of the glycosidic linkages. This chemical reduction affects the carbonyl group involved in a ring formation, which leads to the corresponding sugar alcohol (alditol) as a single compound for each monosaccharide. Acetylation of the free hydroxyl groups affords volatile derivatives, called alditol acetates, which can easily be separated by GLC, yielding a single peak per monosaccharide (Figs. 6.4 and 6.5). Identification is accomplished through retention times obtained from reference monosaccharide alditols and quantification by comparison of peak areas. An exact amount of internal standard (e.g., inositol) is usually added at the beginning of the analysis. Furthermore, identification can be confirmed by a mass spectrum when the gas chromatograph is coupled to a mass spectrometer (GLC-MS) using electron impact (EI) ionization.

A drawback of the alditol acetate method is that it involves multiple manual operation steps, including tedious evaporations, which makes it rather time-consuming and vulnerable to material loss. Starting material >100μg is preferred. Furthermore, the reduction of different aldoses and ketoses can yield the same alditol upon reduction [e.g., glucose and gulose give the same alditol (glucitol), arabinose and lyxose give the same alditol (arabitol)]; at the same time, fructose (a ketose) produces mannitol and glucitol. Identification can be solved by using a deuterated reducing agent ($NaBD_4$). The newly formed alditols will contain a deuterium atom at C1 or C2, respectively (see also Sect. 6.6). Sialic acids cannot be detected, and glycosyluronic acids cannot directly be detected as alditol acetates unless they are prereduced to their neutral sugar counterparts. Care has to be taken to use clean reagents and solvents for hydrolysis, reduction, and acetylation to avoid extraneous contaminating peaks in the gas chromatograms.

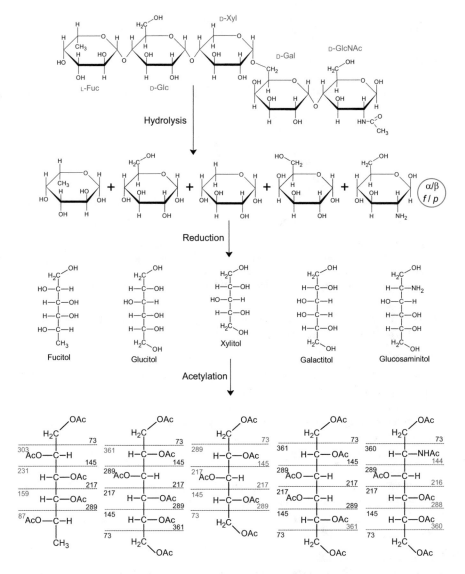

Fig. 6.4 Example of the Alditol Acetate analysis procedure of a hypothetical pentasaccharide. Mass spectral fragmentation (typical red fragments) of the alditol acetate derivatives of pentoses (shown for xylose but same for ribose or arabinose), deoxyhexoses (shown for fucose but same for rhamnose or quinovose), hexoses (shown for glucose and galactose but same for mannose), and *N*-acetyl hexosamines (shown for GlcNAc but is same for GalNAc or ManNAc)

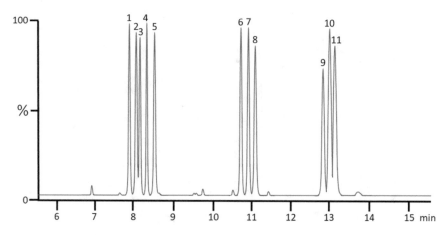

Fig. 6.5 Gas chromatogram of Alditol Acetates on an EC-1 column (30 m). Oven temperature program: 140 °C for 3 min, then 140–250 °C, 8°/min. The peaks represent (1) Rhamnitol, (2) Ribitol, (3) Fucitol, (4) Arabitol, (5) Xylitol, (6) Mannitol, (7) Glucitol, (8) Galactitol, (9) *N*-acetyl glucosaminitol, (10) *N*-acetyl mannosaminitol, and (11) *N*-acetyl galactosaminitol

Protocol 24. Monosaccharide Composition Analysis Using Acid Hydrolysis and GLC(-EIMS) (Alditol Acetates Derivatives)

Materials
Teflon-faced, screw-capped glass tubes (100 × 9 mm; GL 14, Duran Group Mainz)
Pasteur pipettes (long size 230 mm, WU Mainz)
Lyophilizer
Vacuum desiccator with phosphorus pentoxide (P_2O_5)
Heating block (Pierce Reacti-Therm, Rockford, IL)
Evaporation system with Nitrogen (N_2) flow
10-μL syringe (sharp needle)
Gas chromatograph coupled with a flame-ionization detector (FID) or electron-impact mass spectrometer (GLC-EIMS)
GC column EC-1 (30 m × 0.32 mm, Alltech Associates Inc., IL, USA) or DB-1 (30 m × 0.25 mm, J&W scientific)

Chemicals and Solvents
Sodium borohydride ($NaBH_4$ 37.8; Sigma) [eventually, Sodium borodeuteride ($NaBD_4$)]
Prepare 10 mg/mL sodium borohydride (or deuteride) in 1 M NH_4OH
Methanol (Anhydrous CH_3OH, MeOH)
Pyridine (Anhydrous C_5H_5N)
Acetic acid (glacial HOAc; 60.05; Aldrich) Prepare10% acetic acid solution
Acetic anhydride [Ac_2O; $(CH_3CO)_2O$]
Isopropanol (2-propanol)
Phosphorus pentoxide (P_2O_5)
2 M TFA (7.4 mL TFA in 50 mL H_2O) [store at room temperature]

1 M NH$_4$OH (5.5 mL 35% NH$_4$OH in 94.5 mL H$_2$O)

Monosaccharide standard: a mixture of Ara, Rha, Fuc, Xyl, Man, Gal, Glc, GalNAc, and GlcNAc, each 1 mg/mL water (or 1 nmol/μL), but standard mixtures can be prepared relevant to sugars that are expected in the target sample. The standard mixture is carried through the entire procedure in parallel with the samples to be analyzed.

Internal standard (IS): Mannitol or (*myo/meso*)inositol, 1 mg/mL water.

Procedure

Hydrolysis

1. Prepare 0.1–0.5 mg of the target sample (glycoconjugate/polysaccharide) in a screw-capped tube with Teflon lining (screw-top septum vial).
2. Add the exact amount (less than the amount of carbohydrate in the sample) of Inositol (IS).
3. Add 500μL of 2 M TFA, under N$_2$ atmosphere, and screwcap tightly.
4. Incubate in a heating block for 2 h at 120 °C (or 16 h at 100 °C).
   [for samples containing aminosugars, sometimes hydrolysis is performed with 4 N HCl for 6–18 h at 100 °C].
5. Cool to room temperature and dry down by evaporation of the solvent with N$_2$ stream (eventually, use a water bath 40 °C).
6. Add 100μL of isopropanol, vortex, and dry down again.

Reduction

1. Add 250μL of freshly prepared NaBH$_4$ solution (10 mg NaBH$_4$ in 1 mL of 1 M NH$_4$OH).
2. Incubate at room temperature for at least 3 h with intermittent shaking (or overnight at 4 °C).
3. Neutralize dropwise with 4 M (or 10% v/v) acetic acid (sodium borohydride decomposes to sodium borate) until bubbling (release of H$_2$) has ceased; check end pH ~6 with a droplet on pH paper.
4. Add 400μL of MeOH/HOAc (9:1, v/v), vortex, and dry down with N$_2$ stream.
5. Add 200μL of MeOH and dry down. Repeat 4× this step 5 (tetramethyl borate evaporates).

Acetylation

1. Add 300μL of acetic anhydride/pyridine (1:1, v/v), screwcap tightly.
2. Heat for 30 min at 120 °C, with intermittent vortexing (or 2 h at 100 °C).
3. Cool to room temperature, dry down by evaporation with a stream of N$_2$ (water bath 40 °C).
4. Add 100μL of toluene and dry completely with a stream of N$_2$ (do not overdry, as very volatile terminal-stemming pentitols can be lost).
5. Redissolve in DCM (eventually, centrifuge and collect clear supernatant).

GLC

> Inject ~1–3μL clear solvent on GLC-FID/EIMS with an EC-1 (DB-1; AT-1) column (30 m × 0.32 mm i.d.) using a temperature program: 140–250 °C, 8°/ min. The injector temperature is 230 °C and FID 280 °C; in the case of EI-MS, the source temperature is 200 °C, and the electron voltage is 70 eV. He flow rate: 3 mL/min (Fig. 6.5).

Identification of the alditol acetates is based upon the comparison of relative retention times with those of standards run under the same conditions. The relative retention time is the quotient of the retention time of the relevant peak and the IS peak. Quantification can be calculated by the peak areas of each monosaccharide (alditol acetate) compared to the internal standard. In order to obtain robust results, it is recommended to repeat the total analysis three times with identical samples. The EI-mass spectrum of the peaks assists their identification. The EI-mass spectra of the alditol acetates stemming from pentoses, (deoxy)hexoses, and N-acetyl hexosamines will show fragmentation patterns as illustrated in Fig. 6.4.

6.4 Determination of D/L Configuration of Monosaccharides

Monosaccharides are chiral molecules, usually existing in cyclic form, having D- or L-configuration (also called the absolute configuration) (see Sect. 2.1.2). This stereochemistry is based on the D/L configuration of glyceraldehyde, the simplest "monosaccharide" (aldotriose), which has a single (asymmetric) carbon atom (C2). Glyceraldehyde can exist in two nonsuperimposable mirror images called enantiomers.

Higher carbon monosaccharides, depicted in Fisher projection (Chap. 2, Table 2.1) with the most oxidized carbon at the top, follow the glyceraldehyde configuration assignment. The highest numbered asymmetric carbon atom dictates the absolute configuration (–OH on the left = L; –OH on the right = D).

Chiral monosaccharide enantiomers can have divergent biological activities but have identical chemical and physical properties, except for the rotation of polarized light (− left rotation, which agreed by chance with the L-configuration; + right rotation, which agreed by chance with the D-configuration). By consequence, this property is traditionally used to determine the absolute configuration of a monosaccharide, but the drawback is that you need a considerable amount of the monosaccharide and in very pure condition for accurate measuring of specific rotation.

For a precise monosaccharide analysis, the determination of the absolute (D or L) configuration is a prerequisite, especially for monosaccharide constituents of isolated novel polysaccharides. While most natural-occurring sugars are assumed to have the D-configuration, the less common L-forms (L-Rha, L-Ara, L-Gal) have also been found, e.g., in bacterial polysaccharides, marine compounds, and conotoxins [7]. Commonly used chromatographic methods cannot discriminate enantiomers from each other, but several techniques that allow the simultaneous measurement of D/L enantiomeric sugars can be found in the scientific literature.

Two strategies have been successfully adopted: (1) to separate enantiomers on a column containing a chiral stationary phase or (2) to convert enantiomers to their corresponding diastereomeric derivatives by reacting with chiral derivatizing reagents and separation on a common column.

Gas chromatography using columns with a chiral center onto the stationary phase can give separation of enantiomers. For instance, trifluoroacetyl/methyl derivatives of monosaccharides can be separated on a capillary GLC column, containing chiral cyclodextrins as a stationary phase and subsequent detection by FID [8, 9].

Nowadays, HPLC with a chiral column is often used to separate enantiomers. A commercial chiral HPLC column for enantiomer separation is the Chiralpak AD-H column (Daicel, 250 × 4.6 mm) (strong polarity and ability to form multiple hydrogen bonds), which is able to separate the isomers and anomers of several free monosaccharides. Elution with hexane/ethanol/TFA [(7:3):0.1, v/v] at 25 °C allows identification of the monosaccharide and determination of its D- or L-configuration and simultaneously also the determination of the anomeric (α or β) conformation [10].

On the other hand, as mentioned, chromatographic separation of enantiomers can also be accomplished after preparing their corresponding diastereomers. This is accomplished by introducing a special chiral center in the sugar molecule through suitable derivatization, giving the enantiomers different properties. The derivatives can be analyzed on a regular C18 column [11]. Reversed-phase ultra-performance liquid chromatography with UV/MS detection (UPLC-UV/MS) has been used for simultaneous determination of the absolute configuration of monosaccharides derivatized with L-cysteine methyl ester and phenyl isothiocyanate [12].

Another technique, reductive amination using a chiral reagent, such as 2-aminopropanol, has also been employed to resolve D/L enantiomeric monosaccharides. After being aminated with one of enantiomers (R or S), the amino-alditol formed is acetylated and analyzed by GC-MS analysis. By comparison of the retention times with those of known D or L monosaccharides, the absolute configuration of the unknown monosaccharide can be established.

Recently, a D/L separation method for several monosaccharides and uronic acids was established by capillary electrophoresis using aldo-naphthyl imidazole (aldo-NAIM) derivatives [13]. Derivatization with 2,3-naphtalenediamine was performed by direct oxidative condensation of aldose with the aromatic vicinal diamines in the presence of iodine as a catalyst. Also, chiral separation of monosaccharides as ANTS and PMP derivatives can be achieved by ligand-exchange CE (LECE) using borate and a chiral capillary selector [14].

Finally, still, the most popular approach, developed already more than 40 years ago, consists of the reaction of monosaccharides with chiral alcohol, giving diastereomers as the corresponding glycosides, which are then separated as volatile derivatives on a conventional GLC column [15]. The principle has been often repeated later with slight modifications.

Briefly, after hydrolysis or methanolysis of the oligo/polysaccharide or glycoconjugate, the monosaccharides are converted into their corresponding (−)- or (+)-2

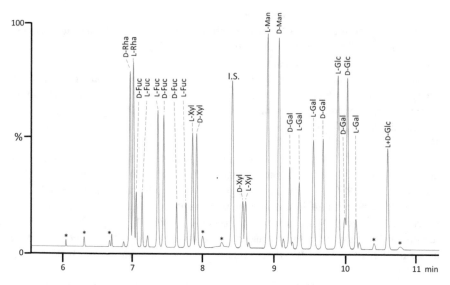

Fig. 6.6 Mixture of six D/L monosaccharides analyzed by GLC as TMS derivatives of their (−)-2 butyl glycosides on a non-chiral EC-1 column. Internal Standard (I.S.) = TMS methyl α-D-galactopyranoside. *Noncarbohydrate contaminants [15]

butyl glycosides and analyzed as TMS derivatives by GLC. The method is applicable for neutral monosaccharides and *N*-acetyl aminosugars as well as for glycosyluronic acids [16]. Although the monosaccharide derivatives give multiple peaks of ring structural and configurational isomers with some of them overlapping in the gas chromatogram, simultaneous determination of different monosaccharide enantiomers is possible for simple mixtures, even slightly contaminated (Fig. 6.6). This method is described in Protocol 25. To view the peak positions on GLC of the D and L forms of an individual monosaccharide, it is also useful to perform a butanolysis with (±)-2-butanol. The (−)-2-butyl D-glycoside and (+)-2-butyl L-glycoside of the same monosaccharide have identical elution/retention times. This also holds for the (−)-2-butyl L-glycoside and (+)-2-butyl D-glycoside of a monosaccharide.

Protocol 25. Monosaccharide D/L Configuration Determination Using (−)-2-Butanolysis and GLC (TMS Derivatives)

Materials

Teflon-faced, screw-capped glass tubes (100 × 9 mm; GL 14, Duran Group Mainz)
Pasteur pipettes (long size 230 mm, WU Mainz)
Vacuum desiccator with phosphorus pentoxide (P_2O_5)
Heating block (Pierce Reacti-Therm, Rockford, IL)
Evaporation system with Nitrogen (N_2) flow
Bench-top centrifuge
10-µL syringe (sharp needle)
Gas chromatograph coupled with a flame-ionization detector (FID) or electron-impact mass spectrometer (GLC-EIMS)

GC column EC-1 (30 m × 0.32 mm, Alltech Associates Inc., IL, USA) or DB-1 (30 m × 0.25 mm, J&W scientific) or ZB-1HT (30 m × 0.25 mm, Phenomenex, Torrance, CA)

Chemicals and Solvents
Silver carbonate (Ag_2CO_3)
Anhydrous methanol (CH_3OH, MeOH)
Anhydrous pyridine (C_5H_5N)
Acetic anhydride [Ac_2O; $(CH_3CO)_2O$]
(R)-(−)-2-butanol (Sigma-Aldrich)
Acetyl chloride (CH_3COCl)
1.0 M (−)-2-butanolic HCl [freshly prepared (R)-(−)-2-butanol/acetyl chloride = 13:1 v/v]
TMS reagent (pyridine-hexamethyldisilazane-trimethylchlorosilane, 5:1:1 v/v/v).

Procedure
Glycoconjugate or oligo-/polysaccharide samples must first be hydrolyzed with TFA or methanolyzed with HCl/MeOH as described in Sects. 5.2.3 and 5.2.4, respectively (eventually including re-*N*-acetylation) and dried, then:

1. Prepare 0.1–0.5 mg of the hydrolyzed/methanolyzed dry target sample in a glass screw-capped tube with Teflon lining (screw-top septum vial).
2. Add 300μL of S-(−)-2-butanol and add 25μL acetyl chloride, under N_2 atmosphere, then cap the tube.
3. Incubate in a heating block overnight (16–18 h) at 80 °C (vortex once after first 15 min).
4. Cool to room temperature.
5. Add 300μL methanol and ~15 mg of solid silver carbonate (Ag_2CO_3) to neutralize and check with a droplet on pH paper.
6. Re-*N*-acetylation: Add 20μL of acetic anhydride to the sample, mix well, and leave at room temperature for at least 4 h in the dark. (If samples do not contain *N*-acetylamino sugars, this step can be skipped).
7. Centrifuge (1 min, ~300 ×g) and collect supernatant in a clean glass tube.
8. Dry up the supernatant with a stream of N_2 and dry further, overnight, in a vacuum desiccator (over P_2O_5).
9. Derivatization: Add 200μL of trimethylsilylation (TMS) reagent, close tube, vortex, and keep the mixture for 30 min at room temperature (or 10 min, 70–80 °C) (then, eventually, centrifuge for 15 s to sediment white precipitate).
10. Analyze about 1–3μL clear sample solvent by GLC-FID/EIMS on an EC-1 (DB-1; AT-1) column (30 m × 0.32 mm i.d.) with a temperature program: 140 °C for 3 min, then 140–250 °C, 8°/min. The injector temperature is 230 °C and FID 280 °C; in the case of EI-MS, the source temperature is 200 °C, and the electron voltage is 70 eV. He flow rate: 3 mL/min.
11. Identification of the TMS butyl glycosides is based upon a comparison of relative retention times with those of standards, run under the same condi-

tions. The relative retention time is the quotient of the retention time of the relevant peak and the IS peak (Fig. 6.6).

6.5 Monosaccharide Analysis by HPLC Methods

6.5.1 High-pH Anion-Exchange Chromatography

HPAEC-PAD (see Sect. 5.3.6) is a particular HPLC system using a special detector. At high pH (>12), monosaccharides are ionized and interact with a nonporous, pellicular, quaternary amine stationary resin in an HPLC column, giving highly selective separation and a high resolution of nonderivatized monosaccharides due to their dissociation constants (pK_a). The very sensitive (<100 nmol/L) system is suitable for the determination of monosaccharides in glycoconjugates and polysaccharides (Protocol 26) [5, 6].

Carbohydrate-containing target samples are first hydrolyzed into the constituent monosaccharides (neutral, amino, and acidic sugars). Optimal conditions for acid hydrolysis are depending on the sample type as discussed in Sect. 5.2. In general, 2 M TFA for 3–6 h at 120 °C or 4 M HCl for 2 h at 100 °C are used. Notably, long-time incubation with TFA can result in partial epimerization of mannose to glucose. Identification of the monosaccharides is based on the correlation of retention times with standard mixtures, eventually, including an internal standard for quantification. Sometimes, co-elution of sugars occurs, depending on the column type and its temperature. For monosaccharide analysis on a CarboPac PA-1 (or PA-20) column, elution is usually performed isocratically with 15 mM NaOH. A typical chromatogram is shown in Fig. 6.7. Temperature control is fundamental for obtaining consistent results. N-acetyl hexosamines and N-acetyl/glycolyl sialic acids can also be analyzed on HPAEC. Sialic acids are released from glycoproteins by hydrolysis in 0.1 M HCl at 80 °C for 1 h (see Chap. 9).

The elution of acidic monosaccharides and oligosaccharides (e.g., N-glycans) requires stronger eluents than those used with neutral sugars. This is accomplished by using sodium acetate, isocratic or as gradient, in the sodium hydroxide eluent. Optimization of the method may be necessary due to the types of columns and commercial instrumentation used. Read the included manuals, protocols, and application notes from the supplying company.

Protocol 26. Monosaccharide Composition Analysis Using Acid Hydrolysis and HPAEC-PAD

Hydrolysis
Materials
Glass vials with Teflon-lined screw caps
Heating block

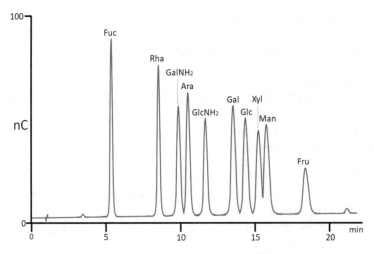

Fig. 6.7 Example of HPAEC-PAD analysis of some underivatized monosaccharides on a CarboPac PA-1 column (250 × 4.6 mm, Dionex), using 15 mM NaOH at a flow rate of 1 mL/min

Evaporation system (N_2)
Trifluoroacetic acid (2 M: 7.6 mL conc. TFA plus 42.4 mL H_2O)
Hydrochloric acid (4 M: 16.6 mL conc. HCl plus 33.5 mL H_2O)
Glc, Gal, Fuc, Man, Xyl, $GlcNH_2$, $GalNH_2$, GlcA, GalA (must all be anhydrous and dry when weight for standards)

Method

1. Start with the exact amount (5–10μg) of dry glycoprotein in a glass Pyrex tube to which, eventually, an exact amount of Internal Standard (IS, usually 2-deoxyrhamnose or 2-deoxyglucose) is added.
2. Add 300μL 4 M HCl (under N_2 atmosphere) and cap the tubes.
3. Incubate for 2 h at 100 °C in a heating block (**or** 0.5 mL of 2 M TFA for 4 h at 100 °C).
4. Cool to room temperature.
5. Evaporate the solution with a stream of N_2 (wash 3× with 100μL water followed by evaporation).
6. Dissolve in 300μL of 15 mM NaOH.

HPAEC-PAD
Materials
Dionex ICS-6000 SP HPLC system fitted with a triple-pulsed amperometric detector (PAD) (Thermo Scientific)
CarboPac PA-1 column (250 × 4.6 mm, Dionex) or PA100 (Dionex Sunnyvale, CA, USA)
Plastic pipettes
Well degassed Milli Q water

50% (w/w) NaOH solution (~19.3 M; carbonate free, do not stir)

200 mM NaOH (prepare 20.8 mL of 50% NaOH in 2 L H_2O, stir 2 min)

15 mM NaOH (prepare 1.6 mL of 50% NaOH in 2 L H_2O, stir 2 min)

Anhydrous sodium acetate (prepare 0.5 M or 1 M in 100 mM NaOH, filtrate through 0.2-μm nylon filter)

All solvents purged for 20 min with helium prior to use (and keep pressurized with He to avoid absorption of CO_2)

The mobile phase should be prepared in plastic bottles

Mono-/oligosaccharide standards

Procedure

1. Samples must be desalted prior to analysis and then dissolved in 15 mM NaOH and filtered through 0.45-μm nylon filter
2. Inject 50μL sample on CarboPac PA-1 (monosaccharides) or PA-100 (oligo/polysaccharides)
3. Elute with 15 mM NaOH at a flow rate of 1 mL/min.
4. Regenerate the column after each run with 200 mM NaOH for 10 min to remove contaminants (eventually peptides and/or amino acids) and to avoid carbonate contamination of the column.
5. Re-equilibrate with 15 mM NaOH.

6.5.2 Ultrahigh-Performance Liquid Chromatography

Recently, another HPLC method with great sensitivity (femtomoles) and speed (~10-min run) for qualitative and quantitative monosaccharide analysis was introduced [17, 18]. Monosaccharides are derivatized with 1-phenyl-3-methyl-5--pyrazolone (PMP) (see Sect. 5.4.2.2, Protocol 21). PMP derivatives of pentoses, deoxyhexoses, hexoses, hexuronic acids, and *N*-acetylhexosamines can be analyzed by ultrahigh-performance liquid chromatography triple quadrupole mass spectrometry (UHPLC/QqQ-MS) using dynamic multiple reaction monitoring (dMRM) (Fig. 6.8). An Agilent 6495 QqQ mass spectrometer coupled with an Agilent 1290 infinity II UHPLC system, equipped with an Agilent ZORBAX Eclipse Plus C18 column, can be used with gradient elution consisting of the aqueous mobile phase A (25 mM ammonium acetate in 5% acetonitrile in water, pH 8.2) and the organic mobile phase B (95% acetonitrile in water). Positive ion-mode electrospray ionization (ESI) and collision-induced dissociation (CID) are applied. In multiple reaction monitoring (MRM), the concentration of an unknown sample is determined by comparing its MS response to that of a known standard, using mass filtering (see Chap. 11).

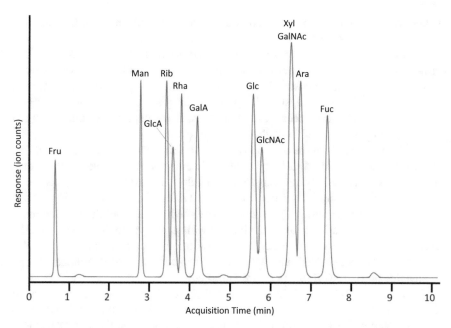

Fig. 6.8 Example of UHPLC/dMRM chromatogram of PMP-labeled monosaccharides on an Agilent ZORBAX Eclipse Plus C18 column, according to [17]

6.6   Determination of Carbohydrate Linkages by Methylation Analysis

6.6.1   Chemical Aspects

A monosaccharide in a complex carbohydrate can be substituted at different positions (see Chap. 2). Monosaccharides can be linked among themself by different linkages, such as (1→2), (1→3), (1→4), (1→5) and (1→6). The so-called methylation analysis is a powerful method, widely used, to obtain structural information, such as linkage patterns, branching points, and monosaccharide ring sizes in oligo-/polysaccharides. These data are obtained by analyzing partially *O*-methylated alditol acetates (PMAAs) derived from the sample. The sample can consist of pure oligo/polysaccharide or glycoconjugate, like glycoprotein or glycolipid.

Methylation analysis will not show the difference between D- and L-sugars and does not reveal the sequence of the constituent monosaccharides (i.e., which residues are attached to each other) and does not indicate the α or β configuration of the linkages. However, methylation analysis remains a standard method in carbohydrate research, in combination with other methods.

The traditional methylation analysis of carbohydrates is performed with the methylsulfinyl carbanion and methyl iodide in dimethyl sulfoxide (DMSO). For many years, the method has been modified and optimized. In 1984, Ciucanu and Kerek published the nowadays regularly applied method which is based on the utilization of methyl iodide, DMSO, and solid sodium hydroxide (Protocol 27). It is critical to achieve complete permethylation for a correct analysis. Consequently, it should be carried out under strictly controlled conditions (e.g., absence of water and salts, an adequate base catalyst, and appropriate reagents) to avoid oxidative degradation [19–21]. Take care, because in some cases, an incomplete permethylation can still occur, in particular, of the hydroxyl group on C3.

The basic principle of the method is to perform a fully O-methylation of all free hydroxyl groups ($-OH \rightarrow OCH_3$) of the constituent monosaccharides in the native oligo-/polysaccharide, using anhydrous dimethyl sulfoxide (DMSO) as polar aprotic solvent. The initial reaction for the methylation is illustrated in Fig. 6.9. A suspension of NaOH/DMSO (as a strong base) deprotonates all the free hydroxyl groups on the saccharide to alkoxide ions, allowing methyl iodide to react with the unstable carbanions to form methyl ethers resulting in a permethylated oligo/polysaccharide. Acetamido and amino groups, if present, are also deprotonated, and methylated and carboxyl functions are esterified.

However, the basic conditions of permethylation lead to the loss of O-linked acetyl groups. Also, phosphate and sulfate can partially be lost if conditions are too rigorous. Better to perform a prior desulfation (50 mM methanolic-HCl, 18 h at RT) because when sulfates are not removed, they are not methylated and remain negatively charged or result in sodium sulfate groups.

During acid hydrolysis of the purified permethylated product, the methyl ether groups are resistant, but the glycosidic linkages (including the methyl glycoside

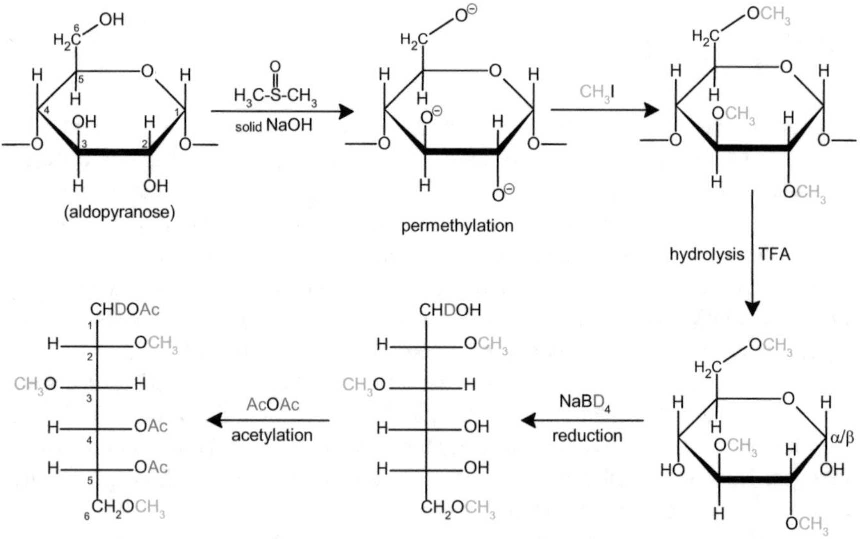

Fig. 6.9 Schematic presentation of permethylation in dimethyl sulfoxide (DMSO), followed by hydrolysis (TFA), reduction (NaBD$_4$), and acetylation (Ac$_2$O/pyridine)

linkage at the reducing end of an oligosaccharide) are cleaved. The resulting mono-saccharides now have free hydroxyl groups at the positions that were originally involved in a glycosidic linkage.

The monomers are then converted to their corresponding alditols (opening of the ring forms) by chemical reduction using $NaBD_4$. The use of sodium borodeuteride in the reduction step introduces a deuterium atom at C1 of the resulting alditol, avoiding the problem of mass symmetry from different partially O-methylated aldoses. The introduction of deuterium allows the C1 to be distinguished in the mass spectrum (Fig. 6.9). The reduction of the monosaccharide obtained by the hydroly-sis also generates a free hydroxyl group at positions that reflect the original ring size. For hexopyranose rings, a hydroxyl will be generated at C5 and for furanoses at C4. However, discrimination between original 4-substituted hexopyranoses and 5-substituted hexofuranoses is problematic.

The subsequent derivatization by acetylation of the remaining free hydroxyls will generate volatile species (partially methylated alditol acetates, PMAAs) for analysis by GLC-electron impact (EI)-MS (Fig. 6.10). The GC retention times of the PMAAs, compared to PMAA standards, will identify the nature of the parent monosaccharide type (e.g., Glc, Man, GlcNAc, etc.), whereas the fragmentation pattern in the mass spectra will identify the original linkage substitutions [22, 23].

Thus, methylation analysis yields the molar ratios of various linkage positions, the branching pattern of (glycosyl) units, and information on ring form but the infor-mation on the anomeric configuration is lost. The percentage (ratio) of PMAAs stemming from the terminal, internal, and branched units provides an indication of the oligo-/polysaccharide structure. Ring size (p or f) for each monosaccharide can be deduced, but methylation analysis does not give information on the α/β-anomeric configuration of the glycosidic linkages nor on the sequence of the monosaccharide residues.

Regarding the methylation analysis, some features need attention. Undermethylation can occur, which would be indicated by high ratios of branching points to terminal sugars in the final analysis. A perfect methylation analysis of branched carbohydrates will have terminal-to-branch point ratios approximating 1:1. To avoid undermethylation, sometimes, double methylation of the material is necessary. High terminal-to-branch point ratios may be indicative of either base-catalyzed degradation or inadequate initial solubilization in DMSO or may result from incomplete hydrolysis.

Protocol 27. Methylation/Linkage Analysis
Prior to methylation analysis, glycan/oligosaccharide (DP > 3) can easily be desalted on a BioGel P-2 column (25 × 1 cm) eluted with water. Lyophilize the void-volume fraction and dry overnight in a desiccator over P_2O_5.

The methylation of hydroxyl groups is performed by two different procedures (*Procedure I* or *Procedure II*), depending on the amount of starting material, the preference of the analyst, and the available equipment. Always work in a fume hood.

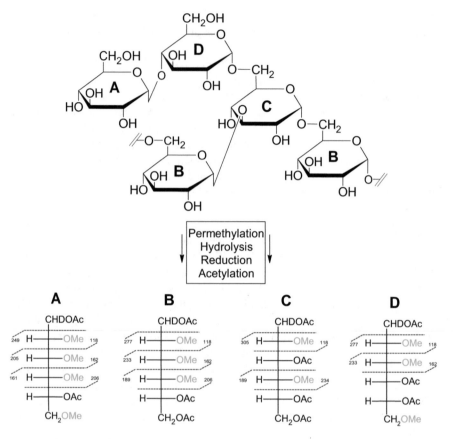

Fig. 6.10 Example of partially methylated alditol acetates (PMAAs) formed by methylation/linkage analysis. Characteristic main MS fragments observable in GC-EIMS are indicated

Materials

Teflon-lined screw-capped glass tubes (100 x 9 mm; GL 14, Duran Group Mainz) or equivalent tubes

Lyophilizer

Vacuum desiccator with phosphorus pentoxide (P_2O_5)

Nitrogen (N_2) evaporation system

Pasteur pipettes (long size 230 mm, WU Mainz)

Small mortar and pestle

Magnetic stirring device

Heating block (VLM or Pierce Reacti-Therm, Rockford, IL)

Water bath

Benchtop centrifuge (suitable for 100×9 mm glass tubes)

Vortex mixer

pH paper

10-μL injection syringe (sharp needle)

Gas chromatograph/electron-impact mass spectrometer system (GLC-EIMS)
Gas chromatography column EC-1, HP5, BPX90 or DB-17 (30 m × 0.32 mm)
Dimethyl sulfoxide, anhydrous (CH_3SOCH_3, DMSO)
Sodium hydroxide (NaOH pellets)
Sodium hydroxide (NaOH) 50% w/w solution
Trifluoroacetic acid (TFA)
Acetic anhydride [Ac_2O; $(CH_3CO)_2O$; Sigma-Aldrich)
Acetic acid (glacial) (CH_3COOH)
Methyl iodide (iodomethane, >99.5%)
Chloroform ($CHCl_3$) Merck
Dichloromethane (CH_2Cl_2, DCM) Merck
Methanol (CH_3OH)
Pyridine (anhydrous, C_5H_5N)
Toluene
Sodium sulfate (anhydrous Na_2SO_4)
Sodium borodeuteride ($NaBD_4$) (Sigma-Aldrich, St. Louis, MO)
Sodium thiosulfate

Procedure I permethylation: This method is usually followed in case of a sufficient amount of carbohydrate (~2–3 mg).

Permethylation: (perform all reactions under N_2 atmosphere, use glass pipettes)

1. Well-dried carbohydrate sample is dissolved in 1 mL anhydrous DMSO, under N_2 atmosphere and stirring with a small Teflon-coated magnetic stir bar for at least 20 min at room temperature (in the case of reluctant polysaccharide, sometimes 1–2 h at 80 °C) in a screw-capped glass tube [note: incomplete solubilization leads to loss of sample and ghost branching points due to undermethylation].
2. Add ~20 mg freshly mortar-powdered NaOH.
3. Stir the mixture for 20 min at room temperature.
4. Add 100μL methyl iodide [CAUTION: methyl iodide is very volatile and highly toxic] and close the tube.
5. Stir for 20 min at room temperature.
6. Cool the tube in an ice bath (DMSO get frozen).
7. Refresh N_2 atmosphere and add again 100μL methyl iodide, close the tube, and wait for room temperature.
8. Stir for 20 min at room temperature.
9. Add 1 mL water containing 4 M sodium thiosulfate (to stop methylation reaction).
10. Add 1 mL chloroform, mix well.
11. Wait for two-layer liquid separation (eventually, centrifuge shortly).
12. Transfer the chloroform layer, using a Pasteur pipette, into a clean glass tube.
13. Add 1 mL chloroform to the residual water layer and mix well.
14. Wait for a two-layer situation (eventually, centrifuge shortly).
15. Take the chloroform layer and combine it with the previous chloroform layer.
16. Add 1 mL of fresh water to the total chloroform fraction and mix well.

17. Wait for a two-layer situation (eventually, centrifuge shortly).
18. Remove upper water layer using a Pasteur pipette and discard.
19. Add 1 mL of freshwater to the chloroform layer and mix well.
20. Wait for two-layer separation (eventually, centrifuge shortly).
21. Remove upper water layer, discard.
22. Repeat steps 19–21.
23. Add ~1 g of anhydrous sodium sulfate to the chloroform fraction to completely remove the remaining water.
24. Filtrate the chloroform fraction through a Pasteur pipette with a piece of chloroform-washed cotton wool to remove sodium sulfate, and collect in a clean Teflon-lined screw-capped glass tube.
25. Evaporate chloroform fraction till dryness with a gentle stream of N_2 (then the sample is ready for **hydrolysis**, **reduction**, and **acetylation**, according to subsequent **Subheadings below**).

Procedure II permethylation: This permethylation is frequently used for a lesser amount of carbohydrate material (0.5–1 mg).
The procedure can be adapted for (micro) permethylation of N/O-glycans for MALDI-MS and/or LC-MS/MS analyses.

First prepare the NaOH/DMSO base reagent

1. Mix 300μL of 50% aqueous NaOH with 600μL of anhydrous methanol in a screw-capped glass tube.
2. Vortex to a clear solution during 30 s.
3. Add 2 mL anhydrous DMSO with Pasteur pipette and vortex for 30 s.
4. Centrifuge (2 min, 3000 ×g), remove. and discard the supernatant (top layer).
5. Add 1.0 mL anhydrous DMSO to bottom layer and vortex.
6. Centrifuge (4 min, 3000 ×g), remove and discard the supernatant (top layer).
7. Repeat 3× steps 5 and 6.
8. Finally, the bottom layer, being the NaOH-containing gel pellet, must be transparent/translucent.
9. Add 500μL anhydrous DMSO to the gel, vortex. and immediately use the suspension (this is the NaOH/DMSO base reagent) for permethylation

Permethylation: (perform all reactions under N_2 atmosphere, use glass pipettes)

1. Dissolve target sample (glycan, free of salts/detergents and dried over P_2O_5) in 500μL anhydrous DMSO (in a screw-capped tube) by stirring with a small Teflon-coated magnetic stir bar for ~20 min at room temperature (take care: incomplete solubilization leads to ghost branching points due undermethylation).
2. Add 200μL of NaOH/DMSO base reagent straight into the sample and close tube.
3. Stir for 15 min at room temperature.
4. Cool the tube in an ice bath (sample gets frozen).

5. Refresh N_2 atmosphere in tube.
6. Add 100μL of methyl iodide (highly toxic!!), close tube, and wait for room temperature.
7. Stir for 15 min at room temperature.
8. Add 200μL of NaOH/DMSO base reagent.
9. Refresh N_2 atmosphere in tube.
10. Add 100μL of methyl iodide (highly toxic!!) and close tube.
11. Stir again for 15 min at room temperature.
12. Add 2 mL pure H_2O (sample gets cloudy).
13. Place a Pasteur pipette into the liquid and gently bubble N_2 through the sample until it gets transparent clear [work in a fume hood].
14. Add 2 mL of DCM and vortex for 30 s.
15. Centrifuge (1 min, 500 ×g), remove and discard the top layer (the water phase).
16. Add 2 mL H_2O, vortex, centrifuge, and remove and discard the aqueous top layer.
17. Add 2 mL H_2O, vortex, centrifuge, and transfer the *bottom* (DCM) layer to a clean Teflon-lined, screw-capped glass tube.
18. Evaporate till dryness with a gentle stream of N_2 at room temperature (then the sample is ready for *hydrolysis*, *reduction*, and *acetylation*, according to subsequent **Subheadings**).

Hydrolysis

1. Add 0.5 mL of 2 M TFA, screwcap the tube, and heat for 2 h at 120 °C with four times intermittent vortexing (In the case of ketoses (fructans) use 0.2 M TFA, 1 h at 100 °C).
2. After cooling down to room temperature, evaporate solvent till dryness with a gentle stream of N_2 (evaporate once more after adding 100μL of methanol).

Reduction

1. Add 0.5 mL water, and add ~8 mg $NaBD_4$.
2. Incubate for 2 h at room temperature with intermediate shaking (bubbling due to release of H_2).
3. Add dropwise 4 M acetic acid (fizzing!) till pH 6 to destroy the excess of reductant (check end pH ~6 with a droplet on pH paper).
4. Add 0.5 mL of methanol and evaporate solvent till dryness with a gentle stream of N_2
5. Repeat 4× step 4 (removal of volatile sodium borate) and dry well finally.

Acetylation

1. Add 0.5 mL of freshly prepared pyridine/acetic anhydride (1:1, v/v) and screw cap well.
2. Heat for 30 min at 120 °C in a heating block with intermittent vortexing.

3. After cooling to room temperature, evaporate solvent till dryness with a gentle stream of N_2 (evaporate once more after adding 50μL toluene).
4. Redissolve in 0.5 mL of DCM.
5. Analyze about 1–3μL clear sample solvent by GLC-EIMS on an EC-1 (DB-1; AT-1) column (30 m × 0.32 mm i.d.) with a temperature program: 140°C for 3 min, then 140–250 °C, 8°/min. Injector temperature 230 °C. For EI-MS, the source temperature is 200 °C, and the electron voltage is 70 eV. He flow rate: 3 mL/min.
6. Identification of the PMAAs is based upon the comparison of relative retention times with those of standards, run under the same conditions. The relative retention time (R_t) is the quotient of the retention time of the relevant peak and the retention time of 2,3,4,6-tetramethyl-1,5-diacetyl glucitol. Keep in mind that co-elution can occur depending on the GC column (polarity) used. The EI-mass spectrum of the peaks affords their ultimate identification as partially methylated alditol acetates, stemming from their corresponding (substituted) aldohexopyranoses/furanoses, aldopentapyranoses/furanoses, and N-acetylhexosamines (Fig. 6.11). Quantification is usually expressed in the percentage ratio of peak areas (or Mol%, which is the peak area divided by the molecular weight). To obtain robust results, it is recommended to repeat the total analysis three times with identical samples.

6.6.2 GLC-EI Mass Spectrometry of PMAAs

Mass spectra are based on the fragmentation of the PMAA molecules under the impact of high-energy electrons (EI), followed by the differentiation of the resulting ion fragments according to their mass-to-charge ratio (m/z) (see also Chap. 11). The presence of deuterium allows for the identification of the C1 location. The fragmentation follows simple rules:

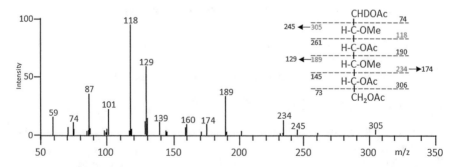

Fig. 6.11 Mass spectrum of PMAA stemming from a 3,6-substituted hexose (see Appendix A for more mass spectra of PMAAs)

1. Primary fragments are generated by the fission of C–C bonds in the alditol chain, whereby the splitting between two carbons with O-methyl groups (methoxylated carbons) is preferred over the splitting between a carbon with an O-methyl group and carbon with an O-acetyl group (acetoxy fragments).
2. Charge resides on the fragment with a methoxy-bearing carbon side (C-OCH₃ group).
3. Cleavage between the carbon carrying an N-methyl-acetylamino group and the carbon with an O-methyl group and then the charge prefers the nitrogen atom.
4. At cleavage between carbons where each carries an O-methyl group, the charge can reside on either fragment but appears most stable on the smaller one.
5. Bonds between carbons both carrying O-acetyl groups (acetoxy-acetoxy fragments) are the most difficult to fragment.
6. m/z of fragment ions containing C1 (deuterium) are even.
7. Secondary fragments are derived by (subsequent) elimination of acetic acid (minus 60) or methanol (minus 32) as well as formaldehyde ($H_2C=O$) (minus 30) and ketene ($H_2C=C=O$) (minus 42) [6].

The typical mass fragments indicate the position of the methyl and acetyl groups in the alditol molecule, which then can be translated to the substitution pattern of the corresponding monosaccharide, demonstrating the linkage types of these monosaccharides in the native material (Fig. 6.10). As an example, the mass spectrum of the partially methylated hexitol acetate stemming from a 3,6-substituted hexose is shown in Fig. 6.11. Monosaccharides with different ring forms (p or f) will also yield characteristic products on methylation analysis. The PMAAs of hexofuranoses will possess an O-methyl group at C5 and not at C4. But note that, a monosaccharide with a pyranose ring, substituted at C4, would yield the same PMAA as a residue with a furanose ring that is substituted at C5. The PMAA mass spectra of hexose, pentose, 6deoxyhexose, and N-acetyl hexosamine (pyranose and furanose ring form), substituted at different places, are depicted in Appendix A.

It has to be noted that the Complex Carbohydrate Research Center at the University of Georgia, USA, has an excellent free-accessible database of PMAA GC-MS spectra (https://www.ccrc.uga.edu/specdb/ms.pmaa/pframe.html). Nowadays, commercial GC-MS systems are equipped with a PC for control and often containing an extended spectra library for PMAAs.

The disadvantage of the methylation analysis method using sodium hydroxide is that native O-acetyl groups are released during the permethylation step and are replaced with a O-methyl group. Also, as already mentioned, phosphate and sulfate can partially be lost. Partial degradation and de-O-methylation at C1 or C3 can occur during the hydrolysis of permethylated N-acetyl hexosamine sugars. Special attention is needed with polymers containing ketoses because their derivatives give two peaks on the chromatogram. Furthermore, it has to be noted that the determination of glycosyluronic acids is not possible as PMAAs by the methylation analysis method of Protocol 27.

In the case of the presence of uronic acids (for example in polysaccharides), more methyl iodide and longer reaction time are needed for the permethylation, and the procedure is modified after the permethylation. Partial saponification of the methyl ester group by NaOH can reduce the yield because sodium salts are not extracted by chloroform. Instead of hydrolysis and reduction, the permethylated product of the uronic acid-containing material is methanolized (3 M HCl/methanol, 18 h at 85°C) yielding partially methylated methyl ester methyl glycosides. After trimethylsilylation of the free hydroxyl groups originally involved in glycosidic linkages, the partially methylated/TMS methyl glycosides are analyzed by GC-MS. Positions of the TMS groups can be identified by EI-mass spectrometry.

6.6.3   Glycosidic Linkage Determination by HPLC-MS/MS

Recently, a modified method to determine and quantitate glycosidic linkages between monosaccharides in glycans was introduced [24]. The method employs ultrahigh performance liquid chromatography coupled with triple quadrupole mass spectrometry (UHPLC/QqQ-MS) in multiple reaction monitoring (MRM) mode. Samples (50µg) are permethylated (50 min) using MeI in DMSO/NaOH. After addition of ice-cold water, the methylated product is extracted using DCM, and this organic fraction is repeatedly washed with ice-cold water to remove excess NaOH and DMSO. Subsequently, dried, hydrolyzed with TFA, and dried.

After the hydrolysis, free hydroxyl groups are present on the positions which were involved in glycosidic bonds. Then, after derivatization of the partially methylated monomers at the reducing ends with 1-phenyl-3-methyl-5-pyrazolone (PMP), analysis is performed by UHPLC/MRM-MS on an Agilent 1290 Infinity II UHPLC system coupled to an Agilent 6495A triple quadrupole (QqQ) mass spectrometer equipped with an Agilent ZORBAX RRHD Eclipse Plus C18 reverse-phase column. Gradient elution is performed with the aqueous mobile phase A (25 mM ammonium acetate in 5% acetonitrile in water, pH 8.2) and the organic mobile phase B (95% acetonitrile in water). Positive ion-mode electrospray ionization (ESI) and collisional-induced dissociation (CID) are applied for multiple reaction monitoring-MS (MRM-MS) (see also Chap. 11). In MRM, the concentration of an unknown sample is determined by comparing its MS response to that of a known standard, using mass filtering [25]. Derivatized terminal, linear, bisecting, and trisecting monosaccharide linkages can be detected by mass spectrometry according to the degree of permethylation (DoPe). The method was successfully employed for polysaccharides in cereal brans [26].

References

1. Cutler P, editor. Protein purification protocols, Methods in molecular biology, vol. 244. 2nd ed. Totowa: Humana Press; 2004.

2. Simpson RJ, editor. Purifying proteins for proteomics: a laboratory manual. New York: Cold Spring Harbor Laboratory Press; 2004.
3. Burgess R, Deutser M, editors. Guide to protein purification, Methods in Enzymology, vol. 463. 2nd ed. Academic; 2009.
4. Mariño K, Bones J, Kattla JJ, Rudd PM. A systematic approach to protein glycosylation analysis: a path through the maze. Nat Chem Biol. 2010;6:713–23.
5. Zhang ZQ, Khan NM, Nunez KM, Chess EK, Szabo CM. Complete monosaccharide analysis by high-performance anion-exchange chromatography with pulsed amperometric detection. Anal Chem. 2012;84:4104–10.
6. Kamerling JP, Gerwig GJ. Strategies for the structural analysis of carbohydrates. In: Kamerling JP, editor. Comprehensive glycoscience-from chemistry to systems biology. Amsterdam: Elsevier; 2007. p. 1–68.
7. Gerwig GJ, Hocking HG, Stöcklin R, Kamerling JP, Boelens R. Glycosylation of conotoxins. Mar Drugs. 2013;11:623–42.
8. König WA, Benecke I, Bretting H. Gas chromatographic separation of carbohydrate enantiomers on a new chiral stationary phase. Angew Chem Int. 1988;20:693.
9. Cooper G, Sant M, Asiyo C. Gas chromatography-mass spectrometry resolution of sugar acid enantiomers on a permethylated beta-cyclodextrin stationary phase. J Chromatogr A. 2009;1216:6838–43.
10. Lopes JF, Gaspar SM. Simultaneous chromatographic separation of enantiomers, anomers, and structural isomers of some biological relevant monosaccharides. J Chromatogr A. 2008;1188:34–42.
11. Tanaka T, Nakashima T, Ueda T, Tomii K, Kouno I. Facile discrimination of aldose enantiomers by reversed-phase HPLC. Chem Pharm Bull. 2007;55:899–901.
12. Wang Y-H, Avula B, Fu X, Wang M, Khan IA. Simultaneous determination of the absolute configuration of twelve monosaccharide enantiomers from natural products in a single injection by a UPLC-UV/MS method. Planta Med. 2012;78:834–7.
13. Lin C, Kuo C-Y, Liao K-S, Yang W-B. Monosaccharide-NAIM derivatives for D-L-configurational analysis. Molecules. 2011;16:652–64.
14. Kodama S, Aizawa S, Taga A, Yamashita T, Kemmei T, Yamamoto A, Hayakawa K. Simultaneous chiral resolution of monosaccharides as 8-aminonaphtalene-1,3,6-trisulfonate derivatives by ligand-exchange CE using borate as a central ion of the chiral selector. Electrophoresis. 2007;28:3930–3.
15. Gerwig GJ, Kamerling JPK, Vliegenthart JFG. Determination of the D and L configuration of neutral monosaccharides by high-resolution capillary G.L.C. Carbohydr Res. 1978;62:349–57.
16. Gerwig GJ, Kamerling JPK, Vliegenthart JFG. Determination of the absolute configuration of monosaccharides in complex carbohydrates by capillary G.L.C. Carbohydr Res. 1979;77:1–7.
17. Xu G, Amicucci MJ, Cheng Z, Galermo AG, Lebrilla CB. Revisiting monosaccharide analysis—Quantitation of a comprehensive set of monosaccharides using dynamic multiple reaction monitoring. Analyst. 2017;143:200–7.
18. Han J, Lin K, Sequria C, Yang J, Borchers CH. Quantitation of low molecular weight sugars by chemical derivatization-liquid chromatography/multiple reaction monitoring/mass spectrometry. Electrophoresis. 2016;37:1851–60.
19. Ciucanu I. Per-O-methylation reaction for structural analysis of carbohydrates by mass spectrometry. Anal Chim Acta. 2006;576:147–55.
20. Price N. Permethylation linkage analysis techniques for residual carbohydrates. Appl Biochem Biotechnol. 2008;148:271–6.
21. Sims IM, Carnachan SM, Bell TJ, Hinkley SFR. Methylation analysis of polysaccharides: technical advice. Carbohydr Polym. 2018;188:1–7.
22. Morelle W, Michalski JC. Analysis of protein glycosylation by mass spectrometry. Nat Protocol. 2007;2:1585–602.
23. Sassaki GL, De Souza LM. Mass spectrometry strategies for structural analysis of carbohydrates and glycoconjugates. Intech; 2013. https://doi.org/10.5772/55221.

24. Galermo AG, Nandita E, Barboza M, Amicucci MJ, Vo TT, Lebrilla CB. Liquid chromatography-tandem mass spectrometry approach for determining glycosidic linkages. Anal Chem. 2018;90:13073–80.
25. Ruhaak LR, Lebrilla CB. Applications of multiple reaction monitoring to clinical glycomics. Chromatographia. 2014;78:335–42.
26. Pasha I, Ahmad F. Monosaccharide composition and carbohydrates linkage identification in cereal brans using UHPLC/QqQ-DMRM-MS. J Food Comp Anal. 2021;96:103732.

Chapter 7
Carbohydrate Analysis of Glycoconjugates

Abstract For studies of glycan structure/function relationships of glycoconjugates, a detailed characterization of the carbohydrate moiety is a prerequisite. To this end, the glycans are usually released and isolated before analysis. This chapter provides protocols for the chemical and enzymatic release of different types of glycans from glycoconjugates and discusses the pro and cons of the different methods. A typical strategy for the analysis of glycoproteins is described. The preparation of glycopeptides and their isolation might form part of this process. Furthermore, a typical scheme for the analysis of proteoglycans and their glycosaminoglycans is provided. Also, attention is paid to the analysis of glycolipids and polysaccharides.

Keywords Glycoproteins · Glycopeptides · Proteoglycans · Glycosaminoglycans (GAG) · Glycolipids · Glycosylphosphatidylinositol anchors · Glycan profiling · Hydrazinolysis · PNGase F digestion · Denaturing · Proteolysis · Reductive amination · Reduction/alkylation · β-elimination · Pronase · Trypsin · Endoglycoceramidase · Lipase · Hydrogen fluoride · Nitrous acid · Fatty acids

7.1 Introduction

Many biological properties and functions of glycoconjugates have been attributed to their carbohydrate parts. The carbohydrates of glycoproteins and glycolipids play critical roles in numerous physiological and pathological processes. This means for studies of structure–function relationship of these molecules, a detailed carbohydrate analysis is a prerequisite. The complete sequencing of oligosaccharides in glycoconjugates is difficult to accomplish by a single method and therefore requires iterative combinations of chemical and physical approaches, which can lead to a final structure. In general, glycoconjugates have to be isolated in pure form from biological materials, such as tissues or cells or body fluids. As mentioned earlier, a detailed discussion of methods for the isolation and purification of individual glycoconjugates (i.e., specific glycoproteins, glycolipids, or proteoglycans) from diverse biological materials is beyond the scope of this book.

7.2   Analysis of Glycoproteins

Glycoproteins are widespread in the living world. They are found inside the cells, on the outside of the cell membranes, and in various extracellular fluids. Many different proteins are present in human blood. Remarkably, all major proteins in human plasma (e.g., transferrin, α_1-acid glycoprotein, fibrinogen, α_1-antitrypsin, haptoglobin, and immunoglobulins) are glycosylated proteins [1, 2]. As shown in Sect. 2.4.1, glycoproteins contain carbohydrates covalently linked to one or more specific amino acids in the polypeptide chain, and the carbohydrate content can range from 1 to 80%. The different carbohydrate moieties (usually called glycans) of these proteins can play a role in the conformational stability of the protein, the solubility (water-binding capacity), protease resistance, and can change the immunogenicity of the protein. The carbohydrate moieties can add charge to the protein. It is known that protein glycosylation is involved in many biological processes, such as cellular signaling, eliciting immune responses, and cancer progression. Most glycoproteins can have multiple glycoforms of glycans present, which means that the same protein molecule can be decorated with different oligosaccharides.

The high prevalence of glycosylated proteins and the huge complexity of their glycans, together with their complicated biosynthetic mechanism, indicate that there must be a significant importance for glycosylation in life. For many years, the deciphering of the glycosylation code of glycoproteins has been the ambition of the glycobiologist. The comprehensive analysis of protein glycosylation is a major requirement for understanding glycoprotein function in biological systems [3].

Apart from that, during the production of recombinant glycoproteins as therapeutics, a detailed analysis of the glycan chains at all stages of development is a requirement [4]. Notwithstanding the use of a distinct human DNA sequence, which provides all of the information necessary for a cell to produce a recombinant glycoprotein with an identical amino acid sequence as the native protein, this does not guarantee that the recombinant protein will be glycosylated in the same way as the native protein. The choice of the expression system and the growth conditions (cell type, pH, temperature, nutrients, etc.) will affect the final glycosylation pattern [5]. This can alter the therapeutic properties and can result in adverse immunological effects. After the detailed study of the glycan structures of the native protein, rapid and robust analytical methods are applied to demonstrate batch-to-batch consistency of glycosylation instead of repeating the detailed and laborious structural analysis.

For the academic study of the biological functions and properties of a specific biological important glycoprotein, the target glycoprotein is frequently isolated from organ tissues, cells, or body liquids, such as blood, urine, and cerebrospinal fluid, where the glycoprotein of interest is characteristically contained in a complex biological mixture (see Sect. 5.1, Fig. 5.1). It must be purified and preconcentrated apart from nonglycosylated proteins, lipids, nucleic acids, and small metabolic molecules. Many methods exist for the isolation and purification of specific glycoproteins from biological samples. For methods for obtaining pure single glycoproteins, the reader is referred to specialized books, for example, *"Purifying Proteins for Proteomics"* [6], *"Protein Analysis and Purification"* [7], and *"Protein Chromatography"* [8]. Due to the multistep processing techniques, such as homogenization, cell lysis, precipitations, centrifugations, dialysis, extractions, filtering, and chromatographic separations, it is a laborious effort to derive a single pure glycoprotein from the biological specimens of interest. Techniques such as solid-phase chromatography, lectin-based glycan capture, immune precipitation, and electrophoresis are among the applied methods. Protocol 28 is just a short protocol as an example of the coarse isolation of glycoprotein material from tissue, but it must be emphasized that more detailed and more adequate methods are necessary for specific important biomolecules.

Protocol 28. Isolation of Glycoprotein from Tissue (Usually Performed After Lipid Extraction)

Procedure

1. Add 1 mL of water to the tissue sections.
2. Homogenize with a sonicator stick.
3. Add 1.75 mL of methanol.
4. Vortex and sonicate in a water bath (25 °C) for ~15 min.
5. Add 3.25 mL of chloroform and vortex.
6. Centrifuge at 15,000 ×g for 15 min.
7. Remove upper-phase layer.
8. Add 3 mL of methanol/water (50/50 v/v) to pellet (lower phase + interphase) and vortex.
9. Homogenize with a sonicator stick.
10. Centrifuge at 15,000 ×g for 15 min.
11. Remove upper-phase layer.
12. Add 1 vol. of ice-cold methanol to pellet (lower phase + interphase) and vortex.
13. Centrifuge at 15,000 ×g for 15 min.
14. Discard the supernatant.
15. Add 1 vol. of ice-cold methanol to pellet and vortex.
16. Centrifuge at 15,000 ×g for 15 min.
17. Discard the supernatant.
18. Repeat 3× steps 15 to 17.
19. Dry the pellet (containing glycoprotein) under a gentle stream of nitrogen.

For carbohydrate analysis, the aim is to get a salt-free material (preferably at least 50–500µg) without extraneous carbohydrates. A typical rapid final desalting of a glycoprotein can be performed by gel filtration on a water-eluted BioGel P-2 column (30 × 0.7 cm). The purity of a glycoprotein and its molecular weight are usually monitored by sodium dodecyl sulfate polyacrylamide gel electrophoresis (SDS-PAGE), although this needs sufficient amounts of material. The carbohydrate moiety contributes to the behavior of a glycoprotein in SDS-PAGE. Deglycosylation will change the mobility of the glycoprotein and is often used to check the presence of glycans. For further studying specific glycoproteins, the purification of the target protein from complex tissue or cell fractions is rather difficult. Thanks to the increased sensitivity of analytical methods needing less material, nowadays, more and more, the protein band of interest is cut from the SDS-PAGE gels and the glycans are enzymatically released by "in-gel" digestion. The advantages are: (1) the protein is already separated from contaminants with minimal sample loss, (2) the release of glycans needs less amount of enzyme, and (3) the liberated glycans are isolated while the protein remains in the gel.

On the other hand, in the case that a pure glycoprotein can be obtained in a higher amount (mg scale), primary structural analysis can easily be followed. This will consist of determining the nature of the protein and the nature and proportion of the linked glycan(s).

The characterization often starts with the amino acid sequence analysis. From the amino acid sequence, potential glycosylation sites for N-linked glycans (–Asn– Xxx–Ser/Thr–) and O-linked glycans (Pro/Ser/Thr-rich regions) can be derived. Unfortunately, glycoproteins are rarely well-defined single compounds. Although the amino acid sequence is mostly homogeneous, heterogeneity may occur in each glycan at each glycosylation site. This feature forms a serious complication in the structure elucidation of the glycans.

In the last 50 years, many chemical and physical techniques have been developed to analyze the oligosaccharides on glycoproteins. In the beginning, relatively large amounts were necessary for chromatographic separations, detection, and identification. For many important glycoproteins, after laborious work, the exact structure of the glycans, including conformations and linkages, has been attained by utilizing chemical analysis methods, together with GLC, MS, and NMR techniques. Today, instrumental analysis methods have become much more sensitive, so less material can be used. In addition, libraries of many known glycan structures also make detailed in-depth analysis less urgent in many cases. Robust high-throughput methods, including microarrays, chip-based HPLC-ESI MS, and automated microchip capillary electrophoresis (CE) techniques, have put the emphasis more on glycan profiles, making use of database comparison with a library of known structures to identify the glycans (see Chaps. 5 and 13). These methods are referred to as glycan profiling, mapping, or fingerprinting [9]. Frequently, the goal is not always to fully characterize all glycan structures in a glycoprotein sample but to compare samples and to focus on the identification of differences.

Already, for some time now, it has been known that the glycans of glycoproteins are not randomly constructed. They show systematic structures and contain a relatively small collection of monosaccharides as building units. Just a few pyranosyl monosaccharides, such as D-galactose (Gal), D-mannose (Man), L-fucose (Fuc), N-acetyl-D-glucosamine GlcNAc), N-acetyl-D-galactosamine (GalNAc), and sialic acid (N-acetylneuraminic acid, Neu5Ac), and in some cases also D-xylose (Xyl), D-glucose (Glc), and D-glucuronic acid (GlcA) participate in the glycan structures of animal/human glycoproteins (for the monosaccharide structures, see Chap. 2, Table 2.1). As shown (see Sect. 2.4.1), there are two main types of glycan linkage to the polypeptide backbone, which can coexist in the same glycoprotein: N-linked and O-linked glycans.

Structurally, the *N-glycan chains*, which are linked to Asn in the protein backbone, have a typical inner core pentasaccharide structure, and they can be classified into three subgroups: (1) high-mannose type, (2) complex type, and (3) hybrid type (Sect. 2.4.1.1, Fig. 2.4). Extensions and branches arise from the varying biosynthetic routes and competing enzymes during this process. This led to a variety of final structures [10]. Structural variations occur, in the first instance, by the presence or absence of Fuc and Xyl, or an extra (intersecting) GlcNAc residue, but more specific outer chain variations of complex-type N-glycans do occur.

The presence of high-mannose N-glycans on proteins is derived from incomplete processing of the N-glycans. Remarkably, contrary to membrane glycoproteins, high-mannose glycans are not commonly attached to serum glycoproteins. The abundance of high-mannose N-glycans is a common feature of cancer tissues (melanoma cells).

The *O-linked glycans* also demonstrate a high variety of structures. These oligosaccharides can be linked to the peptide backbone via different monosaccharides and different amino acids but mainly via GalNAc to Ser/Thr (see Sect. 2.4.1.2). In contrast to N-glycans having one core structure, O-glycans show a selected number of core (root) structures of which the Core 1 structure, Gal(β1–3)GalNAcα1-Ser/Thr, usually extended with one or two sialic acids and is most common and studied (Sect. 2.4.1.2, Fig. 2.5).

Having determined that the protein of interest is glycosylated, for instance by using a colorimetric method (Chap. 4), and having received an indication of the types of glycan chains present, for instance by using monosaccharide analysis (Chap. 6), the next step is to examine these sugar chains in more detail. Several instrumental techniques are employed in the process to elucidate the glycan structure in detail:

1. High-performance liquid chromatography (HPLC)
2. Gas–liquid chromatography (GLC)
3. Capillary electrophoresis (CE)
4. High-performance (pH) anion-exchange chromatography (HPAEC)
5. Lectin array technology (LAT)
6. Mass spectrometry (MS)
7. Nuclear magnetic resonance spectroscopy (NMR)

A thorough glycan structural characterization requires a multidimensional approach, combining different technologies with different levels of complexity. Considering the heterogeneous nature of glycoproteins, different experimental approaches are employed to analyze different features, such as monosaccharide composition, sialylation, glycoforms, carbohydrate chain types, oligosaccharide patterns (antennary profiles), glycosylation site(s), rare substituents, and other specific information that is required. There is no single method available that can be used to completely characterize the glycosylation on a glycoprotein molecule.

The glycan analysis of N- and O-glycans can be tackled in three different ways:

1. Analysis of glycans while still attached to the intact glycoprotein
2. Analysis of glycans attached to peptides after protease digestion
3. Analysis of glycans after release from protein/peptides

In general, nowadays, the most applied analytical methods are:

1. Fluorescently derivatized glycans separated by HPLC and CE
2. Labeled or reduced glycans separated and analyzed by LC-MS/MS
3. Permethylated glycans analyzed by MALDI-MS.

Mass spectrometry (MS)-based glycomics has become one of the most powerful methods for analyzing glycans released from glycoproteins because it benefits from the rapid improvements in sample preparation and chromatographic separation techniques, and increased sensitive instrumentation.

Currently, two major strategies for detailed glycosylation analysis of glycoproteins are mainly followed, the analysis of free glycans and the analysis of glycopeptides (see Sect. 7.2.1, Fig. 7.1) [11–14]. Furthermore, glycosylation analysis can be viewed as glycotyping, which usually includes (1) the determination of the total number of glycoforms present, (2) the characterization of the oligosaccharide structures present site specifically, and (3) in many cases also the determination of the relative abundances of the glycoforms.

HPLC of fluorescently labeled glycans is an advanced technique to obtain glycan profiles based on their abundance in isolated mixtures [15, 16]. Notwithstanding the recent progress in instrumental and analytical techniques, many methodologies in glycan analysis are still far from being routine, reliable, and reproducible [17]. Prior to structural identification techniques (MS or NMR), separation of glycan samples is nowadays achieved by ultra- or high-performance liquid chromatography (UPLC or HPLC) or, in some applications, capillary electrophoresis (CE). Since minimal amounts are generally available, and monosaccharides and oligosaccharides lack strong absorbance of UV light, the detection during separations is a problematic feature.

Consequently, released glycans (oligosaccharides) are often labeled to enhance detection. The labeling sometimes also improves the chromatographic or electrophoretic separation. A chromophore/fluorescent label can be incorporated at the reducing end of the mono-/oligosaccharide by a reductive amination reaction, making use of the ring-open form of the reducing-end monosaccharide (see Sect. 5.4.2.1, Fig. 5.5). In this way, the ultimate detection of carbohydrates can be lowered to quantities as small as pico-moles and even femto-moles. However, note that reductive amination methods to label released glycans result in partial desialylation.

Typically, glycan profiling by mass spectrometry is the method of choice, especially because of the availability of minimal amounts ($<10\mu g$) of material. Nowadays, an elegant method is to release and extract the glycans from immobilized glycoproteins on electrophoresis gels, followed by mass spectrometry (MS), and high-performance anion-exchange chromatography with pulsed amperometric detection (HPAEC-PAD) [18]. Excellent comprehensive protocols have been published for the analysis of N- and O-glycans cleaved from glycoproteins and the determination of site-specific glycan heterogeneity [19, 20]. Accordingly, glycoproteins are immobilized on PVDF membranes and the N-glycans are enzymatically released by peptide-*N*-(*N*-acetyl-β-glucosaminyl)asparagine amidase (PNGase F) and isolated from the membrane. Subsequently, the O-glycans are chemically cleaved from the same protein spot by reductive β-elimination and isolated. After desalting, the glycans are separated and analyzed by porous graphitized carbon liquid chromatography-electrospray ionization tandem mass spectrometry (PGC-LC-ESI-MS/MS), giving the overall set of glycans present in microgram quantities of a protein or a mixture of proteins at femtomole sensitivity. Glycoproteins of interest can also be proteolytically digested (in-gel) and the (glyco)peptides analyzed by capillary LC-ESI-MS/MS with nano-flow technology. However, the described procedures still take 3–5 days and need sophisticated equipment and experience of the analyst, as the data require largely manual interpretation. It should be noted that nonexpert researchers will need detailed practical instructions.

7.2.1   A Typical Approach to the Analysis of a Glycoprotein

First of all, it must be said again that there is no one universal method for the elucidation of the carbohydrate chains of all different glycoproteins. In most cases, protocols have to be adjusted to the specific glycoprotein of interest. In this chapter, a selection of general protocols will be presented that includes the principles and some generally used standard conditions. A strategy for the analysis of a glycoprotein, which might be isolated from cells, tissues, or biological fluids, is depicted in Fig. 7.1.

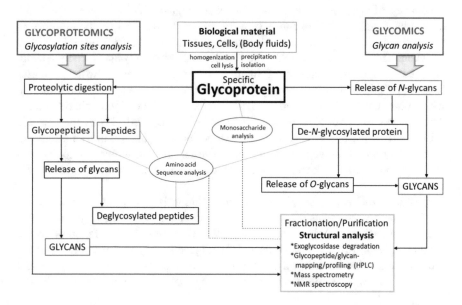

Fig. 7.1 A schematic setup of a glycoprotein analysis strategy, including glycoproteomics and glycomics

A typical strategy for glycoproteins includes that, before releasing N- and O-linked glycans, the glycoprotein is denatured by reduction/alkylation of S–S bridges. Then, eventually, protease digestion will facilitate the enzymatic release of N-glycans. The released intact N-glycans are isolated for further analysis. From the remaining protein/peptide material, the O-glycans are chemically released and also isolated for analysis. Most modern techniques for glycan analysis (MS and NMR) focus primarily on the released glycans. However, the purification of the target sample (glycan and/or glycopeptide) for analysis is a challenging task.

Extraction methods are often used because oligosaccharides are soluble in aqueous-alcoholic solutions, whereas proteins, lipids, and polysaccharides are insoluble. Other interfering molecules (e.g., amino acids, organic acids, and pigments) can be removed by ion-exchange chromatography. Many oligosaccharides are non-charged molecules and can therefore be separated from charged molecules. Nonpolar phase-extraction chromatography can remove nonpolar molecules. Nowadays, to improve the extraction efficiency and save on extraction time, microwave-assisted extraction (MAE) [21] and ultrasound-assisted extraction (UAE) have been used. Also, pressurized liquid extraction increases the yields of carbohydrates [22, 23].

A colorimetric method can be applied to detect sugar bound to a purified protein. The phenol–sulfuric acid assay is the most used method for detecting and quantification of hexoses and pentoses, but other tests can also be used (see Chap. 4).

Furthermore, the issue of whether a protein is modified by N-glycans can be determined by peptidase F digestion (see later) and then examining the changes in molecular weight. Sodium dodecyl sulfate-polyacrylamide gel electrophoresis (SDS-PAGE), ordinarily used for analysis of proteins, can be used to estimate the occupancy of N-glycosylation sites. N-linked oligosaccharides have an average molecular weight of 2 kDa and have a disproportionate effect on the mobility of glycoproteins in SDS-PAGE. In some cases, proteins can be fractionated into distinct bands based on differences in the number of N-linked glycans associated with each band. The time course of incubation with PNGase F (which releases N-glycans) and analysis of products by SDS-PAGE can lead to "laddering," in which consecutive protein bands can differ by an N-linked oligosaccharide chain. Another way to detect glycoproteins in gels during SDS-PAGE is by the periodic acid-Schiff (PAS) reaction, which involves periodate oxidation of vicinal glycan hydroxyl groups followed by Schiff base formation with amine- or hydrazide-based probes. For the PAS method, test kits with protocols are commercially available.

Characterization of both N- and O-glycosylation is somewhat complicated by the fact that glycoproteins often contain more than one N- and/or O-glycosylation site, and at each of the glycosylation sites, there can be several different oligosaccharides (glycoforms). This site-specific heterogeneity may vary in species and tissues and may also be affected by physiological changes. Furthermore, identical oligosaccharides may be found at more than one glycosylation site in the polypeptide chain. The analysis of the glycosylation of a protein is therefore the analysis of a population of glycoforms.

Again, it has to be emphasized that the following items are involved in the glycan analysis:

1. Determination of the primary amino acid sequence of the protein
2. Preliminary monosaccharide analysis of the total glycans of the intact protein
3. Identification of those amino acid residues that carry N- or O-linked oligosaccharides
4. Isolation of glycopeptides of each glycosylated site
5. Determination of the degree of occupancy of each glycosylated site
6. Analysis of the oligosaccharide structures associated with each glycopeptide/glycosylation site
7. A qualitative description of the entire set of glycan structures linked to defined sites
8. A quantitative description of the relative molar distribution of the observed glycans on the individual glycosylation sites (glycoprofiling).

As mentioned earlier, to get a first idea of the type of glycans present on the glycoprotein, with regard to molecular size, charge, and sugars, a qualitative and quantitative monosaccharide compositional analysis is performed on the purified intact glycoprotein. Two methods are commonly used for monosaccharide analysis, HPLC (HPAEC-PAD) and GLC-MS.

The GLC method, using methanolysis and analysis of trimethylsilyl ethers of the methyl glycosides, is the most widely applied (Sect. 6.2). When started with an exact weight of the pure glycoprotein, the molar ratios of the various monosaccharides present can be calculated and the total carbohydrate content can be determined using an internal standard. The results of the monosaccharide composition can be very informative. Typically, for glycoproteins, fucose (Fuc), galactose (Gal), mannose (Man), N-acetylglucosamine (GlcNAc), and N-acetylgalactosamine (GalNAc) are observed since they are the most common. For example, the presence of Man and GlcNAc is an indication for N-linked glycans. If GalNAc is not detected, the presence of mucin-type O-linked glycans can be excluded. When these initial clues regarding the glycan types have been gained, confirmatory evidence can be obtained by release and isolation of the intact glycan(s), preferably in a high yield, nonselectively, and unmodified [24]. Chemical and enzymatic methods are available to liberate glycans as oligosaccharides from the protein. Summarized, the analytical glycomics procedure after extraction/purification of a glycoprotein can consist of:

1. proteolytic digestion: N- and O-linked glycans (→glycopeptides)
2. hydrazinolysis: N- and O-linked glycans (→oligosaccharides)
3. PNGase F digestion: N-linked glycans (→oligosaccharides)
4. alkaline borohydride treatment: O-linked glycans (→oligosaccharide-alditols)

7.2.2   Chemical Release of N- and O-Linked Glycans

There is a chemical method, using anhydrous hydrazine (N_2H_4), to release both N- and O-linked glycans from the protein/peptide. Although using different time and temperature conditions, the hydrazinolysis method will often result in the nonselective release of glycans. The protein part will be destroyed, and there will be some loss of sialic acids (also N-acetyl, N-glycolyl, and O-acyl groups are cleaved). Re-N-acetylation has to be performed leading also to β-acetohydrazide derivatization, which has to be hydrolyzed to produce glycans with reducing ends suitable for labeling. The reaction mechanism of hydrazinolysis is still not fully understood (but involves a non-reductive β-elimination). The major disadvantage of the method is that hydrazine is difficult to obtain, but, moreover, one has to be very careful because anhydrous hydrazine is very hygroscopic, corrosive, toxic, highly inflammable, explosive, and highly toxic. Consequently, the method can only be performed on a small scale. Since the method is typically performed in specialized laboratories due to the relative difficulty and hazards of the process, it will not be discussed in detail here.

In brief, the glycan release by hydrazinolysis involves four steps:

1. Incubation of dry, salt-free glycoprotein with anhydrous hydrazine (~1 mg/100μL, 4–8 h at 65–100 °C) (Fig. 7.2),
2. Removal of the excess hydrazine by evaporation using a rotary evaporator
3. Re-N-acetylation of de-N-acetylated amino groups [usually with saturated sodium bicarbonate solution, containing 8% (v/v) acetic anhydride, 25 min at 4 °C]
4. Hydrolysis to obtain a reducing end at the glycans (usually done by treatment with TFA or Dowex 50W-X2 H⁺).

A GlycoRelease Glycan Hydrazinolysis Kit is commercially available from Prozyme/Agilent, containing chemicals and detailed protocols.

A small advantage of hydrazinolysis is that N- and/or O-glycans, liberated in this way, can subsequently be labeled without any purification (Fig. 7.3). Generally, the reducing ends of the glycans are tagged (e.g., with 2-AB) by reductive amination. The labeled glycans can be separated by chromatographic methods (HPLC) and analyzed by mass spectrometry and/or NMR spectroscopy.

It has to be noted that, in particular with the release of O-glycans, during long reaction time hydrazinolysis, peeling reactions can occur. This is a chemical degradation of the glycan by peeled residues from the reducing ends due to further β-elimination [25]. Particularly, reducing-end residues having a 3-O-substitution are vulnerable to peeling, since the C3 position is in a β-position to the reducing-end carbonyl. This feature together with the required safety precautions gives preference to another chemical method. In that case, the release of O-glycans is performed under carefully controlled mild alkaline conditions. The process is called reductive β-elimination because it is accompanied by a reduction step to give the glycan-alditols (see Sect. 7.2.4).

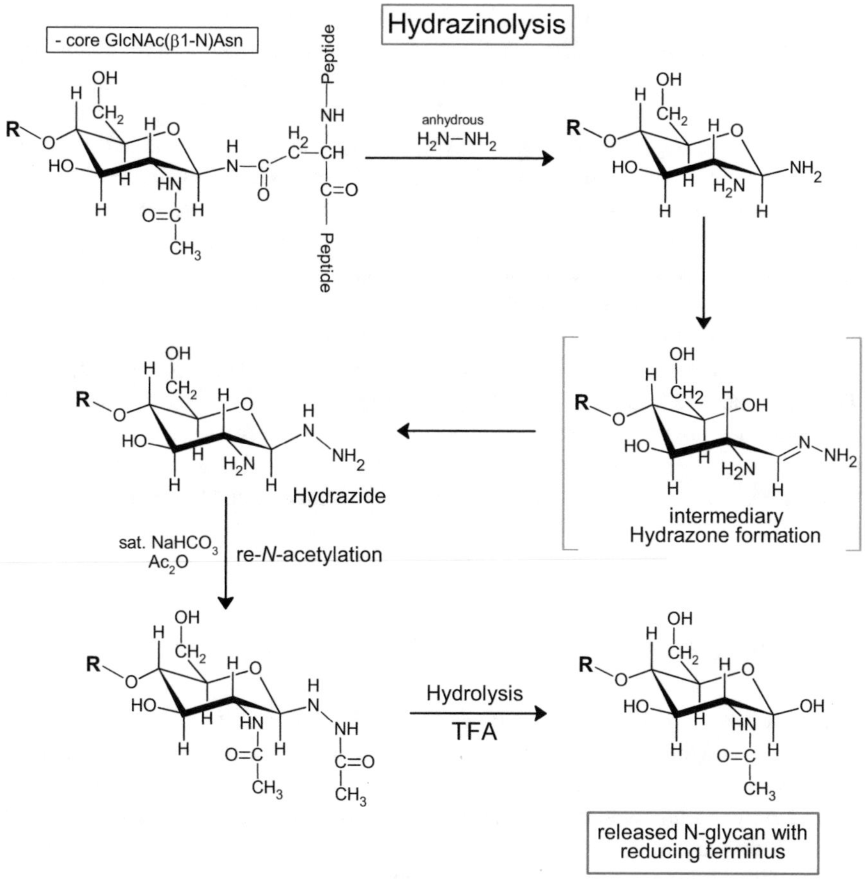

Fig. 7.2 Hydrazinolysis of N-linked glycan

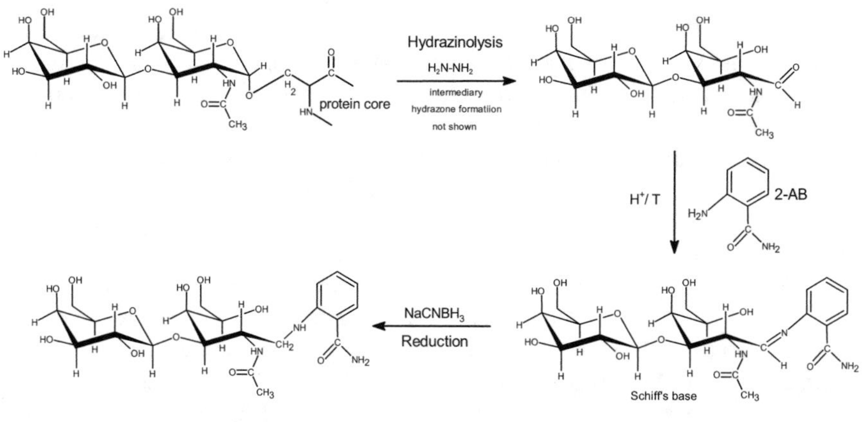

Fig. 7.3 Hydrazinolysis of O-linked glycan directly followed by 2-AB labeling

Researchers are frequently looking for novel chemical methods to release gly-cans [21, 26–29]. For instance, a rapid, rigorous preparative method to obtain large quantities of glycan consists of the treatment of glycoprotein/peptide samples with household bleach, that is, sodium hypochlorite (NaClO), which degrades the pro-tein while releasing intact N- and O-glycans [30]. The method is named oxidative release of natural glycans (ORNG). N-glycans are released as glycosylamines that are then spontaneously converted to nonreduced free N-glycans. However, some loss of reducing-end GlcNAc occurs depending on time and temperature. The release of O-glycans requires high concentrations of NaClO and longer incubation time, yielding different O-glycan products. The reaction mechanisms are still not clear. The method is not suitable for analytical purposes.

Another approach is a "chemoenzymatic" method to release N-glycans, called Threshing and Trimming (TaT), using *N*-bromosuccinimide (NBS) treatment of glycopeptides obtained by pronase digestion [31]. Furthermore, a rapid chemical de-N-glycosylation using hydroxylamine in lithium hydroxide solution can be used to release glycans as oximes, which can be fluorescently tagged by reductive amina-tion [32].

7.2.3 Enzymatic Release of N-Linked Glycans

The preferred method to release, in particular, all common classes (oligomannose, complex, and hybrid) of N-linked glycans is enzymatical, which leaves the peptide backbone intact and obviates possible chemical breakdown of the glycans. Although several enzymes are available for cleaving various oligosaccharide chains from the polypeptide backbone, the principal method of N-glycan release from the glycopro-tein (usually after denaturation) or from glycopeptides is accomplished by treatment with a peptide-*N*-glycosidase enzyme, called *N*-glycanase or peptide-N^4-(*N*-acetyl-β-glucosaminyl) asparagine amidase (EC 3.5.1.52; PNGase F purified from *Flavobacterium meningosepticum*, commercially available). This enzyme is capa-ble of cleaving the linkage between *N*-acetyl-β-glucosamine and asparagine (Asn), liberating, in the first instance, the glycan with a 1-amino-glycosylamine at the reducing end (Fig. 7.4). The aqueous conditions remove the amino group from the reducing terminus of the PNGase F-released glycans. The next reaction might be a subsequent quantitative reduction to an alditol or labeling with a fluorescent tag. The originally glycosylated Asn residue is simultaneously converted by deamina-tion to aspartic acid (Asp), but the peptide chain is kept intact, making it possible to deduce the glycosylation site(s) in the protein/peptide.

Although PNGase F is indifferent to extended structures, also containing sulfate or phosphate groups, it should be noted that PNGase F does not cleave N-glycans containing an $(\alpha 1 \rightarrow 3)$-linked fucose residue (called a "core fucose") attached to the asparagine-linked *N*-acetylglucosamine. This fucose linkage generally occurs in glycoproteins present in insects, parasitic worms, and plants. Furthermore, when

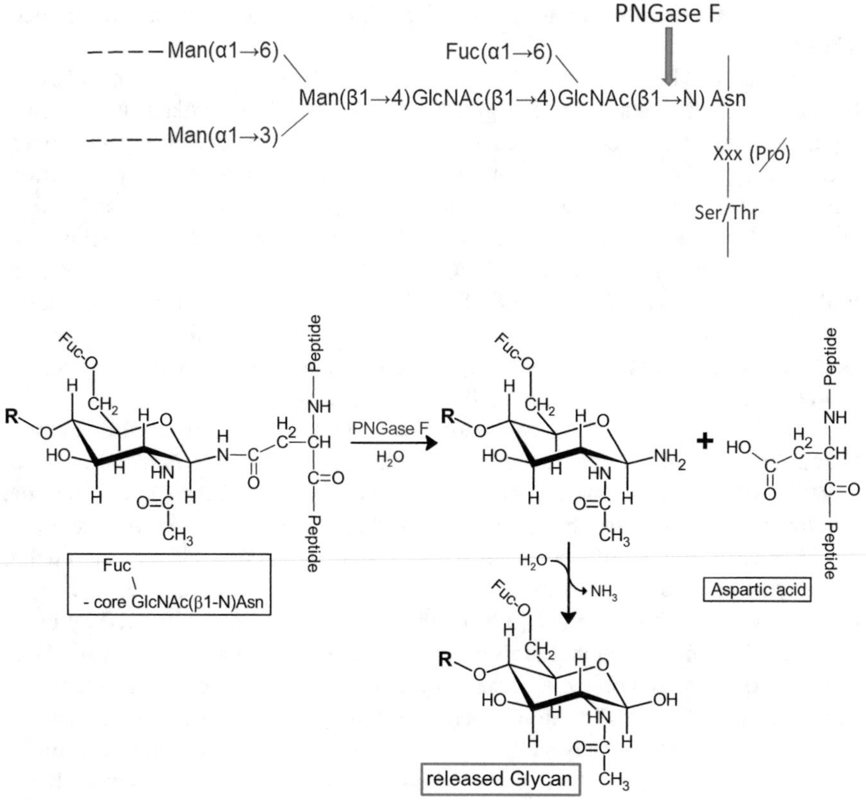

Fig. 7.4 Schematic representation of the enzymatic release of N-linked glycan from a glycoprotein by PNGase F digestion. Release of the C1-amino group from glycosylamine occurs spontaneously in an aqueous solution

used for glycopeptides, PNGase F acts only on glycopeptides containing more than three amino acids. A further complicating factor is that glycoproteins can differ in susceptibility to PNGase F deglycosylation. Sometimes, the crowdedness of glycans on the polypeptide chain, together with the size and conformation (secondary and tertiary structure) of the peptide portion, might hinder the complete release of all N-linked glycans from the glycoprotein. In that case, denaturation of the protein followed by digestion with an appropriate protease (e.g., trypsin) prior to enzymatic deglycosylation might help (see Sect. 7.3). Often, the PNGase F digestion conditions will have to be optimized for each glycoprotein studied to ensure maximum release. Furthermore, in the case of glycopeptides, it has to be noted that PNGase F is ineffective in removing N-linked chains located at the C- or the N-terminus of a peptide because PNGase F requires a minimum peptide length of three amino acids. Usually, after the PNGase digestion, the next step is a reversed-phase chromatographic purification step to separate peptides/proteins and the glycan pool. In general, glycans are desalted using graphite columns before being derivatized with a label [33].

There is another enzyme (from almonds), called peptide-N-glycosidase A (PNGase A), which can release N-glycans having an ($\alpha1\rightarrow3$)-linked fucose to the asparagine-linked N-acetylglucosamine, but it is most efficient on glycopeptides, necessitating proteolytic digestion of the glycoprotein before deglycosylation. Recently, other yeast enzymes able to release N-glycans, named PNGase YI, F-II, and acid-stable PNGase H$^+$, were reported [29]. In addition, there are the endo-β-N-acetylglucosaminidases, named Endo A, Endo C_I and C_{II}, Endo D, Endo F1, F2, F3, Endo H, Endo L, Endo M, and Endo S, all having their own specificities related to the structures of the N-glycans. They cleave the ($\beta1\rightarrow4$)-bond between the two GlcNAc residues (chitobiose) of the N-glycan common core, leaving one GlcNAc residue attached to the asparagine in the protein backbone [34].

It should be noted that a typical in-solution PNGase F digestion requires an incubation time of several hours (usually overnight) and the enzyme cannot be reused. Together with the high cost of commercial PNGase F, these factors have led to the development of alternative protocols. For instance, faster digestions were achieved by microwave irradiation [35] or pressure cycling technology [36]. Immobilizing the enzyme on columns established the reuse of the enzyme and accelerated digestions significantly (to ~30 min) [37]. High throughput was also achieved by using gel blocks in a 96-well setup. In this case, glycoproteins are denatured and alkylated and subsequently immobilized into a polyacrylamide gel and cleaned from detergents. Then, PNGase F digestion is performed in-gel, and the released glycans are washed out for further analysis [38]. Recently, an integrated N- and O-glycomics approach for sequential release of N- and O-glycans from biological samples based on glycoprotein immobilization on polyvinylidene difluoride (PVDF) membrane filter plates in 96-well format was reported [39]. This report from the group of Manfred Wuhrer may serve as an excellent example for glycomics research performed by experts in the field of glycobiology.

For academic studies, when sufficient glycan material is needed for further studies, in-solution digestion is still frequently used and will be discussed in this chapter (Protocol 29). To facilitate the enzymatic release of N-glycans from glycoproteins, a preceding denaturation of the protein is advised, eventually, by disulfide bond reduction and alkylation (see Sect. 7.3). Typically, glycoproteins are denatured by heating in an appropriate detergent (e.g., 0.5% SDS/5% β-mercaptoethanol) and degraded by a protease (e.g., trypsin, pepsin, or chymotrypsin) treatment with or without reductive alkylation. Afterward, a nonionic detergent, for example, Nonidet P-40 (NP-40) or Triton X-100, is added to the reaction medium to counteract the inhibition of PNGase F by SDS. Furthermore, PNGase F digestions should be carried out in solutions that are devoid of buffers containing ammonium salts and urea, when derivatization is planned at the reducing end of the glycan, as the hydroxyl group of the terminal GlcNAc will be substituted by an amine group, making labeling impossible.

For functional analysis studies of N-glycans, it is often necessary to prepare released glycans in sufficient amounts. N-glycan isolation on a larger scale needs adjustment of the PNGase F standard analytical workup procedure. It is possible to substantially increase the amount of glycoprotein and enzyme. Consequently, the volumes and amounts of detergent contaminants will increase, which will complicate the sample handling. In particular, the total removal of detergent is troublesome but can be obtained by using a hydrophobic interaction sorbent (Bio-Beads SM-2, Bio-Rad). The subsequently obtained extract is further purified by a sequence of C18 and graphitized carbon SPE steps [40].

Nowadays, the cleavage enzymes for de-*N*-glycosylation, together with labeling compounds, are commercially available in ready-to-use kits (e.g., GlycoWorks™ *Rapi*Fluor™ Labeling Module from Waters or the Ludger LZ-rPNGaseF-kit or the ProZyme/Agilent kit or Sciex fast glycan kit). Detailed protocols for performing the different steps are included in those kits. More and more, sample cleanup after labeling is performed by HILIC solid-phase extraction (SPE), using, for instance, BioZen™ N-Glycan Clean-Up micro-elution (Phenomenex). Recently, automated systems for de-*N*-glycosylation, including fluorescent labeling, LC separation, and computerized interpretation, became commercially available. For instance, the SmartMS™-enabled BioAccord System (Waters) is an easy-to-use LC-MS system for comprehensive analysis of glycoproteins with built-in analytical workflows for specific analysis such as N-glycan identification and profiling. Highly robust chromatographic separation and accurate mass information can be obtained. Conclusively, an automated workflow from sample preparation to data reporting is possible [41]. It is very instructive to consult the application notes published by the different companies.

Protocol 29. Denaturing of Glycoprotein and Enzymatic Release of N-Linked Glycans by PNGase F Treatment

Materials
Screw-capped Eppendorf cups (2 mL)
Eppendorf centrifuge
Water bath
Water (Milli-Q ultra-high-purity water, Millipore)
DL-dithiothreitol (DTT)
1 M Tris-HCl buffer, pH 7.5
150 mM phosphate buffer, pH 7.5 (207 mg $NaH_2PO_4.H_2O$ in 10 mL H_2O)
Peptide-N4-(acetyl-β-glucosaminyl) asparagine amidase (PNGase F, EC 3.5.1.52,
    Flavobacterium meningosepticum, New England Biolabs, Ipswich, UK)
Sodium dodecyl sulfate (SDS, Sigma-Aldrich Chemie B.V.)
2-mercaptoethanol (β-mercaptoethanol, Sigma-Aldrich Chemie B.V.)
Trifluoroacetic acid (TFA)
Glacial acetic acid
Acetonitrile
Ethanol
Iodoacetamide (IAA)
Centrifugal SpeedVac evaporator
Lyophilizer

Sephadex G-25 column
Eppendorf spin columns with 10 kDa MWCO
CarboGraph Extract-clean cartridges (4 mL, Alltech)
C-18 Extract-clean cartridges (Alltech)

Reduction/Alkylation Procedure

1. Dissolve 250µg glycoprotein in 100µL of 1 M Tris-HCl buffer, pH 8, containing 1% SDS.
2. Add 25µL of 0.5 M dithiothreitol (DTT) in 1 M Tris-HCl buffer, pH 8, mix well.
3. Incubate for 30 min at 60 °C under N_2 atmosphere (to reduce disulfide bonds in the protein).
4. Cool to room temperature.
5. Add 60µL of 0.5 M iodoacetamide (to perform carbamidomethylation).
6. Incubate for 30 min at RT in the dark with intermittent vortexing (Cys residues are irreversibly blocked).
7. Stop reaction by adding dropwise glacial acetic acid to pH 4 (check on pH paper).
8. Remove the excess reagent by SEC on Sephadex G-25 (or with spin columns with a 10 kDa MWCO).
9. Lyophilize the SEC void-volume fraction.

Denaturing procedure

1. Take 250µg of desalted/dry glycoprotein in a screw-capped Eppendorf cup.
2. Add 100µL of 150 mM phosphate buffer, pH 7.5, containing 0.5% SDS and 2% 2-mercaptoethanol.
3. Incubate for 5 min at 95 °C.
4. Cool to room temperature in an ice bath.
5. Centrifuge for 10 min at 1000 ×g.
6. Collect supernatant in clean screw-capped Eppendorf cup.
7. Add 10µL of 25% Nonidet P-40, mix well (to destroy SDS that can inactivate PNGase F during subsequent PNGase F treatment).

PNGase F Treatment in Phosphate Buffer

1. Add 100µL of 150 mM phosphate buffer, pH 7.5, to (denatured/reduced/alkylated) glycoprotein sample (~250µg), mix well.
2. Add 250 mU PNGase F (use at least 1 mU/µg glycoprotein).
3. Incubate overnight at 37 °C with gently shaking.
4. Quench digestion reaction at 100 °C for 3 min.
5. Cool to room temperature in an ice bath.

At this point, the following steps are possible:

1. Dry the sample by SpeedVac and store at −20 °C (ready for reduction or derivatization or labeling later on), or

2. Isolate the N-glycan pool from the reaction mixture and isolate de-*N*-glycosylated protein for release of O-glycans, or
3. Direct labeling (e.g., with 2-AB) for N-glycans analysis.

7.2.3.1 Isolation of N-Glycans and de-N-Glycosylated Protein (Containing O-Glycans)

N-Glycan Pool Isolation and Purification (Desalting) After PNGase F Digestion

1. Add 600μL of ice-cold ethanol or 800μL ice-cold acetone to the PNGase F-treated sample (210μL).
2. Keep the sample for 20 min at −80 °C (or overnight at −20 °C) (precipitation de-*N*-glycosylated protein).
3. Centrifuge at 8000 ×g for 10 min at 4 °C.
4. Collect the supernatant (contains the released N-glycans).

   (the precipitated proteins pellet is dried and stored at −20 °C for O-glycan preparation)

5. Prepare PGC cartridge (nonporous graphitic carbon PGC, Alltech) (graphitized carbon SPE)

   (a) wash PGC cartridge with 3 mL of 85% acetonitrile, containing 0.1% TFA,
   (b) followed by 2 mL of water, containing 0.1% TFA and 1 mL pure water

6. Load supernatant (containing N-glycans) on the cartridge.
7. Wash 3× with 3 mL water to remove salts (do not let the column run dry).
8. Elute N-glycans with 3 mL of 50% acetonitrile in 0.1% TFA.
9. Evaporate acetonitrile with N_2 stream and then lyophilize.
10. Sample can be used for HPAEC-PAD or after permethylation/labeling for (HPLC)MS analysis.

Separation of N-Glycans and de-N-Glycosylated Protein (Still Containing O-Glycans)

Prepare C-18 Extract-clean cartridge (Alltech) (or C18 reversed phase SPE, Strata X-RP cartridge, SepPak C18 SPE column (100 mg/mL; Waters), Bond Elut C18 column (Varian))

1. Acidify the N-glycan sample to pH < 5 with 1% TFA or 5% acetic acid.
2. Apply sample to C-18 SPE cartridge.
3. Elute with 4× 500μL water containing 5% acetic acid.
4. Collect flow-through (which contains N-glycans).
5. Dry by SpeedVac/lyophilization (this sample can be used, for instance, for permethylation and MS analysis).
6. Then, elute with 3 mL of 80% aqueous acetonitrile containing 0.1% formic acid (eluate contains the de-*N*-glycosylated protein/peptides).

7. Elute with 2 mL of 100% 2-propanol and pool with step 6.
8. Dry by SpeedVac/lyophilization (this sample can now be used for O-glycan analysis).

The fractionation and purification of liberated N-glycans after enzyme digestion is a critical item. Even from a pure, single glycoprotein, it is likely that several structurally distinct N-glycans will be obtained. As we have seen in Chap. 2, glycans from proteins can differ in size due to the number and substitution pattern of the monosaccharides and can differ in degree of charge due to difference in sialylation, and possible sulfation or phosphorylation.

It is a challenge to separate different types of glycans in sufficient quantities for structural characterization (see Sect. 8.2). Due to the fact that often only limited amounts of a glycoprotein can be derived from biological samples, the use of high-resolution chromatography coupled with sensitive MS methods has become inevitable, for example, using porous graphitized carbon liquid chromatography with electrospray ionization mass spectrometry (PGC-LC-ESI-MS/MS). Free N-glycans can be analyzed in this way, or, alternatively, the glycans can be reduced before analysis. Increased sensitivity for detection of N-glycans is also achieved via fluorescent/chromophore tagging by reductive amination of the free reducing ends of the released glycan molecules or derivatization of the glycans by permethylation [42]. To enhance the detection of glycans, they are often tagged with a 2-aminobenzamide (2-AB) label directly after enzymatic release by PNGase F (see Sect. 5.4.2.1, Fig. 5.4, Protocol 19). Separation can be performed by HPLC-HILIC [43]. The permethylation of glycans is often used to increase the detection and fragmentation in mass spectrometry. The hydrogens of hydroxyl, amine, and carboxyl groups are converted to methoxy- or methylamine groups (see Sect. 6.6.1).

An initial analysis of the permethylated glycan pool by matrix-assisted laser desorption mass spectrometry (MALDI-TOF) reveals molecular masses of the glycans and can already give information about the glycan heterogeneity and the presence of labile groups that would be removed by later derivatization (see Chap. 11, Fig. 11.2). Furthermore, permethylation of the glycans, giving just a small increase in molecular weight, is often performed for enhancement in sensitivity and more selective fragmentation during advanced MS techniques, for example, ESI-MS/MS. Permethylation improves ionization compared to native glycans and stabilizes sialic acids, which enables the acquisition of complete glycan profiles from complex samples in the positive ion mode. More prevalent cross-ring fragments can be obtained, yielding linkage information with MS/MS. Permethylated glycans are significantly more hydrophobic allowing for easier sample cleanup and chromatography. Very good separations of permethylated N-glycans can be obtained by RP-HPLC on C18 columns (e.g., Spherisorb S5 ODS2 or Sep-Pak C18), using gradient elution with H_2O/acetonitrile 80:20 to 20:80 (v/v).

For the determination of the glycosylation site(s), glycopeptides must be prepared through digestion of the glycoprotein with proteolytic enzymes (see Sect. 7.3). Analysis of the glycopeptides provides information on the identity of the glycans and the position of the glycosylation sites in the polypeptide backbone. In this approach, glycoproteins are digested with proteases such as trypsin (and less commonly used enzymes such as chymotrypsin, LysC, LysN, AspN, GluC, or ArgC) to produce peptides and glycopeptides that are then typically analyzed using MALDI-MS and LC-MS(/MS) methods [44] and less commonly used CE-MS(/MS) methods.

7.2.4 Release of O-Glycans by Reductive β-Elimination

In contrast to N-glycans, it is more difficult to address the various aspects of structural characterization of O-glycans [45–47]. The reasons are:

1. the lack of a consensus amino acid sequence for the glycosylation sites on the polypeptide
2. the high heterogeneity both in the number of glycans and in the extent of their occupancy and
3. the lack of a universal enzyme to release O-glycans from the proteins.

The commercially available O-glycanases only release the unsubstituted Core 1 disaccharide Gal(β1–3)GalNAc(α1-O) and Core 3 disaccharide GlcNAc(β1$\rightarrow$3) GalNAc(α1$\rightarrow$O) from Ser/Thr.

In general, glycoproteins containing O-linked glycans are analyzed in three different ways:

(1) release of O-glycans by chemical treatment, followed by subsequent characterization (O-glycan profiling), (2) digestion of glycoproteins into peptides followed by analysis of the glycopeptides ("bottom-up" approach), or (3) analysis and characterization of O-glycans on intact glycoproteins ("top-down" approach).

A variety of analytical platforms are available for the characterization of the glycans, such as HPLC profiling, lectin affinity chromatography, capillary electrophoresis, mass spectrometry, and NMR spectroscopy. Together with the GC-MS monosaccharide analysis and linkage (methylation) analysis, ^{1}H NMR spectroscopy is the most powerful tool to reveal the sequence and the anomeric linkages of O-glycan constituents.

All O-glycans can be chemically liberated from protein by hydrazinolysis, usually 6 h at 60 °C (Sect. 7.2.2, Fig. 7.3). Then, hydrazine is evaporated and the released glycans are re-N-acetylated and derivatized with 2-AB or 2-AA in the same reaction mixture. Separation of the O-glycans can be achieved by HPLC on an Amide-80 column [48]. However, loss of labile substituents (e.g., O-acetyl) occurs, and the method requires special handling procedures and safety rules, which makes it, in the first instance, not suitable for broad use.

The following method is more convenient. The O-glycans are, most commonly, cleaved from the protein by mild alkaline treatment. A β-elimination reaction using sodium hydroxide cleaves the glycans from the glycoprotein and releases O-glycans with a reducing end. The reaction works only when the carboxy group of Ser/Thr is not free. However, the alkaline conditions have the problem that the protein/peptide

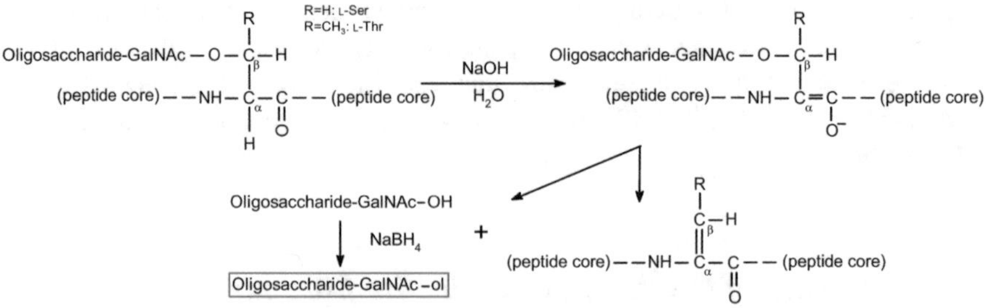

Fig. 7.5 Release of O-glycan from Ser/Thr by reductive β-elimination. During the reaction, the released glycans are converted to alditols, which exclude fluorescent tagging

is degraded during the reaction and also *O*-acetylation is lost. But more severe, the glycan is steadily destroyed under the basic conditions by 'peeling reactions' occurring at the released reducing terminus. To minimize the latter undesirable side reactions, a reduction of the reducing end of the released oligosaccharide is performed simultaneously during the chemical release, yielding the glycans with a terminal alditol [46]. This method is called the reductive β-elimination procedure (Fig. 7.5) (Protocol 30).

Protocol 30. Release of O-Glycans by Reductive β-Elimination

Materials
Pyrex glass tubes equipped with Teflon-faced screw cap
Ice bath
Speed-Vac
SPE PGC cartridges (Grace, IL)
Sodium hydroxide (NaOH)
Sodium borohydride (NaBH$_4$) (reducing agent)
Glacial acetic acid
Methanol
Trifluoroacetic acid (TFA)
Acetonitrile

Procedure

1. Dried (de-*N*-glycosylated) glycoprotein/peptide (~500–1000μg) is suspended in 250μL of water in a screw-capped glass tube.
2. Sonicate for 10 min to maximum dissolving.
3. Add 250μL of 0.5 M NaOH and 500μL of 2 M NaBH$_4$, mix well.
4. Incubate for 16–22 h at 45 °C with intermittent shaking (eventually, for 2 h in an ultrasonic bath at 60 °C) (do not cap tightly to allow the release of H$_2$).
5. Cool on ice, neutralize by adding slowly, dropwise, 25% acetic acid to adjust the pH to 6–7 (avoid loss due to excessive bubbling, vortex repeatedly).
6. Add 500μL methanol/acetic acid (9:1 v/v), mix well.
7. Evaporate to dryness under a stream of N$_2$

8. Add and evaporate 4× 400µL methanol, containing 1% (v/v) acetic acid (removal of borate salts as volatile methyl borate).
9. Add 250µL of water containing 0.05% TFA, vortex to maximum dissolving.
10. Prepare PGC cartridges (equilibrate with water, and 80% acetonitrile containing 0.05% TFA).
11. Load sample onto the column, wash with water containing 0.05% TFA.
12. Elute O-glycans with 40% acetonitrile containing 0.05% TFA.
13. Evaporate acetonitrile and then lyophilize.

In general, the reduced O-glycans are dissolved in 10 mM NH_4HCO_3 and analyzed by PGC-LC-ESI MS/MS. Separation of the glycan-alditols is also possible with HPAEC/PAD under high alkaline conditions. It is important to note that in O-glycosylated proteins, most of the O-linked domains have numerous O-linked sites in great proximity, which often complicates their release and data analysis.

7.3   Analysis of Glycopeptides

7.3.1   Preparation of Glycopeptides from Glycoproteins

An alternative way to obtain glycan samples from glycoproteins for easier further analysis is the preparation of glycopeptides by complete or selective digestion (proteolysis) of the polypeptide chain [49]. The advantage of glycopeptides is also that they preserve information about the glycosylation site in the protein and the site-specific distribution of different glycoforms when multiple glycosylation sites are present [50]. Furthermore, to facilitate the enzymatic release of N-glycans, initial proteolytic digestion to disrupt the polypeptide chain is often performed. Additionally, after the PNGase F treatment of the proteolytic sample, the released N-glycans and the glycopeptides still having O-glycans are isolated separately for analysis [51].

One of the most widely applied analytical techniques for glycopeptide analysis involves the analysis of proteolytic digests by RP-HPLC coupled with ESI-MS/MS [11, 52]. In general, the mass spectrometric analysis experiments on those peptides and glycopeptides are indicated as qualitative/quantitative "bottom-up" glycoproteomics. Glycoproteomics aims at the concomitant identification of not only the composition of the glycan but also the sites of glycosylation and the determination of the protein attached with glycans (see Sect. 7.2.1, Fig. 7.1). Thus, analysis of glycopeptides is used to elucidate the glycan structures at the individual sites. For Asn-linked N-glycans, the site consensus amino acid sequence is known, and only one or two sites are often present on a tryptic-prepared glycopeptide. O-glycans are usually linked to Ser/Thr present in a nonconsensus amino acid sequence (but often an adjacent proline at position +2). The reader is encouraged to consult excellent reviews [53–55] for more detailed information on LC-MS/MS-based glycoproteomics.

It has to be noted that the heavily occupied O-glycan clusters found in mucin-type glycoproteins (see Sect. 2.4.1.2, Fig. 2.6, and Sect. 8.3.2) can dramatically shield the protein from proteolytic digestion. Prior to proteolysis, to maximize the digestion efficiency, glycoprotein denaturation (typically a reduction and alkylation for breakdown of disulfide bridges) is usually performed (see Sect. 7.2.3, Protocol 29).

When the denatured glycoprotein is exposed to extensive proteolysis with pronase (usually mixture of at least ten proteases) or proteinase K, this results in the carbohydrate chain still linked to the single amino acid of the junction. Alternatively, limited proteolysis with specific endoproteinases (e.g., trypsin) to obtain small glycopeptides can also be performed. In most cases, tryptic digestion is most appropriate for generating (glyco)peptides for mass spectrometric analysis, although the nonspecific pronase is still frequently used (Protocol 31). The disadvantage is that glycopeptides prepared with pronase or proteinase K cannot be used for PNGase F digestion because PNGase F requires a peptide linkage on both sides of the glycosylated Asn (see Sect. 7.2.3). A key to successful analysis at the glycopeptide level is to ensure that peptides generated after digestion are not too long.

Trypsin is the better choice of a proteolytic enzyme (Protocols 32 and 33) because it results in peptide fragments (average size of 700–1500 Da) with an amino group at each end, one at the amino-terminus and the other on the side chain of lysine or arginine at the carboxyl-terminus of the peptides. The peptides having C-terminal arginine and lysine are charged, making them easier detectable by MS. However, it is possible that specific glycosylation patterns may confer differential susceptibility to proteolysis. Thus, proteolysis should always be optimized for each glycoprotein. For N- and O-glycopeptides to be analyzed successfully, they should preferably contain only one glycosylation site. The selection of proteases and buffers determines the size of glycopeptides, influencing the hydrophobicity, which subsequently is responsible for the separation of the (glyco)peptides by an HPLC method.

The glycosylation sites are identified by removing the glycans completely by PNGase F digestion, which converts asparagine into aspartic acid in the peptide backbone.

The protease digestion is usually carried out in solution, but can also be done "in-gel" after PAGE. To this end, the gel part containing the Coomassie brilliant blue-colored glycoprotein is cut out of the gel, destained by washing (e.g., with ammonium bicarbonate buffer/acetonitrile), treated with reduction and alkylation reagents, and washed again. Then, the gel part is treated overnight with proteases (e.g., trypsin and pronase). The resulting (glyco)peptides are extracted from the gel matrix with acetonitrile acidified with formic acid or TFA [56]. The mixture of N- and O-linked glycopeptides obtained in the protease digestion can be analyzed directly by LC-MS [57]. Glycopeptides obtained by trypsin digestion are typically analyzed, without derivatization, by mass spectroscopy (MS/MS) after separation through RP-LC, HILIC, or CE.

Protocol 31. Protease Digestion with Pronase in Tris-HCl Buffer

Materials
Eppendorf cups (2 mL)
Eppendorf centrifuge
Water bath 50 °C
Toluene
Sephadex G-50 or Bio-Gel P-2
100 mM Tris/HCl, pH 8, containing 2 mM $CaCl_2$
Pronase (*S. griseus*; Calbiochem) (predigested to destroy glycosidase contamination, but be careful, pronase is generally impure)

Procedure

1. Dissolve 500µg of glycoprotein in 200µL of 100 mM Tris-HCl buffer, pH 8, containing 2 mM $CaCl_2$ in an Eppendorf cup.
2. Add 100µg pronase.
3. Add 1µL toluene, tightly cap the cup.
4. Incubate 2 h at 50 °C with intermittent shaking.
5. Add again 50µg pronase.
6. Incubate overnight at 50 °C.
7. Heat sample for 5 min in boiling water bath.
8. Cool to room temperature.
9. Centrifuge 10 min at 5000×*g*.
10. Collect supernatant in a clean cup.
11. Apply supernatant to a Sephadex G-50 or Bio-Gel P-2 column (50 × 1 cm).
12. Elute with 7% 1-propanol in water.
13. Collect fractions and test for sugar (spot test, Sect. 4.3.1).
14. Pool glycopeptide fractions (for further separation, e.g., by charge).

Protocol 32. Protease Digestion with Trypsin in Bicarbonate Buffer

Materials
Eppendorf cups
Eppendorf centrifuge
Water bath
Toluene
Ammonium bicarbonate (NH_4HCO_3)
TPCK-treated trypsin (Sigma-Aldrich)
Enzyme reagent: dissolve 1 mg of trypsin in 500µL of 100 mM ammonium bicarbonate, pH 7.8 (freshly prepared)

Procedure

1. Dissolve 500µg (reduced and alkylated) glycoprotein in 500µL of 100 mM ammonium bicarbonate, pH 7.8, in an Eppendorf cup.
2. Add 10µL enzyme reagents (20µg trypsin).

3. Add 1μL toluene (to prevent bacterial growth).
4. Incubate for 6 h at 37 °C.
5. Add 10μL enzyme reagents (20μg trypsin).
6. Incubate for 18 h at 37 °C.
7. Add 100μL of water.
8. Lyophilize, which also removes the volatile buffer.
9. Dissolve in 600μL of water and filter through 0.2μm filter ((glyco)peptides can directly be analyzed by LC-MS/MS, Chap. 11).

Protocol 33. Protease Digestion with Trypsin in Phosphate Buffer

Materials

Microvials (1.5 mL or less) with conical bottom
Thermomixer (Eppendorf)
(TPCK-treated) trypsin (Sigma-Aldrich)
Sodium phosphate (Na_2HPO_4/NaH_2PO_4)
Enzyme reagent: dissolve 1 mg of trypsin in 500μL of 200 mM phosphate buffer, pH 7.

Procedure

1. Dissolve 200μg (reduced and alkylated) glycoprotein in 50μL of 200 mM phosphate buffer, pH 7.
2. Add 5μL of enzyme reagent, mix well.
3. Incubate for 5 h at 37 °C under gently shaking.
4. Add 10μL of enzyme reagent, mix well.
5. Incubate for 18 h at 37 °C under gently shaking.
6. Stop digestion by heating for 5 min at 95 °C.
    (Sample can now directly be used for PNGase F digestion in phosphate buffer (Protocol 29)).

7.3.2 Separation and Isolation of Glycopeptides

Protease digestion of a glycoprotein results in a complex mixture of nonglycosylated peptides and glycopeptides having N- and O-linked glycans. In some cases, HPAEC can be used to separate glycopeptides but that is highly dependent on the structures. Take note that, in this case, further peptide cleavage can occur at high pH.

Glycopeptides are usually separated by RP-HPLC and subsequently analyzed by coupled mass spectrometry. However, in the LC-MS/MS analysis, the proteolytic glycopeptides often exhibit poor ionization efficiency and will often be masked by the unmodified peptides originating from the nonglycosylated regions of the glycoprotein. Hence, prior glycopeptide enrichment must be performed, possible with lectin-affinity chromatography, HILIC SPE, or (PGC)-RP SPE. The isolation and purification procedures of the glycopeptides demand expertise to a certain extent from the performer. Several detailed discussions of the methodologies with relevant protocols can be found in the glycobiology literature [18, 58–64].

In brief, glycopeptides can be enriched by solid-phase extraction (SPE) techniques. For instance, by using C18-reversed-phase (RP) cartridges, first conditioned with acetonitrile, followed by water containing 0.1% (v/v) TFA or 0.1% (v/v) formic acid. The peptide mixture in the same solvent is loaded and the column is washed with the same solvent. Glycopeptides are eluted with acetonitrile containing 0.1% (v/v) TFA or 0.1% (v/v) formic acid.

Further purification can be performed by passing the glycopeptide fraction through a preconditioned PGC cartridge column. Porous graphitized carbon has an affinity for substrates with hydrophilic properties, such as glycopeptides with relatively short amino acid sequences. Glycopeptides with long amino acid sequences (usually from tryptic digests), however, are less hydrophilic.

After washing with 0.05% TFA, the retained glycopeptides are eluted with 25% acetonitrile containing 0.05% TFA, followed by drying with a SpeedVac. After redissolving in water, MALDI-TOF-MS can give an overview of the glycopeptide masses in the mixture.

The obtained glycopeptide fraction is often prefractionated (e.g., by HILIC) prior to LC-MS/MS analysis [44, 65–69]. Many different commercially available RP-HPLC columns are used, and acetonitrile-gradient elution conditions will have to be optimized to ensure maximum separation for analysis by mass spectrometry. The separation of glycopeptides with RP-HPLC is mainly based on the binding interaction with the amino acid sequence of the peptide backbone. Mass spectrometric methods that are generally used are MALDI-MS, ESI-MS, CID MS/MS, and ion mobility MS [44, 52, 70–73] (see Chap. 11). Mass spectrometric (MS/MS) glycoproteomics experiments are preferably performed in the positive ion mode. In recent years, much emphasis has been placed on developing bioinformatic tools to simplify the interpretation of glycopeptide data (see Chap. 13).

7.4 Analysis of Proteoglycans and Their Glycosaminoglycans

Proteoglycans (PGs) are conjugates of protein and glycosaminoglycan (GAG) (see Sect. 2.4.2). They are considered among the most structurally complex glycoconjugates. Several PGs have been analyzed and given intricate names, such as aggrecan, versican, biglycan, decorin, fibromodulin, and lumican. These biological macromolecules are present in high levels in matrices on the extracellular surface of connective tissue cells. Although many of the PG activities depend on the core protein, numerous physiological functions and interactions are defined by the GAG sugar sequences. Therefore, structural elucidation of GAG chains is a prerequisite for exploring the mechanisms of PG–ligand interactions. The general strategy for structural analysis of GAGs involves several steps that begin with the isolation of PGs from tissues or cultured cells (Fig. 7.6).

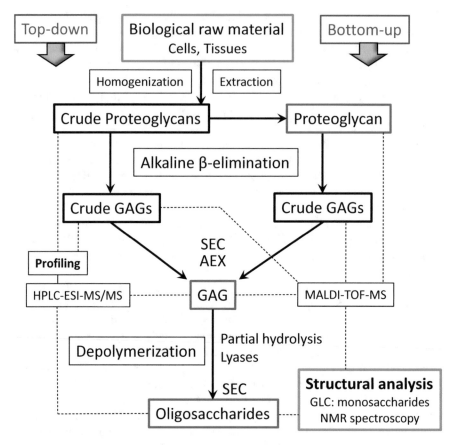

Fig. 7.6 A typical scheme for proteoglycan and GAGs analysis

Isolation and purification of proteoglycans are difficult and need a certain level of expertise because proteoglycans can form multimolecular aggregates. Dissociative extraction techniques under specific conditions are required as well as density gradient centrifugation techniques in some cases. For a detailed discussion of the preparation of biological tissue and isolation/purification procedures of PGs and GAGs, the reader is referred to the following adequate papers [74–78].

Briefly, extraction, dialysis, and ultrafiltration are used for the isolation of PGs from physiological fluids and tissue, followed by different chromatographic fractionations. The procedures are very time-consuming and labor-intensive, needing some expertise. Some practical tuition would be desirable for nonexperts. Tissues or cells have to be defatted, sliced, ground, and homogenized, followed by extraction of PGs using buffered solutions of chaotropic agents. As an example, a simplified procedure for the isolation of PG from connective tissue such as cartilage is shown (Protocol 34).

In the first instance, crude proteoglycan mixtures are often separated and purified by gel permeation chromatography (GPC) using long columns with small diameter (e.g., 100 × 1 cm) filled with Sepharose-2B or 4B. Notably, proteoglycans with the same core polypeptide will display considerable size heterogeneity due to the GAGs. This will result in separation by size-exclusion chromatography (SEC). However, the large hydrodynamic size, high water solubility, presence of many charge groups (sulfate), and the high polarity of proteoglycans make them more suitable for separation and analysis by anion-exchange chromatography (AEX) (Protocol 35). Hence, after dialysis, proteoglycans may be first separated by anion-exchange chromatography based on the charge conferred by the sulfate groups. This also implicates that separation of PGs is possible by capillary electrophoresis, according to their charge/mass ratio.

Protocol 34. Crude Preparation of Proteoglycans from Connective Tissue

Materials
Eppendorf tubes
Pasteur pipettes
Eppendorf centrifuge
Lyophilizer
Phosphate buffered saline (PBS, Ca and Mg free, Sigma-Aldrich)
Papain solution (1 mg papain/mL PBS buffer)
Ethanol

Procedure

1. Immerse 500μg dried tissue (e.g., cartilage, muscle) in 500μL digestion PBS buffer [137 mM NaCl, 10 mM phosphate, 2.7 mM KCl, pH 7.4].
2. Add 50μL papain solution, mix well.
3. Incubate for 24 h at 60 °C with end over end mixing.
4. Add again 50μL papain solution, mix well.
5. Incubate for 24 h at 60 °C with end over end mixing.
6. Centrifuge for a short time and collect the supernatant.
7. Lyophilize the supernatant.
8. Dissolve lyophilized supernatant in 500μL water, vortex.
9. Add 2 mL ice-cold ethanol and vortex.
10. Keep for 4 h at −80 °C (or 18 h at −20 °C) (precipitation of PG).
11. Centrifuge at 12,000×g for 10 min.
12. Discard the ethanol supernatant.
13. Add 100μL water, mix well.
14. Add 2 mL ice-cold ethanol and vortex.
15. Keep for 4 h at −80 °C (or 18 h at −20 °C) (precipitation of PG).
16. Centrifuge at 12,000 ×g for 10 min.
17. Discard the ethanol supernatant (complete removal of artifacts).
18. Lyophilize the pellet which contains proteoglycans (PGs).
19. Dissolve dry PG sample in 50 mM Tris-HCl buffer, pH 8.0, containing 6 M urea.

20. Filter through a 0.22µM membrane filter (*PG sample*).

The obtained filtered *PG sample* can be separated further by gel-filtration chromatography (SEC) on a Superose 6 HR10/30 column, equilibrated with 50 mM Tris-HCl buffer, pH 8.0, containing 6 M urea, and eluted with the same buffer. The fractionation can be monitored by 280 nm detection or carbohydrate spot test (Sect. 4.3.1, Protocol 1) and profiled for uronic acid-containing proteoglycans by the carbazole assay (Sect. 4.4.4, Protocol 13). Further separations can be obtained by anion-exchange chromatography of the fractions (Protocol 35).

Protocol 35. Anion-Exchange Chromatographic Separation of Proteoglycans

Materials
DEAE ion-exchange cartridge (Millipore/GE Healthcare)
Peristaltic pump
Detector 280 nm
50 mM Tris-HCl buffer, pH 8.0, containing 6 M urea
50 mM Tris-HCl buffer, pH 8.0, containing 6 M urea and 2 M NaCl

Procedure

1. Flush the cartridge (column) with 10 mL of water, flow rate ~1 mL/min.
2. Equilibrate with 10 mL of 50 mM Tris-HCl buffer, pH 8.0, containing 6 M urea.
3. Apply the filtered *PG sample* to the cartridge.
4. Elute with 20 mL of 50 mM Tris-HCl buffer, pH 8.0, containing 6 M urea (highly anionic molecules bind to column).
5. Elute until the 280-nm absorbance (stemming from protein) has fallen back to almost the zero plateau.
6. Then, elute the bound proteoglycans with 40 mL linear-elevated salt gradient from 0.15 M to 2 M NaCl (monitor by sugar spot test) and collect fractions. Proteoglycan solutions are desalted by dialysis and lyophilized. Maybe, further ion-exchange separations under similar conditions will be required, for example, on Mono-Q (Pharmacia) or Q-Sepharose.

After isolation and purification of a proteoglycan, the presence of uronic acid can be demonstrated colorimetrically by the carbazole reaction (see Sect. 4.4.4, Protocol 13). A more sensitive colorimetric method (<5µg/mL GAG) is the dimethyl-methylene blue dye-binding assay, which can also be used in microtiter plates (Protocol 36). Then, a monosaccharide analysis, for instance after methanolysis (see Sect. 6.2, Protocol 23), will give qualitative and quantitative information on the constituents of the GAG carbohydrate chains.

Protocol 36. Colorimetric Determination of GAGs

Materials
ELISA plate reader with 525 nm detector
96-well optical grade, flat-bottomed microtiter plate
NaCl p.a.

0.1 M HCl (prepare: 1 M HCl is 0.82 mL 37% HCl plus 9.18 mL water)
Glycine, free acid (>99%)
1,9-dimethyl methylene blue (Taylor's Blue) Aldrich, UK

Prepare dye solution:

1. Dissolve 3.04 g glycine and 2.37 g NaCl in 95 mL 0.1 M HCl at 25 °C under stirring.
2. Add slowly 16 mg dimethyl methylene blue gradually (2 min), under vigorously stirring.
3. After 2 h stirring, add water to a total of 1 L.
4. Stir overnight (then, store in the dark at room temperature).

Procedure

1. Make GAG standards (0–5μg in final vol. of 25μL).
2. Divide in microtiter plate wells (determine sample in triplicate).
3. Put target sample in 25μL in wells (different dilutions and at least duplicates).
4. Add 200μL of dye solution, mix.
5. Read absorbance at 525 nm immediately.

Detailed investigation of the GAGs demands the release of the GAGs from the protein core [79]. The high heterogeneity of the sulfation patterns of the GAG chains makes the isolation of homogeneous GAG oligomers and structural characterization a challenging task. The enormous structural diversity of GAG requires a laborious strategy for investigation, which encompasses:

1. The global evaluation of the number of glycoforms obtained after detachment from the core protein
2. The determination of chain lengths
3. Collection of data upon single-chain epimerization
4. Assessment of the overall sulfate content
5. Determination of the sulfation distribution along each chain
6. The identification of over- and under-sulfated regions and
7. The accurate identification of sulfation sites within the monomer ring.

This strategy is usually indicated as glycotyping of proteoglycans [80].

The protein core of proteoglycan is routinely sequenced either directly using proteomics or indirectly using molecular biology through the encoding DNA. The structural characterization of GAGs of PGs can be achieved by two strategies, referred to as "top-down," where a purified intact GAG chain is directly analyzed without pre-depolymerization or "bottom-up," where GAG chains are depolymerized using controlled chemical or enzymatic methods (Fig. 7.6). Mass spectrometry has become the key methodology in the structural elucidation of GAGs.

For the top-down study, the intact GAG chains are liberated from the polypeptide core. The GAG components are released by either chemical or enzymatic methods. GAGs in proteoglycans are O-linked to Ser/Thr, via the tetrasaccharide linker HexA($\beta1\rightarrow3$)Gal($\beta1\rightarrow3$)Gal($\beta1\rightarrow4$)Xyl($\beta1\rightarrow$O-Ser/Thr (see Sect. 2.4.2). This means that the release of GAGs from the protein can be affected by alkaline treatment with NaOH (β-elimination; Fig. 7.5). The serine residues to which the GAGs were O-linked are converted to dehydro-alanine, but the core protein stays intact. Direct reduction with NaBH$_4$ of the released carbohydrate chains after cleavage is included to prevent peeling reactions. A protocol for the release of GAG from PG by alkaline β-elimination is given below (Protocol 37).

Often, isolated PGs are treated with nonspecific proteases to degrade the protein core and facilitate the release of GAGs from the core peptides by β-elimination. Some endo-β-xylosidases are able to enzymatically release intact GAG chains by splitting the Xyl-Ser linkage. The resulting crude GAGs are collected by ultrafiltration or through precipitation. To obtain detailed sequence information, oligosaccharides are analyzed after depolymerization of the GAG with specific lyase enzymes, such as chondroitinases, keratanases, and hyaluronidases. Lyases will produce oligosaccharide fragments with a 4,5-unsaturated monosaccharide on the non-reducing end.

Protocol 37. Release of O-Linked GAGs from Proteoglycans by Reductive β-Elimination

Materials
Pyrex glass tubes equipped with Teflon-faced screw cap
Ice bath
Speed-Vac or Lyophilizer
SPE PGC cartridges (Grace, IL)
Sodium hydroxide (NaOH) (prepare: 1 M NaOH is 5.3 mL 50% NaOH solution plus 94.7 mL water)
Sodium borohydride (NaBH$_4$)
Glacial acetic acid (HOAc)
Methanol (CH$_3$OH)
Trifluoroacetic acid (TFA)
Acetonitrile (ACN)

Procedure
1. Dissolve lyophilized PG sample (~500µg) in 500µL 0.1 M NaOH in a screw-capped glass tube.
2. Add 19 mg solid sodium borohydride, vortex.
3. Incubate for 18 h at 37–45 °C (do not cap tightly to allow the release of H$_2$) with intermittent shaking, in the dark (eventually, for 2 h in ultrasonic bath at 60 °C).
4. Place sample on ice.
5. Neutralize slowly by adding (droplets) glacial acetic acid (to pH 5).
6. Add 500µL methanol/acetic acid (9:1 v/v), mix well.
7. Evaporate to dryness under a stream of N$_2$
8. Add and evaporate 4× 400µL methanol (removal of borate salts as volatile methyl borate).

9. GAGs sample ready for RP-SPE.

RP-Solid Phase Extraction (RP-SPE)

1. Place a C18 microspin column (PepClean, Thermo-Pierce) in a 1.5 mL microcentrifuge tube.
2. Wash 3× with 0.1% acetic acid by centrifugation 1500 ×g for 15 s.
3. Place C18 microspin column in new 0.5 mL Eppendorf tube.
4. Apply GAGs sample solution onto the column, wait 5 min.
5. Add 30μL 0.1% acetic acid.
6. Centrifuge 1500 ×g for 15 s.
7. Repeat 2× steps 5 and 6.
8. Combine eluants and dry by SpeedVac.
9. Dissolve in 50μL water and vortex.
10. Add 450μL chilled ethanol and vortex.
11. Store at −20 °C for 18 h (4 h at −80 °C).
12. Centrifuge 12,000 ×g for 10 min (ink-mark position of pellet).
13. Discard ethanol supernatant.
14. Repeat 1× steps 10–13.
15. Collect the *pellet* (containing GAGs).

Strong Anion-Exchange-SPE

1. Place a SAX microspin (Harvard) column into 1.5-mL microcentrifuge tube.
2. Wash 2× with 200μL water by centrifugation 1500 ×g for 15 s.
3. Wash 4× 200μL of 300 mM Na_2HPO_4 by centrifugation 1500 ×g 15 s.
4. Wash 3× 100μL 50 mM phosphate buffer, pH 3.5.
5. Apply GAG *pellet* in 50μL 50 mM phosphate buffer, pH 3.5, onto the column.
6. Centrifuge at 500 ×g for 15 s.
7. Collect the eluant.
8. Apply eluant again to column.
9. Centrifuge at 500 ×g for 15 s.
10. Collect the eluant.
11. Repeat 2× steps 8–10 with 5 min wait in last step (multiple steps to achieve complete binding).
12. 3× wash the column with 100μL 50 mM phosphate buffer, pH 3.5.
13. Centrifuge 1500 ×g for 15 s.
14. Place the spin column in a new 0.5 mL Eppendorf tube.
15. 2× apply 1 M NaCl solution/centrifuge 1500 ×g for 15 s.
16. Collect eluants (containing glycans) and dry.
17. Perform 1× ice-cold ethanol precipitation as before.

The high-molecular-weight GAG chains obtained after β-elimination can be analyzed in their native form by MALDI-TOF-MS (mostly in negative-ion mode) to evaluate the chain lengths and overall sulfate content. However, the application of MALDI-MS in GAG analysis has been limited due to the propensity for labile sulfate groups to undergo decomposition. ESI-MS is more frequently applied to GAG analysis. ESI is generally recognized as a mild ionization method useful for polar compounds and is known to result in less sulfate group cleavage than MALDI [81].

Subsequently, GAG samples are usually prepared through extensive separation processes, such as SEC, AEX, RP-LC, RP-IPC, and HILIC. In selected cases, direct analysis is performed with mass spectrometry (HPLC-MS/MS) [82]. For a more thorough structural characterization of a GAG chain, in most cases, the combined application of several analytical methods is necessary [83]. For instance, NMR spectroscopy (Chap. 12) can assist in the analysis and quantification of the constituent monosaccharide and is a good option for establishing the GAG structure [84].

In the bottom-up study, GAGs sequencing predominantly relies on tandem mass spectrometry of short GAG chains resulting from enzymatic or chemical depolymerization. GAGs are typically depolymerized using polysaccharide lyase enzymes to produce a distribution of overlapping oligosaccharides of different sizes [85]. The most widely used enzymes to depolymerize GAGs into defined oligosaccharides are bacterial or sheep-testicular hyaluronidases as well as chondroitinases [86]. When depolymerization is performed by lyases, complex GAG mixtures containing chains of variable length and degree of sulfation are formed. The enzymes heparinases and heparitinases (for heparin and heparan sulfate), hyaluronidases (for hyaluronic acid), and chondroitinases (for chondroitin and dermatan sulfates) cleave the carbohydrate chain after the HexNAc residue, thereby creating oligosaccharides, which have a 4,5-unsaturated uronic acid residue (ΔUA) at the nonreducing end. Exception, there is a *Bacillus* sp. keratanase, cleaving keratan sulfate mainly into saturated Gal($\beta1 \rightarrow 4$)GlcNAc disaccharides. Furthermore, treatment with endo-β-galactosidases for keratan sulfate is possible.

The compositional and structural analysis of GAG after enzymatic depolymerization is done either by direct ESI or MALDI-MS screening (mostly in negative-ion mode) or separation with on-/off-line ESI detection and multistage MS sequencing of the single components after desalting of the by SEC or HPLC obtained fractions [87–90]. CZE-ESI-MS^n is also used, giving separation not only according to the carbohydrate chain length but also to the degree of sulfation [80, 91]. To convert the complicated and massive MS datasets in GAG analysis into meaningful structural information, sophisticated bioinformatics tools and databases have been developed. For instance, GlycReSoft or GlyCompSoft for automated LC/MS data [92–95] (see Chap. 13).

Removal of O-sulfate esters from GAGs can be afforded by treatment with 0.5 M methanolic-HCl for 4 h at 32 °C. Then, cool on ice and neutralize with 1 M NaOH. Next, dialyze the sample trice against freshwater for 48 h at 4 °C and lyophilize. The size of the liberated GAG oligosaccharides may be determined by gel filtration chromatography. NMR spectroscopy and gas-chromatographic monosaccharide analysis (methanolysis, see Sect. 6.2) reveal the disaccharide compositions and structures of the GAG fractions.

7.5   Analysis of Glycolipids

7.5.1   General Aspects

A glycolipid is an association of an oligosaccharide with a lipid (see Sect. 2.4.3). Glycolipids are abundant in diverse tissues of humans and higher animals, but they also occur in plant and microbial cells. Their role is to provide energy and also to serve as markers for cellular recognition, being responsible for attaching cells to form tissues. To this end, glycolipids appear, in particular, on the exoplasmic surface of the cell membrane. Their carbohydrates are found on the outer surface of all eukaryotic cell membranes, extending from the phospholipid bilayer into the aqueous environment outside the cell. The associated oligosaccharides are mostly branched and frequently contain acetate, phosphate, and sulfate groups. The precise carbohydrate structure mainly confers the biological function of a particular glycolipid. The human glycolipids are known to carry the blood group-specific epitopes (Sect. 2.4.1.3). Two glycolipid families are defined, the *glyco-sphingosine* type and the *glyco-glycerol* type (Sect. 2.4.3, Fig. 2.11).

In the *glycosphingolipids* (GSLs), a great variety of oligosaccharides is glycosidically linked (via glucose or galactose) to ceramide (Cer) that anchors the glycan onto the cell membrane. GSLs can be neutral as well as acidic. In the case of the presence of terminal sialic acid residues on GSLs, they are indicated as *gangliosides*.

In the *glyco-glycerol* type, the oligosaccharide is linked via a glycosidic bond to the primary alcohol function of a glycerol molecule, which is further esterified by two (invariably different) fatty acid molecules. The glycerol-type glycolipids predominantly occur in plants. Furthermore, there is the glycosyl-phosphatidylinositol (GPI) membrane anchor, which connects an extracellular protein via carbohydrate and lipid to the cell membrane (see Sect. 2.4.4, Fig. 2.12). A detailed discussion of the methods for the isolation and purification of specific glycolipids is beyond the scope of this book. The reader is kindly referred to other sources [96, 97].

As an example, a typical procedure (Protocol 38) is given for rough isolation/purification of GSL from the membrane fraction of cells or tissue. Glycolipid isolation/purification involves three steps:

1. extraction of lipids from biological matrices
2. bulk separation from major lipid and non-lipid contaminants and
3. chromatographic resolution of individual species.

In the first instance, the total lipid fraction (including GSL) is usually obtained from tissue or cells by multiple solid–liquid extraction (SLE) with organic solvents (e.g., chloroform/methanol mixtures in different combinations).

Protocol 38. Extraction of Crude Glycolipids (GSLs) from Tissues

Materials
Chloroform
Methanol
Sodium chloride (NaCl)
Ultrasonic water bath
SpeedVac
Centrifuge

Procedure

1. Homogenize (sonicate) tissue or cells with chloroform/methanol (2:1 v/v) for 2 min at 4 °C (recommended tissue-to-solvent ratio: 1:20).
2. Centrifuge at 1000 ×g for 5 min and collect the supernatant.
3. Add 1 vol. of water, containing 0.5% NaCl, to the sediment, mix well by vortexing.
4. Add 9 vol. of chloroform/methanol (1:1 v/v), mix and sonicate again for 2 min at 4 °C.
5. Centrifuge at 1000 ×g for 5 min and collect supernatant and pool this with the previous supernatant.
6. Add chloroform/methanol (1:2 v/v) to sediment, mix well by vortexing.
7. Incubate for 30 min at 45 °C with intermediate mixing.
8. Centrifuge at 1000 ×g for 10 min and collect supernatant and pool with previous supernatants.
9. Dry the supernatant pool in SpeedVac.

This extract contains total lipids and glycolipids (but also some glycopeptides/proteins, sterols, and triglycerides depending on the biological source), so further fractionation/purification is necessary.

Extracts are freed of low molecular weight hydrophilic contaminants (salts, sugars, amino acids, etc.) by dialysis (dialysis tubing MW cutoff ~10 kDa) or by reversed-phase chromatography (RP-SPE).

Reversed-Phase Chromatography
Fractionation using Sep-Pak C18 cartridges (SPE)

Procedure

1. Wash column by injection of 5 mL of water, 15 mL of methanol/water (1:1, v/v), 5 mL of methanol, and finally 5 mL of chloroform/methanol/water (2:43:55, v/v/v).
2. Inject max. 5 mL GSL in chloroform/methanol/water (2:43:55, v/v/v).
3. Collect the eluate.
4. Reinject this eluate.
5. Again elute with 5 mL of chloroform/methanol/water (2:43:55, v/v/v).
6. Reinject this eluate.

7. Now elute successively with 5 mL of methanol/water (1:1, v/v), 5 mL methanol/ water (4:1, v/v), 5 mL methanol, 5 mL chloroform/methanol (1:1, v/v), and 5 mL chloroform, and collect the fractions.
8. Check the eluates by TLC for GSLs (see Sect. 7.5.2).

Then, other chromatographic methods (ion-exchange and gel filtration) are used for separation and purification. During the isolation and separation procedures, care must be taken to ensure that the removal or migration of labile substituents does not occur. For further analysis, glycolipids can be separated into neutral GSLs and acidic glycolipids (gangliosides) by anion-exchange chromatography, using columns like DEAE or Sepharose-Q.

A typical procedure is as follows:

1. Generate DEAE-Sephadex A-25 into the acetate form by washing with 0.8 M ammonium acetate in methanol.
2. Pack a glass column (25 × 1 cm).
3. Wash the packed column with chloroform/methanol/water 15:30:4 v/v/v.
4. Apply the glycolipid sample in the same solvent.
5. Elute with the same solvent the neutral glycolipids.
6. Then, elute with 0.8 M ammonium acetate in methanol the gangliosides.

Eventually, a separation of gangliosides according to sialic acid content (mono, di, tri) can be obtained with gradient elution 0–1.0 M ammonium acetate in methanol.

7.5.2   Thin-Layer Chromatography of Glycolipids

Mixtures of glycolipids can be separated by (HP)TLC, and different detection reagents (usually those for nonreducing sugars) allow recognition of individual glycolipids bands. For instance, orcinol–sulfuric acid detects all sugars and resorcinol–HCl detects gangliosides by their sialic acids. TLC (Protocol 39) remains widely used as it provides good reproducibility and high sensitivity. For TLC, the more polar the glycolipid the more polar the running solvent should be. However, a single TLC band, even in multiple solvent systems, does not indicate the absence of multiple species because closely related GSLs can comigrate. Small ceramide differences or saccharide variations can occur.

Protocol 39. TLC of Glycolipids and GSLs

Materials
HPTLC–silica gel$_{60}$ plates (10 × 10 cm, 0.2 mm, Merck 5633 Darmstadt).
or commercial aluminum-backed, high-performance TLC plates
Oven at 110 °C
Developing tank lined with filter paper (TLC developing chamber)
5 µL micro-syringe (Hamilton)
Hair-dry blower or heat gun

Procedure

1. Apply samples (lipids in methanol/chloroform 2:1, v/v) in 5-mm bands at the origin of the TLC plate and dry by blowing warm air from a hair dryer (include original GSL as reference).
2. Place plate in an equilibrated tank with the appropriate running solvent, close the tank.
3. Generally used TLC running solvents are:

   (a) for small di/tri/tetrasaccharide lipids: chloroform/methanol/water (60:35:8, v/v/v)
   (b) for neutral glycolipids (GSLs): chloroform/methanol/water (65:35:8, v/v/v)
   (c) for large acidic glycolipids (GSLs): chloroform/methanol/water (50:55:18, v/v/v)
   (d) for gangliosides (globo-, ganglio-, and lacto-series GSLs): chloroform/methanol/0.02% aqueous calcium chloride or KCl (50:40:10, v/v/v)

4. Allow undisturbed ascending of running solvent front up to 1 cm below the top of the plate.
5. Take out the plate and allow the solvent to evaporate in a closed fume hood at room temperature (overnight), followed by gentle heat blowing with hair dryer.
6. Dip the plate quickly in staining solvent and dry horizontally under heating.

Staining Solvents

For all GSLs: 0.2% Orcinol in 2 M H_2SO_4. Prepare: 200 mg of orcinol in 11.4 mL of H_2SO_4, and make up to 100 mL with water (CAUTION) (store at 4 °C). This is the most commonly used for detection of free and lipid-bound carbohydrates. Pinkish–violet coloration after heating 10 min at 100 °C.

For sialic acid-containing GSLs: Resorcinol/HCl-Cu^{2+} reagent. Prepare: 200 mg of resorcinol in 10 mL of water, then add 80 mL of HCl and 0.25 mL of 0.1 M $CuSO_4$, fill to 100 mL with water (store at 4 °C in the dark). Gangliosides (sialic acid) give blue–violet coloration after heating for ~20 min at 110 °C in an oven with a ventilator (cover TLC with glass plate before heating). Neutral GSLs pinkish violet after ~10 min.

Glycolipids can often be characterized by high-performance column chromatography coupled with MS and/or NMR without the need for the release of glycans. For gangliosides, negative ionization with MS/MS has been shown to produce informative fragment ions.

Subsequently, the oligosaccharide moieties of glycolipids can be released by enzymes, such as lipases that remove the lipid head groups or by endoglycoceramidase (EGCase). The latter enzyme does not cleave the linkage between monosaccharide and ceramide (in cerebrosides) or between oligosaccharides and diacylglycerol (in glycoglycerolipids). The commercial *Rhodococcus* sp. endoglycoceramidase reflects different specificities, being EGCase I mainly for globo-, EGCase II mainly for lacto- and ganglio-, and EGCase III (EGALC) for 6-gala series GSLs (Sect. 2.4.3, Table 2.3). As an example, Section 7.5.3 gives a procedure using EGCase to release glycans from glycolipids by endoglycoceramidase (EGCase) (Protocol 40). Frequently, small glycolipids can also be analyzed as entire structures.

7.5.3   Carbohydrates of Glyco(Sphingo)Lipids

The characterization of oligosaccharides of glycolipids is carried out mostly using the analytical techniques described in previous chapters. The analysis generally starts with qualitative and quantitative monosaccharide determination. To release the monosaccharide components from glycolipids, methanolysis (1 M HCl–methanol, 18 h, 85 °C) is used, which separates the sphingosine and the fatty acids from the oligosaccharide and simultaneously cleaves the oligosaccharide into the constituent monosaccharides as methyl glycosides. After re-N-acetylation of any hexosamines or sialic acids, the methyl glycosides are analyzed by GLC-MS as TMS derivatives (see Sect. 6.2, Protocol 23). During the methanolysis of GSLs for monosaccharide analysis, also the fatty acid amides in the ceramide are released as fatty acid methyl esters. These fatty acids can be isolated by extraction of the methanolysate before re-N-acetylation, just by adding hexane, mixing, and centrifugation. The isolated and dried hexane fraction can be trimethylsilylated and analyzed by GLC-MS. The remaining methanol fraction is used for further monosaccharide analysis.

Alternatively, monosaccharide analysis can be performed after acid hydrolysis (4 M TFA, 4 h, 100 °C) and GLC/MS of the alditol acetates (Sect. 6.3, Protocol 24).

Furthermore, methylation analysis is performed to obtain information about the carbohydrate substitution pattern (see Sect. 6.6, Protocol 27). However, in the case of glycolipids containing hexosamines, it is recommended to hydrolyze the permethylated glycolipid, as follows:

1. Dissolve the permethylated glycolipid in glacial acetic acid/5 M sulfuric acid (19:1 v/v, 0.3 mL) and heat in a capped vial at 80 °C for 18 h,
2. Add water (0.3 mL) and continue heating for a further 5 h,
3. Pass the cooled hydrolysate through a Dowex 50W-X8 ion-exchange resin column (~200 mg, acetate form), in Pasteur pipette, and wash with methanol (4 mL) (sulfate ions are removed).
4. Collect total eluate and dry with a stream of N_2.

Subsequent reduction and acetylation give partially methylated alditol acetates (PMAAs), which are analyzed by GLC-MS (see for PMAA mass spectra, Sect. 6.6.2, Fig. 6.11).

Nowadays, HPLC coupled with mass spectrometry (HPLC-ESI-MS/MS) is generally used in the analysis of glycolipids and the oligosaccharide analysis also relies mostly on mass spectrometry (MALDI and MS/MS techniques) using, for instance, advanced nano-LC/ESI QTOF MS systems (Agilent, CA). Simple glycolipids can often be characterized by MS and, as an example, the reader is referred to the analysis of GSLs by HILIC/ESI-MS/MS [98]. Also, NMR spectroscopy can suffice sometimes without the need for the release of the glycans. Direct NMR analysis of GSL of moderate size can yield much information about sequence, linkage positions, and anomeric carbon configuration of the oligosaccharide (see Chap. 11). However, often for structural analysis of the carbohydrates of glycolipids, selective cleavage of the carbohydrate moiety is preferred. The release of sugar chains of GSLs can be achieved by ozonolysis or periodate oxidation [99], but these methods are not suitable for alkaline-sensitive GSLs. These methods will not be discussed here.

Preferably, intact oligosaccharides are enzymatically released from neutral and sialylated glycosphingolipids (GSLs), using a glycosyl-N-acyl sphingosine 1,1-β-D-glucanohydrolase, generally referred to as endoglycoceramidase (EGCase, EC3.2.1.123) (Protocol 40). The enzyme catalyzes the hydrolysis of Galβ1→and Glcβ1→Cer linkages of neutral, sialylated, and sulfated glycosphingolipids. Hence, the release of a glycan from the lipid sphingosine or ceramide is usually performed enzymatically with commercially available (recombinant) endoglycoceramidases. Endoglycoceramidases release entire glycans from GSLs but are specific for certain GSL structures [100, 101]. Optimal conditions often have to be determined by using different incubation times and/or enzyme amounts, followed by checking the reaction mixture by TLC (Protocol 41) (disappearance of glycolipids bands from starting material indicates complete release of glycans).

Finally, released glycans are usually derivatized with a fluorescent tag and chromatographically analyzed. The oligosaccharide chains of glyco(sphingo)lipids, derived by enzymatic liberation with endoglycoceramidase, are usually labeled with 2-aminopyridine.

Possible methods for the analysis of the released oligosaccharides are:

1. ion-exchange HPLC of anionic oligosaccharides
2. size-exclusion HPLC of neutral oligosaccharides and
3. HPAEC-PAD for both anionic and neutral oligosaccharide

Additionally, sequential exoglycosidase degradation of the glycan of glyco(sphingo) lipids can be used for monosaccharide sequence determination, allowing deduction of the position and stereochemistry of the linkages; however, this method is complex and requires expertise to a certain extent from the analyst due to the necessity of different buffer conditions. As a consequence, it is not immediately appropriate for the nonexpert. At this moment, mass spectrometry is also intensively used for the analysis of glycolipids [102, 103].

Protocol 40. Release of Glycans from Glycolipids/Glycosphingolipids by Endoglycoceramidase (EGCase)

Materials
Eppendorf cups (2 mL)
Eppendorf centrifuge

Sonicator water bath
Heating water bath
Sodium acetate
Triton X-100
Chloroform
Methanol
Endoglycoceramidase (EGCase, EC3.2.1.123)

Procedure

1. Take a dry (~25µg) GSL in an Eppendorf cup.
2. Dissolve in 100µL of 25 mM sodium acetate buffer, pH 5.0–6.0, containing 0.2% (w/v) Triton X-100.
3. Sonicate for 1 min (material must be completely dissolved).
4. Add 50µL of 10 mM of EGCase enzyme in the same buffer.
5. Incubate overnight at 37 °C.
6. Stop the reaction by heating in a boiling water bath for 5 min.
7. Add 2 vol of chloroform/methanol (2:1 v/v), vortex well.
8. Centrifuge in Eppendorf centrifuge for 10 min at 2000 rpm.
9. Take the upper aqueous phase (contains the glycans; glycolipids and ceramides remain in lower organic phase).
10. *Desalting:* apply the aqueous phase sample on Sephadex G-10 gel-filtration column (20 × 0.6 cm).
11. Elute with water, collect the oligosaccharide fractions (check with sugar spot test).
12. Lyophilize the fractions (store at −20 °C).
13. Dissolve fractions in 50µL of methanol, use 1µL for TLC analysis.

Protocol 41. TLC of Glycans from Glycolipids and Glycosphingolipids (GSLs)

Materials
HPTLC-silica gel$_{60}$ plates (10 × 10 cm, 0.2 mm, Merck 5633 Darmstadt)
or commercial aluminum-backed, high-performance TLC plates
Oven at 110 °C
Developing tank lined with filter paper (TLC developing chamber)
5µL microsyringe (Hamilton)
Hair-dry blower or heat gun

Procedure

1. Apply the samples in 5-mm bands at the origin of the TLC plate and dry by blowing warm air from a hair dryer (include original GSL as reference).
2. Place the plate in an equilibrated tank with the appropriate running solvent, close the tank.
3. Running solvent for separation of oligosaccharides: 1-butanol/acetic acid/water (2:1:1, v/v/v)

running solvent for separation of glycolipids: chloroform/methanol/0.02% aqueous calcium chloride (60:40:9, v/v/v)

4. Allow undisturbed ascending of running solvent front up to 1 cm below the top of the plate.
5. Take out the plate and allow the solvent to evaporate in a closed fume hood at room temperature (overnight), followed by gentle heat blowing with hair dryer.
6. Dip the plate quickly in staining solvent 0.2% Orcinol in 2 M H_2SO_4. Prepare: 200 mg of orcinol in 11.4 mL of H_2SO_4, and made up to 100 mL with water (CAUTION) (store at 4 °C).
7. Dry horizontally under heating for ~10 min at 110 °C in a ventilated oven. Pinkish-violet coloration appears for carbohydrates.

7.5.4   Carbohydrates of Glycosylphosphatidylinositol (GPI) Anchors

To analyze GPI anchors, a combination of chemical and enzymatic methods must be used depending on the structural information desired. A minimal GPI-anchor precursor consists of a core glycan [ethanolamine (EtN)-PO_4→6)Man(α1→2) Man(α1→6)Man(α1→4)GlcNH$_2$], which is α-linked to D-*myo*-inositol of phosphatidylinositol (see Sect. 2.4.4, Fig. 2.12). This glycan core GPI structure may be modified. Extra substituents can be present on the glycan core, being Man, Gal, GalNAc, and sialic acid residues. Also, the precise nature and composition of the lipid moiety (diacylglycerol, monoacylglycerol, alkylacylglycerol, and ceramide) may vary with cell type. Mammalian GPIs contain a glycerophospholipid moiety, in particular, 1-alkyl-2-acylglycerol. Furthermore, a protein is often fused to the complex structure of a GPI. In this way, such a protein is present at the outside of the cell membrane, where the GPI is anchored into the phospholipid bilayer. The proteins attached to GPI anchors may be glycoproteins, containing N-glycans and/or O-glycans. GPI-anchored proteins are abundant constituents of the cell membranes in parasites. Some parasites are responsible for human diseases such as malaria (*Plasmodium sp.*) and toxoplasmosis (*Toxoplasma sp.*) (see Chap. 3).

   The detailed discussion of the biology, isolation, and purification of GPIs and their proteins is beyond the scope of this book but can be found, for instance, in [104] and a recent review [105]. In brief, to facilitate isolation and purification, the GPIs are initially metabolically labeled with [^{3}H]-radiolabeled ethanolamine or monosaccharides (mannose, glucosamine, inositol). GPI-anchored membrane proteins (GPI-APs) can be isolated by treatment of membrane fractions with nonionic detergents (e.g., Triton X-114) at low temperature (4 °C), which extracts soluble and integral membrane proteins, including GPI-anchored proteins. At room temperature, two phases separate, and the amphiphilic proteins remain associated with the detergent-enriched phase.

The GPIs may be recovered from the cells by differential solid–liquid extractions with chloroform/methanol and separated by anion exchange and/or hydrophobic interaction chromatography, monitored by liquid scintillation counting.

Specific chemical and/or enzymatic treatments, as indicated in Fig. 7.7, are used to produce GPI fragments for structural analysis. To obtain GPI without the protein, rigorous proteolysis using proteases (pronase, trypsin, papain) can be performed to degrade the protein. However, a general way to produce a GPI glycan fragment involves cleaving the phosphodiester bonds between mannose and the protein–ethanolamine and between inositol and the lipid moiety. This is achieved by chemical dephosphorylation using ice-cold aqueous hydrofluoric acid. Hydrogen fluoride (HF) cleaves the phosphodiester bonds, yielding the glycan core free from protein and lipid (Protocol 42).

The glycan fragment is then converted to a neutral glycan by N-acetylating or deaminating the glucosamine residue. The resulting neutral glycan products are usually analyzed by chromatographic techniques, like SEC (BioGel P-4), HPLC (HPAEC), and TLC, in conjunction with monosaccharide/linkage analysis and NMR spectroscopy. NMR spectroscopy is always a good addition for complete structural characterization [106]. Structural studies have shown that the complexity and diversity of GPIs structures in protozoa cells are greater than in mammalian cells [107].

Isolation of the protein, still bound via the ethanolamine to four sugar residues, can be obtained by nitrous acid deamination (Protocol 43). Treatment with dilute nitrous acid (HONO) cleaves the linkage between the glucosamine residue and the phosphatidylinositol, converting glucosamine into anhydro-mannose (AHM). A GPI-anchored protein can also be selectively released from the lipid moiety with phospholipase enzymes. GPI-APs are usually treated with phosphatidylinositol-specific phospholipase C (PI-PLC) and/or D (GPI-PLD). Lipase C cannot cleave a GPI anchor in which the inositol is acylated. This kind of GPI especially appears in mammalian cells. Prior treatment with mild alkali removes the fatty acid on the inositol ring. Alkaline treatment also removes the lipid-anchor moiety. The released (glyco)protein part can be further characterized by LC-MS/MS. Techniques such as HPLC, HPAEC, and MS are frequently used to analyze pure GPIs.

Structural characterization of the GPI anchors and identification of the carbohydrate components have been historically a laborious analytical task. Compositional monosaccharide analysis can be performed by using methanolysis for the liberation of the monosaccharide residues as methyl glycosides, followed by GLC-MS of the trimethylsilyl derivatives (see Chap. 6). A summarized procedure for GPI is presented in Protocol 44. Simultaneously, a qualitative lipid/fatty acids analysis is possible. A quantitative fatty acid analysis with internal standard heptadecanoic acid is described in Protocol 45. Hydroxyester-linked fatty acids are extremely labile to alkali. Alkaline released fatty acids are acidified to protonate the fatty acids, then partition them into an organic solvent, and derivatize them to the fatty acid methyl esters (FAME) for GLC-MS analysis.

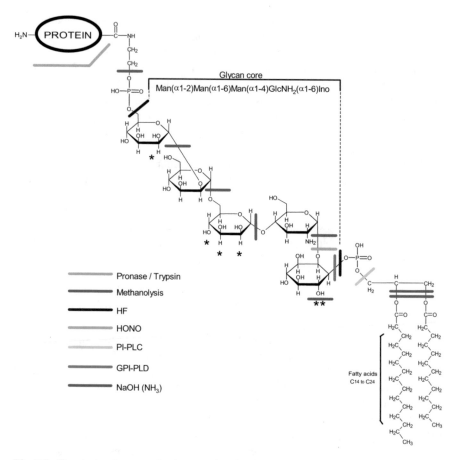

Fig. 7.7 Chemical and enzymatic cleavage sites for the production of GPI fragments, using the colored-indicated specific techniques. *Additional sugar modification (see text), ** possible acylation (palmitic acid)

Protocol 42. Preparation of the Neutral Glycan from Whole-Protein-GPI by HF Treatment

Materials
Teflon-coated screw-capped glass tubes
Screw-capped 2-mL Eppendorf tubes
Eppendorf centrifuge
Lyophilizer
Heating block (Pierce Reacti-Therm, Rockford, IL)
N_2 evaporation system
Trifluoroacetic acid (TFA)
Ammonia (30%)
Methanol

48% aqueous HF (store at −20 °C)
LiOH (saturated solution)
Solid NaHCO$_3$
Acetic anhydride [(CH$_3$CO)$_2$O]
Dowex AG50-X12 (H$^+$) [Bio-Rad]

Procedure

1. Take 150μg of lyophilized/dry protein–GPI in a small screw-cap glass tube with Teflon lining.
2. Add 200μL of 20 mM TFA.
3. Heat for 1 h at 80 °C (mild acid hydrolysis to remove sialic acid if present).
4. Cool to room temperature and dry by lyophilization.
5. Add 200μL of ammonia/methanol (1:1, v/v).
6. Heat for 1 h at 45 °C (to remove inositol palmitoylation if present).
7. Dry under a stream of N$_2$
8. Dissolve residue in 300μL water.
9. Transfer the solution into a screw-capped 2-mL Eppendorf tube.
10. Dry by lyophilization.
11. Add 100μL of ice-cold 48% aqueous HF.
12. Incubate for 48–72 h at 0 °C (dephosphorylation to yield the glycan core, de-N-acetylation occurs in the case of the presence of N-acetyl sugars).
13. Transfer the incubation mixture into a screw-capped 2-mL Eppendorf tube filled with 400μL frozen saturated LiOH solution (prepared on dry ice).
14. Keep in ice bath for 15 min and mix well by vortex.
15. Centrifuge for 30 s at 2500 rpm to sediment the LiF precipitate.
16. Collect the supernatant.
17. Wash/centrifuge the LiF pellet twice with 100μL ice-cold water.
18. Pool the supernatants.
19. Add 40 mg of solid NaHCO$_3$ to supernatants pool, mix well and cool on ice bath.
20. Add slowly 3× 30μL acetic anhydride at 10 min intervals (N-acetylation, do not vortex).
21. Allow the sample to come to room temperature and keep 20 min at RT.
22. Load the solution on a column of 5 mL Dowex AG50-X12 (H$^+$).
23. Elute with 20 mL of water.
24. Dry the eluate by lyophilization.
25. Redissolve in 100μL water.
26. Sample can be used for separation by HPAEC, as well as on a BioGel P-4 column.

Protocol 43. Preparation of Protein-EtN-Glycan Moiety and the Phosphatidylinositol (PI) Moiety by Nitrous Acid Deamination

Materials

2-mL Eppendorf tube with screw cap
Eppendorf centrifuge
Solution A: 0.2 M sodium acetate (0.81 g CH_3COONa/50 mL H_2O)
Solution B: 0.2 M acetic acid (1.15 mL glacial acetic acid/100 mL H_2O)
Prepare 100 mL buffer: Add 10 mL of solution A to 90 mL of solution B, pH ~4.
10% (w/v) Nonidet P-40 in water
0.4 M sodium nitrite ($NaNO_2$) in buffer (freshly prepare: 0.27 g/10 mL)
1 M sodium nitrite in buffer (freshly prepare: 0.68 g/10 mL)
Water-saturated n-butanol

Procedure

1. Dissolve 100µg dried protein-EtN-GPI in 200µL of acetate buffer in 2-mL Eppendorf tube.
2. Add 4µL of Nonidet P-40 solution (to solubilize the sample).
3. Add 200µL of 0.4 M sodium nitrite, check pH is ~4.
4. Incubate for 3 h at room temperature.
5. Add 100µL of 1 M sodium nitrite.
6. Keep for 6 h at room temperature (HONO is generated in situ by the action of acid on sodium nitrite; pale yellow/brown color appears).
7. Add 500µL of water-saturated n-butanol, vortex.
8. Centrifuge for 10 s in Eppendorf centrifuge.
9. Collect the butanol phase.
10. Repeat steps 7, 8, and 9, pool the butanol extracts (containing the phosphatidylinositol (PI) moiety; take an aliquot for TLC analysis or analyze by (negative ion) ES-MS and ES-MS-CID-MS).
11. The aqueous phase contains the protein linked to the glycan with the 2,5-anhydromannose (AHM). The AHM can be reduced with $NaBH_4$ to 2,5-anhydromannitol.

TLC Analysis of Butanol Extract

Silica 60 plates, solvent: chloroform/methanol/acetic acid/water 25:15:4:2 (v/v), or chloroform/methanol/water 4:1:1 or 10:10:3, develop 3x with intermediate air drying.

Protocol 44. Monosaccharide Composition Analysis

Materials

Teflon-faced screw-capped glass tubes (100 × 13 mm; Reacti-Vials, Duran Group Mainz)
Lyophilizer
Vacuum desiccator with phosphorus pentoxide (P_2O_5)
Heating block (Pierce Reacti-Therm, Rockford, IL)
SpeedVac Evaporation system

Gas chromatography coupled with a flame-ionization detector (FID) or an electron-impact mass spectrometer (GLC-EIMS)

Chemicals and Solvents

1.0 M methanolic HCl (prepared by dilution of commercial ampulla containing 3 M methanolic HCl (Supelco, USA))

Pyridine, anhydrous (C_5H_5N; Sigma Aldrich)

Acetic anhydride [Ac_2O; $(CH_3CO)_2O$]

Anhydrous methanol (CH_3OH, MeOH)

Trimethylsilylation (TMS) reagent is a freshly prepared mixture of anhydrous pyridine-hexamethyldisilazane-trimethylchlorosilane, 5:1:1 (v/v/v). (CAUTION: corrosive)

Procedure

1. Take 100µg dry protein–GPI (salt and detergent free) in a screw-cap tube with Teflon lining.
2. Add an exact amount of Internal Standard solution (mannitol 10 nmol).
3. Lyophilize and then dry, overnight, in a vacuum desiccator (over P_2O_5).
4. Add 500µL of 1.0 M HCl–methanol under nitrogen atmosphere (tightly close tubes).
5. Heat for 8 h at 85 °C (vortex once after the first 15 min).
6. Cool to room temperature.
7. Add ~100µL of pyridine to neutralize HCl.
8. Add 100µL of acetic anhydride, mix well.
9. Leave for 2 h at room temperature.
10. Dry in SpeedVac and 1× with methanol.
11. Add 200µL of trimethylsilylation (TMS) reagent, close tube, vortex.
12. Keep the mixture for 30 min at room temperature.
13. Inject 1µL on GLC, column SE-54 or EC-1, temp 140–250 °C with 8°/min, then, 250–300 °C with 15°/min, then, 300 °C for 20 min.
14. Identification of the TMS methyl glycosides is based upon the comparison of relative retention times with those of standards run under the same conditions.

During the methanolysis of GPI for monosaccharide analysis, also fatty acids are released as methyl esters and are visual during GLC analysis.

Protocol 45. Fatty Acid Analysis

Materials

Teflon-faced screw-capped glass tubes (100 × 13 mm; Reacti-Vials, Duran Group Mainz)

Lyophilizer

Vacuum desiccator with phosphorus pentoxide (P_2O_5)

Heating block (Pierce Reacti-Therm, Rockford, IL)

SpeedVac Evaporation system

Gas chromatography coupled with a flame-ionization detector (FID) or electron-impact mass spectrometer (GLC-EIMS)

Ammonia (30%)

Heptadecanoic acid

Di-ethyl ether

Dichloromethane (DCM)

Hydrochloric acid (HCl)

1.0 M methanolic HCl (prepared by dilution of commercial ampulla containing 3 M methanolic HCl (Supelco, USA))

Pyridine, anhydrous (C_5H_5N; Sigma Aldrich)

Acetic anhydride [Ac_2O; $(CH_3CO)_2O$]

Anhydrous methanol (CH_3OH, MeOH)

Saturated diazomethane in ether (has yellow color)

Trimethylsilylation (TMS) reagent is a freshly prepared mixture of anhydrous pyridine-hexamethyldisilazane-trimethylchlorosilane, 5:1:1 (v/v/v) (CAUTION: corrosive)

Procedure

1. Take the exact amount of dry GPI plus heptadecanoic acid (exact amount as IS) in 2-mL Eppendorf tube with screw cap.
2. Add 400μL of ammonia/methanol (1:1, v/v), close well.
3. Incubate 6 h at 50 °C.
4. Dry in SpeedVac, 3× with 100μL 50% methanol (to remove ammonia).
5. Dissolve in 300μL of 10 mM HCl in water.
6. Add 300μL of di-ethyl ether, mix well, and wait 5 min.
7. Centrifuge for 15 s in Eppendorf centrifuge.
8. Transfer the ether phase to a clean tube.
9. Repeat 2× steps 6, 7, and 8; combine the ether phases.
10. Dry under a gentle stream of N_2
11. Put the tube in an ice bath.
12. Add 50μL aliquots of ether saturated with diazomethane until yellow color remains.
13. Leave for 20 min on ice.
14. Evaporate with a gentle stream of N_2 to dryness (in fume hood).
15. Dissolve residue in dichloromethane.
16. Analyze 1μL on GC-MS, temp 140–250 °C with 8°/min, then, 250–300 °C with 15°/min, then, 300 °C for 20 min. Compare retention times and mass spectra with commercial fatty acid methyl esters (FAMEs) (e.g., characteristic ions are m/z 74 and 87; m/z $[M]^+$; m/z $[M-31]^+$).

7.6 Analysis of Polysaccharides

The determination of the primary structure of polysaccharides is rather difficult due to the fact that polysaccharides often consist of mixtures of compounds differing in molecular mass and may be built up from irregular elements that are unevenly distributed over the chain (see Sect. 2.3).

Polysaccharides are employed in the food, cosmetic, and pharmaceutical industries. Structural analysis of the concerning polysaccharide may say something about the function in these applications. The characterization of polysaccharides is a labor-intensive and time-consuming endeavor. Multiple chemical and physical analytical methods are often needed to obtain a complete picture of the structure [108, 109].

The primary structure of a polysaccharide is defined by the monosaccharide composition, the configuration of the glycosidic linkages, the position of glycosidic linkages, sequence of the monosaccharides, as well as the nature, number, and location of appended noncarbohydrate groups (Sect. 5.1, Table 5.1). The spatial structure is furthermore responsible for the properties.

In general, the analysis starts with the isolation of polysaccharides from a natural source (e.g., "medicinal" plants/fruits, human/animal tissues, and bacterial cultures), followed by a thorough purification. There are many different procedures for the isolation and purification of polysaccharides, which are highly dependent on the composition and structure [110]. The determination of the molecular mass (M_w) is usually performed by SEC with laser light scattering (LLS) or refractive index (RI) detection, although it is difficult to obtain suitable standards that possess the same hydrodynamic volume to prepare a calibration curve.

Polysaccharides vary greatly in molecular size (10 to >100 kDa) and can contain all kinds of ketoses, aldoses, anhydrosugars, aminosugars, and sugar acids. Homopolysaccharides are composed of one type of monosaccharides, while heteropolysaccharides contain different types of monosaccharides. The polymer may be linear, branched, or occasionally a cyclic polymer.

Furthermore, polysaccharide can be linked to lipid, as in the case of lipopolysaccharide (LPS), expressed on the cell wall of Gram-negative bacteria. These polysaccharide chains are structurally very diverse, containing monosaccharides not found elsewhere in nature.

In the first instance, the total carbohydrate content of the target polysaccharide is determined by the phenol–sulfuric acid colorimetric method (see Sect. 4.3.2, Protocol 2). For a quantitative/qualitative analysis of the constituent monosaccharides, the polysaccharide must be depolymerized. The glycosidic bonds between the monosaccharide residues are cleaved by strong acid and elevated temperature. Optimal conditions have to be found for maximal hydrolytic cleavage with minimal degradation. The yields of monosaccharides released depend on the nature of the polysaccharide (homo/hetero), type of glycosidic bonds, degree of cross-linking, and number of hydrogen bridges between sugars. Furthermore, the type of acid used, acid strength (pH), reaction time, and temperature (see also Sect. 5.2). Especially, polysaccharides containing acidic sugars (e.g., pectins and alginates) or having β-linkages (e.g., cellulose and chitin) can be difficult to hydrolyze quantitatively. Sulfuric acid and trifluoroacetic acid are commonly used for hydrolysis. After hydrolysis, underivatized monosaccharides are determined by HPAEC-PAD (Sect. 6.5.1). Gas–liquid chromatography (GLC) is used to analyze derivatized monosaccharides (Sects. 6.2 and 6.3). In the latter case, solvolysis is usually performed by methanolysis (1–3 M HCl/methanol, 18 h, 85 °C). Additionally, the glycosidic-linkage pattern of the monosaccharides of the polysaccharide is determined by a "methylation analysis" (Sect. 6.6).

An indispensable, powerful technique to study the structure of polysaccharides is NMR spectroscopy (see Sect. 12.5). Proton and carbon NMR spectroscopic experiments can give information on the type of constituent monosaccharides, their anomeric configuration, ring size, the position and type of glycosidic linkages, and the presence of noncarbohydrate substituents [108]. To complete and confirm the ultimate structure, frequently, all these chemical and physical analyses are also performed on fragments obtained by specific degradation of the polysaccharide (Sect. 5.2.7). Using the results of oligosaccharides consisting of overlapping parts of the polysaccharide, the sequence of residues and glycosidic linkages can be deduced.

References

1. Mariño K, Saldova R, Adamczyk B, Rudd PM. Changes in serum *N*-glycosylation profiles: functional significance and potential for diagnostics. Carbohydr Chem. 2012;37:57–93.
2. Stumpo KA, Reinhold VN. The N-glycome of human plasma. J Proteome Res. 2010;9:4823–30.
3. Defaus S, Gupta P, Andreu D, Gutiérrez-Gallego R. Mammalian protein glycosylation—structure versus function. Analyst. 2014;139:2944–67.
4. Zhang L, Luo S, Zhang B. Glycan analysis of therapeutic glycoproteins. MAbs. 2016;8:205–15.
5. Yang X, Bartlett MG. Glycan analysis for protein therapeutics. J Chromatogr B 2019;1120:29–40.
6. Simpson RJ, editor. Purifying proteins for proteomics: a laboratory manual. New York: Cold Spring Harbor Laboratory Press; 2004.
7. Rosenberg IM. Protein analysis and purification. Basel: Birkhauser; 2005.
8. Walls D, Loughran S, editors. Protein chromatography. Totowa: Humana Press; 2017.
9. Hong Q, Ruhaak LR, Stroble C, Parker E, Huang J, Maverakis E, Lebrilla CB. A method for comprehensive glycosite-mapping and direct quantitation of serum glycoproteins. J Proteome Res. 2015;14:5179–92.
10. Moremen KW, Tiemeyer M, Nairn AV. Vertebrate protein glycosylation: diversity, synthesis and function. Nat Rev Mol Cell Biol. 2012;13:448–62.
11. Mariño K, Bones J, Kattla JJ, Rudd PM. A systematic approach to protein glycosylation analysis: a path through the maze. Nat Chem Biol. 2010;6:713–23.
12. Krishnamoorthy L, Mahal LK. Glycomic analysis: an array of technologies. ACS Chem Biol. 2009;4:715–32.
13. Mechref Y, Hu Y, Desantos-Garcia JI, Hussein A, Tang H. Quantitative glycomics strategies. Mol Cell Proteomics. 2013;12:874–84.
14. Gaunitz S, Nagy G, Pohl NLB, Novotny MV. Recent advances in the analysis of complex glycoproteins. Anal Chem. 2017;89:389–413.
15. Domann PJ, Pardos-Pardos AC, Fernandes DL, Spencer DI, Radcliffe CM, et al. Separation-based glycoprofiling approaches using fluorescent labels. Proteomics. 2007;7(suppl. 1):70–6.
16. Zhang Y, Peng Y, Yang L, Lu H. Advances in sample preparation strategies for MS-based qualitative and quantitative N-glycomics. TrAC Trends Anal Chem. 2018;99:34–46.
17. DeLeoz MLA, Duewer DL, Fung A, Liu L, Yau HK, et al. NIST interlaboratory study on glycosylation analysis of monoclonal antibodies: comparison of results from diverse analytical methods. Mol Cell Proteomics. 2020;19:11–30.

18. Vreeker GCM, Wuhrer M. Reversed-phase separation methods for glycan analysis. Anal Bioanal Chem. 2017;409:359–78.
19. Jensen PH, Karlsson NG, Kolarich D, Packer NH. Structural analysis of N- and O-glycans released from glycoproteins. Nat Protocol. 2012;7:1299–310.
20. Anumula KR. Analysis of Ser/Thr-linked sugar chains. In: Post-translational modification of proteins: tools for functional proteomics, Methods in molecular biology, vol. 1934. Springer Science + Business Media; 2019. p. 33–42.
21. Yuan JB, Wang CJ, Sun YJ, Huang LJ, Wang ZF. Nonreductive chemical release of intact N-glycans for subsequent labeling and analysis by mass spectrometry. Anal Biochem. 2014;462:1–9.
22. Lv GP, Hu DJ, Cheong KL, Li ZY, Qing XM, Zhao J, Li SP. Decoding glycome of *Astragalus membranaceus* based on pressurized liquid extraction, microwave-assisted hydrolysis and chromatographic analysis. J Chromatogr A. 2015;1409:19–29.
23. Cai K, Hu D, Lei B, Zhao H, Pan W, Song B. Determination of carbohydrates in tobacco by pressurized liquid extraction combined with a novel ultrasound-assisted dispersive liquid-liquid microextraction method. Anal Chim Acta. 2015;882:90–100.
24. Narimatsu H, Kaji H, Vakhrushev SY, Clausen H, Zhang H, Noro E, Togayachi A, Nagai-Okatani C, Kuno A, Zou X, Cheng L, Tao S-C, Sun Y. Current technologies for complex glycoproteomics and their applications to biology/disease-driven glycoproteomics. J Proteome Res. 2018;17:4097–112.
25. Goso Y. Malonic acid suppresses mucin-type O-glycan degradation during hydrazine treatment of glycoproteins. Anal Biochem. 2016;496:35–42.
26. Chen WX, Smeekens JM, Wu RH. Comprehensive analysis of protein N-glycosylation sites by combining chemical deglycosylation with LC-MS. J Proteome Res. 2014;13:1466–73.
27. Goso Y, Sugaya T, Ishihara K, Kurihara M. Comparison of methods to release mucin-type O-glycans for glycomic analysis. Anal Chem. 2017;89:8870–6.
28. Kameyama A, Dissanayake SK, Thet Tin WW. Rapid chemical de-N-glycosylation and derivatization for liquid chromatography of immunoglobulin N-linked glycans. PLoS One. 2018;13:e019800.
29. Zhang Q, Li Z, Song X. Preparation of complex glycans from natural sources for functional study. Front Chem. 2020;8(508):1–12.
30. Song X, Ju H, Lasanajak Y, Kudelka MR, Smith DF, Cummings RD. Oxidative release of natural glycans for functional glycomics. Nat Methods. 2016;13:528–34.
31. Song X, Ju H, Zhao C, Lasanajak Y. Novel strategy to release and tag N-glycans for functional glycomics. Bioconjug Chem. 2014;25:1881–7.
32. Kameyama A, Thet Tin WW, Toyoda M, Sakaguchi M. A practical method of liberating *O*-linked glycans from glycoproteins using hydroxylamine and an organic superbase. Biochem Biophys Res Commun. 2019;513:186–92.
33. Blanchard V, Gadkari RA, Gerwig GJ, Leeflang BR, Dighe RR, Kamerling JP. Characterization of N-linked oligosaccharides from human chorionic gonadotropin expressed in the methylotrophic yeast Pichia pastoris. Glycoconj J. 2007;24:33–47.
34. Kobata A. Exo- and endoglycosidases revisited. Proc Jpn Acad Ser B. 2013;89:97–118.
35. Sandoval W, Arellano F, Arnott D, Raab H. Rapid removal of N-linked oligosaccharides using microwave assisted enzyme catalyzed deglycosylation. Int J Mass Spectrom. 2007;259:117–23.
36. Szabo Z, Guttman A, Karger BL. Rapid release of *N*-linked glycans from glycoproteins by pressure-cycling technology. Anal Chem. 2010;82:2588–93.
37. Szigeti M, Bondar J, Gjerde D, Keresztessy Z, Szekrenyes A, Guttman A. Rapid N-glycan release from glycoproteins using immobilized PNGase F microcolumns. J Chromatogr B. 2016;1032:139–43.
38. Royle L, Campbell MP, Radcliffe CM, White DW, Harvey DJ, Abrahams JL, Kim YG, Henry GW, Shadick NA, Weinblatt ME, Lee DM, Rudd PM, Dwek RA. HPLC-based analy-

sis of serum N-glycans on a 96-well plate platform with dedicated database software. Anal Biochem. 2008;376:1–12.

39. Zhang T, Madunić K, Holst S, Zhang J, Jin C, Ten Dijke P, Karlsson NG, Stavenhagen K, Wuhrer M. Development of a 96-well plate sample preparation method for integrated N- and O-glycomics using porous graphitized carbon liquid chromatography-mass spectrometry. Mol Omics. 2020;16:355–63.

40. Valk-Weeber RL, Dijkhuizen L, Van Leeuwen SS. Large-scale quantitative isolation of pure protein N-linked glycans. Carbohydr Res. 2019;479:13–22.

41. Sun X, Tao L, Yi L, Ouyang Y, Xu N, Li D, Linhardt RJ, Zhang Z. N-glycans released from glycoproteins using a commercial kit and comprehensively analyzed with a hypothetical database. J Pharmaceut Anal. 2017;7:87–94.

42. Ruhaak LR, Huhn C, Waterreus WJ, De Boer AR, Neusüss C, Hokke CH, Deelder AM, Wuhrer M. Hydrophilic interaction chromatography-based high-throughput sample preparation method for N-glycan analysis from total human plasma glycoproteins. Anal Chem. 2008;80:6119–26.

43. Ruhaak LR, Zauner G, Huhn C, Bruggink C, Deelder AM, Wuhrer M. Glycan labeling strategies and their use in identification and quantification. Anal Bioanal Chem. 2010;397:3457–81.

44. Wuhrer M, Catalina MI, Deelder AM, Hokke CH. Glycoproteomics based on tandem mass spectrometry of glycopeptides. J Chromatogr B. 2007;849:115–28.

45. Wada Y, Dell A, Haslam SM, Tissot B, Canis K, et al. Comparison of methods for profiling O-glycosylation. Mol Cell Proteomics. 2010;9:719–27.

46. Zauner G, Kozak RP, Gardner RA, Fernandes DI, Deelder AM, Wuhrer M. Protein O-glycosylation analysis. Biol Chem. 2012;393:687–708.

47. You X, Quin H, Ye M. Recent advances in methods for the analysis of protein O-glycosylation at proteome level. J Sep Sci. 2018;41:248–61.

48. Anumula KR. Single tag for total carbohydrate analysis. Anal Biochem. 2014;457:31–7.

49. Gerwig GJ, Vliegenthart JFG. Analysis of glycoprotein-derived glycopeptides. In: Jollès P, Jörnvall H, editors. Proteomics in functional genomics. Basel: Birkhäuser Verlag; 2000. p. 159–86.

50. Zhu Z, Desaire H. Carbohydrates on proteins: site-specific glycosylation analysis by mass spectrometry. Annu Rev Anal Chem. 2015;8:463–83.

51. Morelle W, Michalski JC. Analysis of protein glycosylation by mass spectrometry. Nat Protocol. 2007;2:1585–602.

52. Dalpathado DS, Desaire H. Glycopeptide analysis by mass spectrometry. Analyst. 2008;133:731–8.

53. Dallas DC, Martin WF, Hua S, German JB. Automated glycopeptide analysis—review of current state and future direction. Brief Bioinform. 2013;14:361–74.

54. Thaysen-Andersen M, Packer NH. Advances in LC-MS/MS-based glycoproteomics: getting closer to system-wide site-specific mapping of the N- and O-glycoproteome. Biochim Biophys Acta. 2014;1844:1437–52.

55. Yang Y, Franc V, Heck AJR. Glycoproteomics: a balance between high-throughput and in-depth analysis. Trends Biotechnol. 2017;35:598–609.

56. Nwosu CC, Huang J, Aldredge DL, Strum JS, Hua S, Seipert RR, Lebrilla CB. In-gel non-specific proteolysis for elucidating glycoproteins (INPEG)—a method for targeted protein-specific glycosylation analysis in complex protein mixtures. Anal Chem. 2013;85:956–63.

57. Stavenhagen K, Plomp R, Wuhrer M. Site-specific protein N- and O-glycosylation analysis by a C18-porous graphitized carbon-liquid chromatography-electrospray ionization mass spectrometry approach using pronase treated glycopeptides. Anal Chem. 2015;87:11691–9.

58. Goldberg D, Bern M, Parry S, Sutton-Smith M, Panico M, Morris HR, Dell A. Automated N-glycopeptide identification using a combination of single- and tandem-MS. J Proteome Res. 2007;6:3995–4004.

59. Ueda K, Takami S, Saichi N, Daigo Y, Ishikawa N, Kohno N, Katsumata M, Yamane A, Ota M, Sato TA, Nakamura Y, Nakagawa H. Development of serum glycoproteomic profiling

technique: Simultaneous identification of glycosylation sites and site-specific quantification of glycan structure changes. Mol Cell Proteomics. 2010;9:1819–28.

60. Gilar M, Yu YQ, Ahn J, Xie HW, Han HH, Ying WT, Qian XH. Characterization of glycoprotein digests with hydrophilic interaction chromatography and mass spectrometry. Anal Biochem. 2011;417:80–8.

61. Parker BI, Thaysen-Andersen M, Solis N, Scott NE, Larsen MR, Graham ME, Packer NH, Cordwell SJ. Site-specific glycan-peptide analysis for determination of N-glycoproteome heterogeneity. J Proteome Res. 2013;12:5791–800.

62. Goldman R, Sanda M. Targeted methods for quantitative analysis of protein glycosylation. Proteomics Clin Appl. 2015;9:17–32.

63. Ji ES, Lee HK, Park GW, Kim KH, Kim JY, Yoo JS. Isomer separation of sialylated O- and N-linked glycopeptides using reversed-phase LC-MS/MS at high temperature. J Chromatogr B. 2019;1110–1111:101–7.

64. Qing G, Yan J, He X, Li X, Liang X. Recent advances in hydrophilic interaction liquid interaction chromatography materials for glycopeptide enrichment and glycan separation. Trends Anal Chem. 2020;124:115570.

65. Selman MHJ, Hemayatkar M, Deelder AM, Wuhrer M. Cotton HILIC SPE microtips for microscale purification and enrichment of glycans and glycopeptides. Anal Chem. 2011;83:2492–9.

66. Pasing Y, Sickman A, Lewandrowski U. N-glycoproteomics: mass spectrometry-based glycosylation site annotation. Biol Chem. 2012;393:249–58.

67. Ongay S, Boichenko A, Govorukhina N, Bischoff R. Glycopeptide enrichment and separation for protein glycosylation analysis. J Sep Sci. 2012;35:2341–72.

68. Alley WR, Mann BF, Novotny MV. High-sensitivity analytical approaches for the structural characterization of glycoproteins. Chem Rev. 2013;113:2668–732.

69. Chen CC, Su WC, Huang BY, Chen YJ, Tai HC, Obena RP. Interaction modes and approaches to glycopeptide and glycoprotein enrichment. Analyst. 2014;139:688–704.

70. Desaire H. Glycopeptide analysis, recent developments and applications. Mol Cell Proteomics. 2013;12:893–901.

71. Kolli V, Schumacher KN, Dodds ED. Engaging challenges in glycoproteomics: recent advances in MS-based glycopeptide analysis. Bioanalysis. 2015;7:113–31.

72. Nilsson J. Liquid chromatography-tandem mass spectrometry-based fragmentation analysis of glycopeptides. Glycoconj J. 2016;33:261–72.

73. Jin C, Harvey DJ, Struwe WB, Karlsson NG. Separation of isomeric O-glycans by ion mobility and liquid chromatography-mass spectrometry. Anal Chem. 2019;91:10604–13.

74. Fu L, Suflita M, Linhardt RJ. Bioengineered heparins and heparan sulfates. Adv Drug Deliv Rev. 2016;97:237–49.

75. Whitelock JM, Iozzo RV. Isolation and purification of proteoglycans. Methods Cell Biol. 2002;69:53–67.

76. Ly M, Laremore TN, Linhardt RJ. Proteoglycomics: recent progress and future challenges. OMICS. 2010;14:389–99.

77. Fasciano JM, Danielson ND. Ion chromatography for the separation of heparin and structurally related glycosaminoglycans: a review. J Sep Sci. 2016;39:1118–29.

78. Woods A, Couchman JR. Proteoglycan isolation and analysis. Curr Protoc Cell Biol. 2018;80:e59.

79. Prabhakar V, Capila I, Sasisekharan R. The structural elucidation of glycosaminoglycans. Methods Mol Biol. 2009;534:147–56.

80. Amon S, Zamfir AD, Rizzi A. Glycosylation analysis of glycoproteins and proteoglycans using capillary electrophoresis-mass spectrometry strategies. Electrophoresis. 2008;29:2485–507.

81. Zaia J. Glycosaminoglycan glycomics using mass spectrometry. Mol Cell Proteomics. 2013;12:885–92.

82. Solakyildirim K. Recent advances in glycosaminoglycan analysis by various mass spectrometry techniques. Anal Bioanal Chem. 2019;411:3731–41.

83. Beccati D, Lech M, Ozug J, Gunay NS, Wang J, Sun EY, Pradines JR, Farutin V, Shriver Z, Kaundinya GV, Capila I. An integrated approach using orthogonal analytical techniques to characterize heparan sulfate structure. Glycoconj J. 2017;34:107–17.
84. Liu X, St Ange K, Wang X, Lin L, Zhang F, Chi L, et al. Parent heparin and daughter LMW heparin correlation analysis using LC-MS and NMR. Anal Chim Acta. 2017;961:91–9.
85. Li L, Ly M, Linhardt RJ. Proteoglycan sequence. Mol Biosyst. 2012;8:1613–25.
86. Wang W, Wang J, Li F. Hyaluronidase and chondroitinase. Adv Exp Med Biol. 2017;925:75–87.
87. Zaia J. On-line separations combined with MS for analysis of glycosaminoglycans. Mass Spectrom Rev. 2009;28:254–72.
88. Volpi N, Linhardt RJ. High-performance liquid chromatography-mass spectrometry for mapping and sequencing glycosaminoglycan-derived oligosaccharides. Nat Protoc. 2010;5:993–1004.
89. Li G, Li L, Tian F, Zhang L, Xue C, Linhardt RJ. Glycosaminoglycanomics of cultured cells using a rapid and sensitive LC-MS/MS approach. ACS Chem Biol. 2015;10:1303–10.
90. Yu Y, Zhang F, Colón W, Linhardt RJ, Xia K. Glycosaminoglycans in human cerebrospinal fluid determined by LC-MS/MS MRM. Anal Biochem. 2019;567:82–4.
91. Zamfir AD. Applications of capillary electrophoresis electrospray ionization mass spectrometry in glycosaminoglycan analysis. Electrophoresis. 2016;37:973–86.
92. Wang X, Liu X, Li L, Zhang F, Hu M, Ren F, et al. GlycCompSoft: software for automated comparison of low molecular weight heparins using top-down LC/MS data. PLoS One. 2016;11:1–13.
93. Hu H, Khatri K, Zaia J. Algorithms and design strategies towards automated glycoproteomics analysis. Mass Spectrom Rev. 2017;36(4):475–98. https://doi.org/10.1002/mas.21487.
94. Duan J, Amster IJ. An automated, high-throughput method for interpreting the tandem mass spectra of glycosaminoglycans. J Am Soc Mass Spectrom. 2018;29:1802–11.
95. Hogan JD, Klein JA, Wu J, Chopra P, Boons G-J, Carvalho L, et al. Software for peak finding and elemental composition assignment for glycosaminoglycan tandem mass spectra. Mol Cell Proteomics. 2018;17:1448–56.
96. Christie WW, Han X. Lipid analysis: isolation, separation, identification and lipidomic analysis. Cambridge: Woodhead Publishing; 2010.
97. Owen DM, editor. Methods in membrane lipids. Totowa: Humana Press; 2014.
98. Akiyama H, Ide M, Yamaji T, Mizutani Y, Niimi Y, Mutoh T, Kamiguchi H, Hirabayashi Y. Galabiosylceramide is present in human cerebrospinal fluid. Biochem Biophys Res Commun. 2021;536:73–9.
99. Song X, Smith DF, Cummings RD. Nonenzymatic release of free reducing glycans from glycosphingolipids. Anal Biochem. 2012;429:82–7.
100. Li Y-T, Chou C-W, Li S-C, Kobayashi U, Ishibashi Y-H, Ito M. Preparation of homogenous oligosaccharide chains from glycosphingolipids. Glycoconj J. 2009;26:929.
101. Albrecht S, Vainauskas S, Stöckmann H, McManus C, Taron CH, Rudd PM. Comprehensive profiling of glycosphingolipid glycans using a novel broad specificity endoglycoceramidase in a high-throughput workflow. Anal Chem. 2016;88:4795–802.
102. Sarbu M, Zamfir AD. Modern separation techniques coupled to high performance mass spectrometry for glycolipid analysis. Electrophoresis. 2018;39:1155–70.
103. Barrientos RC, Zhang Q. Recent advances in the mass spectrometric analysis of glycosphingolipidome—a review. Anal Chim Acta. 2020;1132:134–55.
104. Dangerfield JA, Metzner C. GPI membrane anchors: the much-needed link. Sharjah: Bentham Science Publishers; 2010.
105. Kinoshita T. Biosynthesis and biology of mammalian GPI-anchored proteins. Open Biol. 2020;10:190290.
106. Striepen B, Zinecker CF, Damm JLB, Melgers PAT, Gerwig GJ, Koolen M, Vliegenthart JFG, Dubremetz JF, Schwarz RT. Molecular structure of the "low molecular weight anti-

gen" of *Toxoplasma gondii*: a glucose alpha1-4 N-acetylgalactosamine makes free glycosyl-phosphatidylinositols highly immunogenic. J Mol Biol. 1997;266:797–813.

107. Orlean P, Menon AK. GPI anchoring of protein in yeast and mammalian cells, or: how we learned to stop worrying and love glycophospholipids. J Lipid Res. 2007;48:993–1011.

108. Gerwig GJ. Structural analysis of exopolysaccharides from lactic acid bacteria. In: Kanauchi M, editor. Lactic acid bacteria: methods and protocols, Methods in molecular biology, vol. 1887. Springer; 2019.

109. Song E, Shang J, Ratner D. Polysaccharides, polymer science: a comprehensive reference, vol. 10. Set: Elsevier; 2012.

110. Shi L. Bioactivities, isolation and purification methods of polysaccharides from natural products: a review. Int J Biol Macromol. 2016;92:37–48.

Chapter 8
Structural Characterization of Released Glycans

Abstract This chapter will highlight techniques for the analysis of released glycans in their native, reduced, or labeled form. Nowadays, standardized chromatographic separations followed by mass spectrometric analysis and retention time comparison using on-line databases are the usual methods. Minimal amounts of material can be investigated due to fluorescent/chromophore labeling and the increased sensitivity of the analytical equipment. Fractionation and enzymatic sequence analysis will be discussed, in particular for N-glycans. The release and fractionation of O-glycans, as abundant components of mucins, will also be discussed. Extensive literature references are provided.

Keywords N-glycans · O-glycans · Fluorophore labeling · 2-AA · 2-AB · 2-AP · NP-HILIC · RP-HPLC · Dextran ladder · Glucose units (GU) · Exoglycosidases · Mucins · Reductive β-elimination

8.1 Introduction

In the case of glycoproteins and glycolipids, the analysis of the sugar chains is important to fully understand the function of the entire glycoconjugate [1]. As described already in previous chapters, glycan analysis is frequently performed after the release of the glycans in the native form, the reduced form, or labeled with a fluorophore or chromophore. Subsequently, the glycans are usually separated and isolated by chromatographic techniques. Different methods can be used, such as SEC-type separations, lectin-affinity-based trapping, anion-exchange, and reversed-phase or hydrophilic-interaction (HILIC) chromatography (see Sect. 5.3). Methods to release glycans from glycoconjugates have been discussed in Chap. 7.

After isolation of pure glycans, various methods are on hand for the structural analysis of the oligosaccharides. Before applying chemical methods, it is advised to first record NMR spectra, although this is dependent on the available amount of the glycan (at least 10–50 nmol) (see Chap. 12). NMR spectroscopy is nondestructive, so the samples can be reused for further investigation. As a next step, a monosaccharide analysis and linkage analysis can be performed (see Chap. 6). Another approach, in particular when only minimal amounts are available, is separation by

© Springer Nature Switzerland AG 2021

G. J. Gerwig, *The Art of Carbohydrate Analysis*, Techniques in Life Science and Biomedicine for the Non-Expert, https://doi.org/10.1007/978-3-030-77791-3_8

HPLC and comparison of elution times of the individual glycans to a dextran ladder (expressed in glucose units (GU)), which can predict structures by using web-available software/databases (Chap. 13). Furthermore, the use of mass spectrometric techniques (e.g., ESI-MS/MS) in the primary structural analysis of glycans has grown enormously due to the possibility of direct coupling of mass spectrometers to advance separation equipment (UHPLC-MS and CE-MS). Also, here, on-line software/databases are excellent aids for the interpretation of the data (e.g., GlycoWorkBench).

Another method to obtain information about the monosaccharide types and the linkages between them is the sequential degradation of the oligosaccharide with specific exoglycosidases followed by NP-HPLC (see Sect. 8.2.3).

8.2   Analysis of N-Glycans

As mentioned earlier, to perform structural analysis, N-glycans are commonly released from denatured glycoproteins (Sect. 7.2.1). A supplementary protocol for the release of N-glycans from glycoproteins, including the use of detergents prior to PNGase F digestion, is given in Protocol 46. N-glycans, enzymatically liberated from glycoproteins, are usually directly labeled with a fluorophore, be it 2-aminobenzamide (2-AB), 2-aminobenzoic acid (2-AA), or 2-aminopyridine (2-AP or PA) [2–4]. The 2-AA label is preferred for capillary electrophoresis, due to its negatively charged acidic group, but it also enables HPLC separations, such as hydrophilic interaction liquid chromatography (HILIC), mixed-mode HILIC/anion exchange, and weak anion-exchange chromatography [5]. The labeling procedures are discussed in Sect. 5.4.2 (Protocols 19 and 20) (see also Figs. 5.4 and 5.5). The fluorescent tagging of glycans at the reducing end, through a quantitative reductive amination method, is used to amplify the detection limit but at the same time to increase the hydrophobicity of the oligosaccharide, improving their chromatographic separation.

Protocol 46. Denaturing of Glycoprotein and Release of N-Glycans by PNGase F Treatment

Materials
Glass screw-capped microvials (2 mL)
Phosphate buffer (200 mM, pH 6.5)
Sodium dodecyl sulfate (SDS)
2-mercaptoethanol
NP-40 (Merck, Darmstadt, Germany)
PNGase F recombinant (Roche)
Microcrystalline cellulose (Merck)
Calbiosorb beads (Merck) for detergent removal (wash 3× with Milli-Q water and then resuspended in water 1:1 (v/v))
Thermomixer (Eppendorf)

Centrifugal evaporator (Speed-Vac)

SepPak C18 SPE / C18 Extract-Clean cartridges (Alltech) (washed 3× with 400µL of 80% ACN containing 0.1% TFA and equilibrated 3× with 400µL of 0.1% aqueous TFA)

CarboGraph Extract-Clean cartridges (Alltech) (washed 3× with 400µL of 80% ACN containing 0.1% TFA and equilibrated 3× with 400µL of 5% ACN containing 0.1% TFA)

Procedure

1. Dissolve glycoprotein in 200 mM phosphate buffer (5–10µg/µL).
2. Add 1% SDS and 10% 2-mercaptoethanol and incubate 5 min at 95 °C (to denature glycoprotein).
3. Cool down! Then add NP-40 to a concentration of 10% and dilute sample 10× with 200 mM phosphate buffer, pH 7.5.
4. Add 50 U of PNGase F and incubate for 4 h at 37 °C under gently shaking.
5. Add again 50 U of PNGase F and incubate for 16 h at 37 °C under gently shaking.
6. Add Calbiosorb beads (50µL/100µL sample volume) and incubate for 16–20 h at RT under gently shaking.
7. Centrifuge gently for 5 s and pipet the supernatant (containing N-glycans) into a new screw-capped microvial tube.

N-Glycan Purification

1. Acidify sample to pH <4 with 1% TFA.
2. Apply the sample to a C18 cartridge and collect the flow-through (contains N-glycans).
3. Elute 3× with 400µL of 0.1% aqueous TFA and pool with flow-through of step 2.
4. Apply the flow-through/pool to a CarboGraph cartridge.
5. Wash 3× with 400µL of 0.1% aqueous TFA (to remove salts).
6. Elute N-glycans with 400µL of 25% acetonitrile containing 0.1% TFA.
7. Repeat step 6 twice.
8. Dry eluted N-glycan pool by centrifugal evaporation (or lyophilize).
9. Concentrate the N-glycans by subsequently adding 200, 100, and 50µL Milli-Q water with intermediate centrifugal evaporation.

8.2.1   Fractionation of N-Glycans

Oligosaccharides can be fractionated by traditional gel-filtration chromatography (e.g., SEC on Bio-Gel P-2, P-4, P-6), but HPLC methods are now preferred. As a first step, the pool of released (labeled) N-glycans is often separated by ion-exchange chromatography into charged and neutral glycans and, if possible, on a preparative scale (Protocol 47). The negatively charged glycans, containing sialic acid and/or sulfate groups, bind to the anion-exchange column, while neutral glycans pass through. Usually, separation is achieved into mono-, di-, tri-, and tetra-sialylated glycans. Further glycan characterization can be achieved, for instance, by analysis of the collected peaks on reversed-phase and normal-phase HPLC. RP-HPLC directly coupled with electrospray ionization mass spectrometry (LC/ESI-MS/MS) offers a good method to analyze N-glycans.

In principle, 2-AB labeled glycans are suitable for chromatographic separations by HILIC, WAX, and RP-HPLC and detection/analysis by ESI-MS. At the same time, MALDI-TOF-MS in both positive and negative ion modes, directly on the N-glycan pool, reveals an overview of the distribution of N-glycan structures (see Chap. 11, Fig. 11.2). Anion-exchange (WAX) and C18 reversed-phase chromatography are preferred for the separation of substantial amounts of natural glycans due to the higher solubility of the glycans in the water content of the mobile phase. Nevertheless, 2-AB-labeled N-glycans are typically fractionated/analyzed on a normal-phase (HILIC) HPLC (glycan fingerprinting) (Protocol 48). The structures of the individual glycans can be predicted from the elution time that is expressed in glucose units (GU) with reference to a dextran ladder (Sect. 8.2.2), and quantification is achieved by detector response. An extensive database (GlycoBase and autoGU) has been developed which uses the standardized elution positions of 2-AB-labeled glycans in hydrophilic interaction liquid chromatography (HILIC TSK-Amide-80 column) with fluorescence detection for structural assignment (see Sect. 5.3.3, Fig. 5.2 and Chap. 13). The modern monochromatic-based detectors show utmost sensitivity (in the femtomolar range) [5, 6]. The glycan structures can often be inferred by comparison to similar data from known N-glycans in several reference databases [7, 8] (see Chap. 13).

As said, separation of labeled glycans can also be achieved by reversed-phase liquid chromatography and is often used for the separation and isolation of substantial amounts of glycans (Protocol 49). RP-HPLC on C18 columns is frequently used for the analysis of N-glycans labeled with 2-PA (see Sect. 5.4.2.1, Protocol 20) (see also Fig. 5.5). Furthermore, porous graphitized carbon chromatography (PGC) is very suitable for fractionation/analysis of N-glycans, labeled or as oligosaccharide alditols (Protocol 50). An HPLC system consisting of a PGC column directly connected with a mass spectrometer is extremely suitable for the analysis of N-glycans, especially with the MS^n technique [9, 10]. The use of mass spectrometry in the primary structural analysis of N-glycans has grown enormously, due to the limited amounts often available. Retention times in combination with glycan mass analyses are often sufficient for structural assignment of the glycans [11, 12] (see Sect. 5.3.4).

In the case of isolated and purified N-glycans in sufficient amounts, the analysis of the structure is rather straightforward. A very efficient method for determining the complete primary structure of glycans is by $^1H/^{13}C$ NMR spectroscopy [13]. In many cases, the typical NMR spectra can already reveal the structure of the glycan by comparison to library spectra. NMR spectroscopy of carbohydrates is described in Chap. 12. Confirmation of the composition may be obtained by the determination of the constituent monosaccharides, using the methanolysis/GLC method (Sect. 6.2) and by methylation analysis to confirm the linkage patterns of the monosaccharides (Sect. 6.6). Furthermore, the sequential arrangement of the monosaccharide units can be determined by two-dimensional NMR spectroscopy and by enzymatic sequence analysis (Sect. 8.2.3), when dealing with novel structures.

Protocol 47. Anion-Exchange Chromatography of N-Glycans
Column: Mono-Q HR 5/5 (50 × 5 mm) (Merck 54807, GE Healthcare)

1. Wash column with 2 mL of water at a flow rate of 1 mL/min.
2. Inject sample dissolved in 50µL of water.
3. Elute with a linear gradient from 0 to 100 mM NaCl in water (10 mL).
4. Monitor eluate at 214 nm (neutral oligosaccharides are in void volume, monosialylated are eluted at a ~15 mM NaCl, disialyl at ~40 mM, trisialyl at ~50 mM, tetrasialyl at ~70 mM NaCl).
5. Clean the column by washing with 2 mL of 100 mM NaCl.

Protocol 48. NP-HILIC-HPLC of N-Glycans Labeled with 2-AB or 2-AP (PA)
HPLC: UltiMate 3000 SD HPLC system (Thermo Fisher Scientific, Waltham, MA) equipped with a Jasco FP-920 fluorescence detector.
    Column: Acquity UPLC Glycan BEH Amide (1.7µm, 100 × 2.1 mm) (Waters)

1. Wash the column with 80% acetonitrile in 250 mM formic acid, pH 3.0 (adjusted with ammonia) at a flow rate of 0.5 mL/min.
2. Inject the sample dissolved in the same solvent.
3. Elute with the same solvent, adding a linear gradient of 22–40% water in 65 min.
4. Monitor eluate at (λ_{ex} 330 nm, λ_{em} 420 nm).
5. Clean the column by washing with final gradient conditions for 10 min.

    Column: TSK-gel Amide-80 (5µm, 250 × 4.6 mm) (Merck, Anachem) or GlycoSep N (5µm, 250 × 4.6 mm) (Prozyme/Agilent)

Solvent A: 80% acetonitrile in 50 mM ammonium formate, pH 4.4
Solvent B: 50 mM ammonium formate. pH 4.4
Elution with A, adding a linear gradient of 3% B to 43% B in 50 min at a flow rate of 1 mL/min.
(See Sect. 5.3.3, Fig. 5.2 for the HPLC profile of 2-AB-labeled N-glycans released from human IgG).

    Column: Supelcosyl LC-NH$_2$ (5µm, 250 × 4.6 mm) (Merck) for free (neutral) N-glycans

Elution with Acetonitrile/H$_2$O = 70:30 for 15 min, then linear gradient to 50:50 (v/v) in 60 min at a flow rate of 1 mL/min.

Protocol 49. RP-(C18) HPLC of N-Glycans Labeled with 2-AP (PA)
Column: Shim-Pak CLC-ODS (250 × 6 mm) (column temperature 55 °C) (Shimadzu)

Solvent A: 10 mM phosphate buffer, pH 4
Solvent B: 0.5% 1-butanol in solvent A

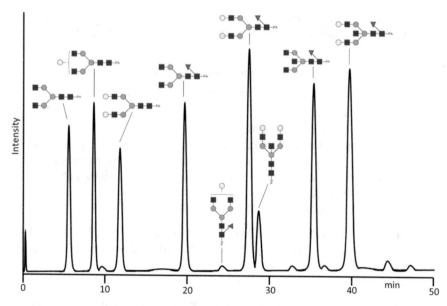

Fig. 8.1 Example of RP-HPLC analysis of PA-oligosaccharides (neutral N-glycans)

Elution with linear gradient from A/B = 80:20 to A/B = 45:55 in 45 min at a flow
   rate of 1 mL/min.
Monitor eluate at (λ_{ex} 320 nm, λ_{em} 400 nm) (Fig. 8.1).
Lyophilize fractions and desalt on a Sephadex G-15 column.

Column: Cosmosil 5C18-P (150 × 4.6 mm) (Nacalai) or Amino AS-5A (5μm,
250 × 4 mm) or Hypersil ODS (5μm, 50 × 4.6 mm) (ThermoFisher Scientific)

Solvent A: 50 mM acetic acid, pH 5 (adjusted with trimethylamine)
Solvent B: 0.5% 1-butanol in solvent A
Elution with linear gradient from A to B in 50 min at a flow rate of 1 mL/min, or
Solvent A: 100 mM ammonium acetate, pH 4 (Prepare: 0.1 M acetic acid and add
   drops of 4 M aqueous ammonia till pH 4), Solvent B: 0.5% 1-butanol in solvent A
Elution with linear gradient from A/B = 95:5 to A/B = 5:95 in 45 min at a flow rate
   of 1.5 mL/min.
Monitor eluate at (λ_{ex} 315 nm, λ_{em} 400 nm).

Protocol 50. HPLC of N-Glycans on a Graphitized-Carbon Column
Column: HyperCarb 3–5-μm particles graphitized carbon (75 × 2 mm) (Thermo
Fisher Scientific)

Elution with linear gradient from (A) 3% acetonitrile in aqueous 1% (v/v) formic
   acid to (B) 90% acetonitrile in aqueous 1% (v/v) formic acid in 60 min at a flow
   rate of 2 mL/min. Or elution with (A) 5 mM NH_4HCO_3/2% acetonitrile, for
   30 min. and then, a linear gradient to (B) 5 mM NH_4HCO_3/80% acetonitrile in

50 min at a flow rate of 2 mL/min. PGC columns require a long equilibration time and are highly susceptible to contamination. So, use clean samples.

8.2.2   Preparation of Glucose Oligomers Standard (Dextran Ladder)

During HPLC separations, the structures of the individual glycans can be deduced from the elution time that is expressed in glucose units (GU) with reference to a dextran ladder [14]. Reference data can be found in databases online. The recipe for the preparation of a Dextran ladder solvent is as follows.

1 g Dextran (MW 200 kDa) is heated in 10 mL of 0.1 M HCl for 4 h at 100 °C. The solution is passed through a column containing 2 mL of Bio-Rad AG3-X4A (OH⁻ form, 100–200 mesh) to remove hydrochloric acid. The eluent (mixture of mono- to oligosaccharides) is diluted with water. The HPLC runs of (fluorescently labeled) aliquots of this solution (DP ~1–20 glucose units) are used as an external standard to calibrate the HPLC system for structural analysis of (identically labeled) N-glycans in GU values (see Chap. 13).

Alternatively, dissolve 10 mg dextran in 1 mL of 1 M TFA and heat at 80 °C for 30 min. After this partial acid hydrolysis, the sample is dried by evaporation with an N_2 stream. The oligosaccharides are reduced by adding 500µL of 1 M $NaBH_4$ in 50 mM NaOH and gently stirring for 2 h. After dropwise neutralization with 4 M acetic acid, desalting is performed by five times co-evaporating with methanol. HPLC runs of aliquots of this solution (containing gluco-oligosaccharide-alditols; DP ~3–13 glucose units) are used as an external standard to calibrate the HPLC system (PGC-LC-ESI-MS/MS) in GU values for structural analysis of reduced N-glycans.

8.2.3   Enzymatic Sequence Analysis of N-Glycans

As discussed earlier, the features that make structural analysis of glycans complicated include:

1. the variety of positional glycosidic bonds that link individual monosaccharide residues
2. the anomeric configuration of each monosaccharide and
3. the presence of branched structures.

When a single, pure (labeled) glycan or glycopeptide has been obtained from a mixture, one way to get information about the oligosaccharide structure can be achieved by enzymatic sequencing, in particular with respect to N-glycans. This means sequential digestion with glycan-specific exoglycosidases to remove terminal monosaccharides one by one from the glycan's nonreducing end. This method for structural characterization of glycans (denoted as exoglycosidase array method) is not the easiest one, and the interpretation of the results needs some skills and practice of the analyst for the following reasons.

Reliable results depend on complete digestion and, in particular, on the purity of the enzymes. They must have strict specificity toward particular monosaccharide species and their anomeric configuration. Several exoglycosidases also have an aglycone specificity or discriminate between positional isomers, and their actions are often sterically affected. At present, more than 40 exoglycosidases are available, specific for a particular type of monosaccharide, the anomeric configuration (α or β), and the position of attachment to an adjacent monosaccharide in a larger oligosaccharide sequence. An overview of commonly used enzymes for the structural elucidation of N-glycans is provided in Fig. 8.2.

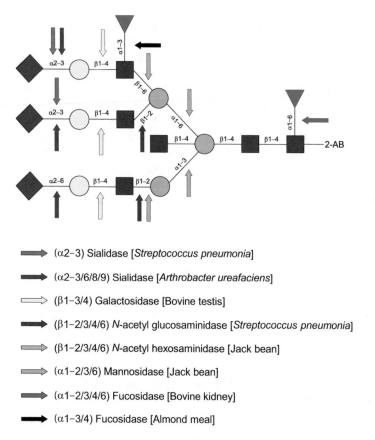

⟹ (α2–3) Sialidase [*Streptococcus pneumonia*]

⟹ (α2–3/6/8/9) Sialidase [*Arthrobacter ureafaciens*]

⟹ (β1–3/4) Galactosidase [Bovine testis]

⟹ (β1–2/3/4/6) N-acetyl glucosaminidase [*Streptococcus pneumonia*]

⟹ (β1–2/3/4/6) N-acetyl hexosaminidase [Jack bean]

⟹ (α1–2/3/6) Mannosidase [Jack bean]

⟹ (α1–2/3/4/6) Fucosidase [Bovine kidney]

⟹ (α1–3/4) Fucosidase [Almond meal]

Fig. 8.2 Overview of commonly used enzymes for exoglycosidase sequence analysis, demonstrated on a complex-type N-glycan. The arrows indicate the cleavage sites, removing the terminal residue if present. Symbol and color notation of the monosaccharides are according to Chap. 2, Table 2.1

Typically, an (e.g., 2-AB-labeled) oligosaccharide (~100 pmol) is incubated overnight (to allow the reaction to go to completion) with a relatively high concentration of the particular enzyme in 50 mM sodium acetate buffer, pH 5.5, at 37 °C. Then, the reaction products are subjected to chromatographic analysis (e.g., BioGel P-4 or NP/RP-HPLC). Comparison of chromatographic glycan profiles before and after an enzyme digest will show to what extent glycans have been digested. The shift of the product peak compared to the initial peak of the starting material is indicative of the release of monosaccharide residues.

For instance, if sialic acids are present in a terminal position, first, they can be removed with specific sialidases (also called neuraminidases), differentiating between (α2→3)- and (α2→6)-linked sialic acids. Removal of galactose with β-galactosidase would indicate the presence of terminal β-galactose. The release of N-acetylglucosamine from de-galactosylated glycan but not from the intact glycan by specific β-N-acetyl hexosaminidases indicates the presence of galactose linked to N-acetylglucosamine. In the same way, a shift after fucosidase digestion confirms the presence of fucose. By using more exoglycosidases with strict specificities, like mannosidases, more structural information can be obtained [15, 16].

For evaluation, the sizes of the analyzed glycan as well as the respective fragments resulting from the exoglycosidase treatment have to be determined. The size of oligosaccharide fragments can be determined by MALDI-TOF mass spectrometry. HPLC-MS analysis of the reaction products is an ideal method for monitoring the progress and confirmation of the results.

The digestion process with subsequent enzymes is repeated until the oligosaccharide has been successfully cleaved completely. In most cases, comparison with known standards treated in the same manner allows tentative glycan identification. In this way, information can be obtained concerning the sequence and type of the monosaccharides and the type of linkages and a tentative proposal of the structure of the starting material can be made.

However, in the case of complex branched structures having several nonreducing termini accessible to an enzyme, a sufficiently accurate method to measure the precise number of monosaccharides cleaved must be used. It should be warned that the use of enzymes for determining structures requires great care on the part of the researcher. The practical work with glycosidases is not that simple. The buffer and enzyme concentrations are critical to obtaining maximum enzyme efficiency. The often quite concentrated salt suspensions tend to dry, the enzymes are not indefinitely stable, and pipetting of the rather small volumes bears the risk of misdosing the enzyme. Thus, failure to degrade a sample may be the result of activity loss of the enzyme. For accurate and reliable structural assignment to be made, the purity of the enzyme reagents and starting oligosaccharides must be very high. Confirmative evidence for a structure by other methods is often essential.

Since the exoglycosidase sequential degradation method requires specialized expertise and facilities, and is relatively slow and labor-intensive, it will not be further discussed here in this book as a technique for the nonexpert. The interested reader is referred to [17].

This also holds for an alternative sequencing strategy, using optimized mixtures of exoglycosidases rather than single enzymes, called the Reagent Array Analysis Method (RAAM) [18]. However, it must be noted that much progress is observed in the development of computational methods to automate exoglycosidase data interpretation in conjunction with modern, highly sensitive analytical workflows to improve glycan profiling [19, 20]. The rapidly growing field of recombinant glycotherapeutics has provided the impetus for the development of reproducible robust high-throughput analysis methods. Exoglycosidase digestion is regularly used combined in a rapid-throughput analytical HPLC (HILIC) platform called GlycoBase and autoGU (see Chap. 13).

Over the past few years, several companies (e.g., GlycoWorks, Ludger, Waters, GlycoPrep, Prozyme, and Oxford Glycosciences Ltd.) have been offering kits with clear protocols that simplify and standardize glycan profiling workflows, improving the speed and sensitivity of N- and O-glycan sample preparation. Furthermore, computational methods for exoglycosidase array data interpretation (GlycanAnalyzer/GlycoStore) have appeared [8, 21]. GlycanAnalyzer has software that allows for the confirmation of glycan structures by analyzing data derived from HILIC-UPLC-MS after the application of exoglycosidases. Exoglycosidase removal of monosaccharides results in signature peak shifts, in both UPLC and MS, yielding an effective way to pattern match N-glycan structures with high detail. Available is also a Hamilton STAR robotics platform (Reno, Nevada) coupled to LC for high-throughput analysis of N-glycans, using the 2-aminobenzamide (2-AB)-labeled glycans in the database GlycoBase [6]. It is possible to identify specific glycosylation features of recombinant glycoproteins using combinatorial exoglycosidase digestions coupled with HILIC analysis.

In the coming years, it is expected that this will become the norm in software packages associated with commercial glyco-analytical platforms. Recently, there is also a drive in the direction of analysis at the glycopeptide level by RP-HPLC-MS, specifically when assessed as part of a multi-attribute method (MAM) workflow [22, 23]. Individual recovered fractions of released oligosaccharides/glycopeptides can be further analyzed using the general analytical techniques (see Chap. 5, Table 5.2), including mass spectrometry and NMR spectroscopy.

8.3 Analysis of O-Glycans

8.3.1 General Aspects

Characterization of O-linked glycans is still considered a difficult task [24, 25]. First, because they include a wide variety of structures (see Sect. 2.4.1.2), and second, because O-linked glycans must be released from the glycoconjugates by chemical methods due to the lack of a generic cleavage enzyme. The only available enzyme (known as *O*-glycanase or endo-α-*N*-acetylgalactosaminidase) has a very limited specificity (only for the release of the nonextended Core 1 disaccharide Gal(β1→3)GalNAc(α1→O) and the Core 3 disaccharide GlcNAc(β1→3) GalNAc(α1→O)), so not suitable for general application to remove all the different O-linked carbohydrates. For the structural analysis, it is essential that the O-glycans are released in a nonselective manner. Two chemical techniques are used to obtain O-glycans.

The most generally applicable cleavage reaction to release O-glycans from Ser and Thr in the protein backbone is by mild alkaline treatment under reductive conditions, called reductive β-elimination (see Sect. 7.2.4, Protocol 30, Fig. 7.5, see also Sect. 7.4, Protocol 37, and Sect. 8.3.2, Protocol 52). However, the method does not work for amino acids, such as Tyr, HyPro, and HyLys. For the elimination reaction to occur, a hydrogen atom must be present in α-position to the electron-withdrawing group present in the β-hydroxyamino acids, like serine and threonine. The released O-glycans are, during the elimination, simultaneously converted to their corresponding oligosaccharide-alditols. This means that the terminal *N*-acetylgalactosamine (GalNAc) is reduced to an alditol (GalNAc-ol) using sodium borohydride. This is done to avoid the undesirable "peeling" side reaction, in which further β-elimination of the released O-glycan chain can occur [25]. However, it has to be noted that some loss of alkali-labile *O*-sulfate esters, if present, can occur. To avoid further complications, it is better to first release N-glycans by PNGase F treatment, as they can also be partially released under alkaline conditions. The alkaline-borohydride hydrolysis does not allow direct labeling with fluorophores or UV chromophores to amplify detection during subsequent separation. If labeling is needed, reoxidation has to be performed. But the underivatized oligosaccharide-alditols can be separated by HPAEC.

The other chemical method to obtain O-glycans is hydrazinolysis (see Sect. 7.2.2). This method needs safety precautions. A mild hydrazinolysis releases O-glycans with a free reducing terminus in high yield and with minimal degradation [26]. However, loss of substitutions, in particular *O*-acyl groups, at various positions on sialic acids, or even desialylation and substantial peptide-bond cleavage, occurs during hydrazinolysis. Re-*N*-acetylation is necessary. Subsequently, labeling with 2-aminopyridine (2-AP/PA) can be performed by reductive amination without any purification (Fig. 8.3).

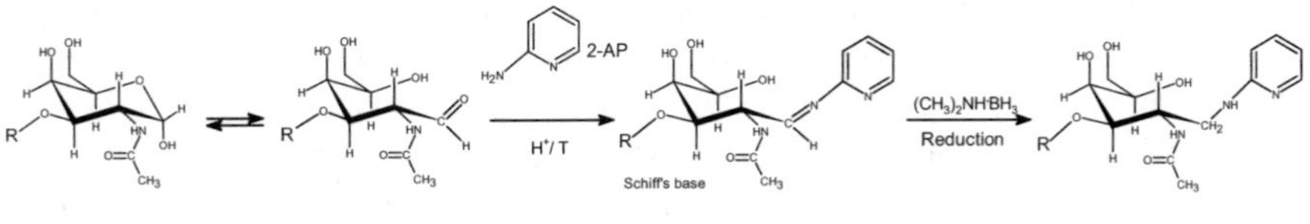

Fig. 8.3 Labeling of released O-glycan with 2-AP

The O-glycans liberated by hydrazinolysis can also be reduced, usually performed in 50 mM NaOH (adjusted to pH 11 with saturated boric acid) with NaBH$_4$ (4 h at room temperature).

Subsequent desalting is performed by gel-filtration chromatography on Bio-Gel P-2 (15 × 1 cm), eluted with water.

8.3.2　Mucin Glycoproteins

The O-glycosylation via GalNAcα1→Ser/Thr extensively occurs in secreted and membrane glycoproteins. This type of O-glycosylation is defined as mucin-type glycosylation because it is abundantly found in mucus glycoproteins, which are thus called mucins. Mucins contain large numbers of clustered O-glycans due to the presence of a large number of serine and threonine residues (usually in tandem repeats) in an uncharged and often proline-rich peptide context (apomucin) (see Sect. 2.4.1.2, and Fig. 2.6).

The extensive glycan clustering has a profound effect on the secondary structure of the glycoprotein. Many O-glycans in mucins carry sulfate groups. The complex structure makes the characterization of the structure of mucins a big challenge. Most mucins are made up of native macromolecules composed of subunits linked through disulfide bridges. Separation and isolation of individual mucin subunits need reduction and alkylation of these disulfide bridges.

During aqueous preparation of mucin from mucus, glycolipids are removed by extraction with chloroform/methanol (2:1, v/v). Subsequently, the lyophilized mucin fraction is dissolved in 0.1 M Tris-HCl buffer, pH 8.0, at a concentration of ~5–10 mg/mL. This pool of mucin proteins is purified via gel filtration chromatography (e.g., Sepharose CL-4B or CL-2B column, 40 × 1 cm), eluted with the same buffer, and collected in the void volume (sugar spot test detection). Dialysis 2–3× against water for 48 h at 4 °C is performed to further desalt the mucin fraction. Then, samples are dried by lyophilization.

As we already know, O-linked glycans can be released from the protein by two chemical methods, hydrazinolysis, and reductive β-elimination. So, these methods are also used for the deglycosylation of mucins. Hydrazine initiates a β-elimination reaction, generating a glycan hydrazine derivative (Sect. 7.2.2). However, the use of the hydrazine method (Protocol 51) for the release of O-linked sugars from high-molecular-weight mucins is complicated. Since there should be no salts, dialysis is performed first. Anhydrous conditions are necessary. Large-scale application is not possible due to the toxic and explosive nature of anhydrous hydrazine. For optimal results, different reaction times must be tested, and the reaction temperature must be found around 60 °C. As a consequence, a high level of expertise is required.

Alternatively, the O-linked glycans can be released by reductive β-elimination as discussed in Sect. 7.2.4. For mucins, Protocol 52 can be used. Most mucins will yield at least 20 different oligosaccharide-alditols after reductive β-elimination, which makes identification not easy. The released reduced O-glycans from a glycoprotein may exist in a multitude of different structures. Frequently, different chromatographic techniques are necessary to separate them (Sect. 8.3.3). The detailed analysis of the very different O-glycans is still a challenging endeavor [27, 28].

Protocol 51. Release of O-Glycans from Mucin by Hydrazinolysis

Materials
Polypropylene vials (1.6 mL) with Teflon screw caps (100 × 13 mm)
Heating block
3- or 5-mL plastic syringe with a Luer lock
Nylon Acrodisc syringe filter, 0.45μm (Gelman, no. 4438, Fisherbrand, 09-719-5)
Dowex 50W-X2 (200–400) cation exchanger H$^+$ form
Acetonitrile
Ammonium bicarbonate
Acetic anhydride [Ac$_2$O; (CH$_3$CO)$_2$O]
Anhydrous hydrazine (high quality)
Acetic acid, glacial
2% boric acid (w/v, granular) in methanol (store at RT)

Procedure

1. Dissolve salt-free glycoprotein to about 10–20 mg/mL in Milli-Q water.
2. Take 10–20μL of the sample (<2 mg) into the vial (screw-cap test tube). Also, Blank: 10μL Milli-Q water in a separate vial.
3. Dry samples very well by lyophilization or in a vacuum centrifuge (Savant) and again at 80 °C for 10 min (samples must be completely dry!).
4. Add 50–300μL anhydrous hydrazine (under dry argon/nitrogen atmosphere, use glass pipettes, avoid metal and plastic) and cap vials immediately, shake gently (CAUTION: work in fume hood, wear skin/eye protection, because hydrazine is a strong base, highly toxic, flammable, and corrosive).
5. Heat vials in a heating block for 3–6 h at 60 °C.
6. Cool to RT and evaporate hydrazine with a gentle stream of dry argon/nitrogen (2× toluene) (under fume hood!!).

Re-N-Acetylation

7. Add 200μL of saturated NaHCO$_3$ aqueous solution (freshly prepared).
8. Add 8μL of acetic anhydride, vortex, let stand for 5 min.
9. Add again 8μL of acetic anhydride, vortex, let stand for 30 min.

Desalting

10. Cool on ice and add Dowex 50W-X2 (H$^+$ form) up to pH 3.
11. Filter through a small glass column (10 × 0.5 cm) and collect the eluent.
12. Wash the column with 5 vol. water and collect eluate and combine with the previous eluent.
13. Lyophilize (ready for labeling, e.g., PA or 2-AA or ANTS).

Although mild hydrazinolysis conditions have been used, peeling is often observed. To minimize the degree of peeling, some improvements have been reported [26, 29]. However, since all the amide groups in the glycans are still deacylated and then re-acetylated, sialic acid-containing O-glycans will only show Neu5Ac. Recently, a new method for liberating O-linked glycans from glycoproteins was introduced, using hydroxylamine and 1,8-diazobicyclo[5,4,0]undec-7-ene (DBU) with 2-AA labeling [30]. O-glycans showed no diacylation. The method looks promising but needs extended application.

Protocol 52. Release of O-Glycans from Mucin by Reductive β-Elimination

Materials
Glass 3-mL Reacti-vials with Teflon-lined screw caps (Pierce)
Sodium borohydride
50% glacial acetic acid
Dowex 50X-12 H+ form resin (Sigma)
Methanol
0.2μm Anotop-IC sample filters (Whatman)

Procedure

1. Dissolve 500μg mucin in 1 mL water in a 3-mL Reacti-vial.
2. Add an equal volume of freshly prepared 2 M NaBH$_4$ in 0.05 M NaOH.
3. Incubate for 16 h at 45 °C (do not close tube, H$_2$ gas is produced).
4. Cool on ice bath.
5. Degrade excess borohydride by dropwise addition of 50% glacial acetic acid till pH 6.
6. Load solution on Dowex 50X-12 H$^+$ column (3 × 0.5 cm, Pasteur pipette).
7. Elute with 3 volumes of water.
8. Lyophilize.
9. Evaporate 4× with methanol to remove boric acid as methyl borate.
10. Redissolve in 500μL water and filter through 0.2μm Anotop-IC filter (Ready for HPAEC-profiling).

8.3.3   Fractionation of O-Glycans

Oligosaccharide-alditols can be readily separated according to size by fractionation on acrylamide-based gel filtration media, such as Bio-Gel P-4 or Sephadex G-25 (Protocol 53). However, to obtain a satisfactory resolution, the columns have to be long and the flow rate has to be slow, resulting in long separation times. Preferably, HPLC is used, for example, RP-HPLC (Protocol 54). Ion-exchange chromatography is used when the separation of neutral and acidic glycans is desired. Subsequent separation can be performed on primary amine-bond silica and by graphitized carbon columns. Permethylated O-glycans can satisfactorily be separated on PGC columns [31]. HPLC is used for rapid and sensitive analysis of hydrazine-released, fluorescently labeled O-glycans [32, 33]. The isolated O-glycans are usually characterized by mass spectrometry with MALDI and ESI ionization [34].

Protocol 53. Separation of Reduced O-Glycans by Gel Filtration Chromatography

Materials
Bio-Gel P-4 column (<400 mesh, 120 × 1 cm) or Sephadex G-25 column (medium grade, 70 × 5 cm) (column temperature 55 °C will improve separation)
Peristaltic pump
0.1 M pyridine acetate, pH 5.0

Procedure

1. Dried alkaline β-eliminated/reduced O-glycans are dissolved in 0.1 M pyridine acetate, pH 5.0.
2. Apply sample to the column and elute with 0.1 M pyridine acetate, pH 5.0 (30 mL/h).
3. Collect fractions, monitored by RI detection and test for carbohydrate with resorcinol/H_2SO_4 (spot test) method.
4. Lyophilize isolated carbohydrate-containing fractions.
5. Fractions can be further separated by HPLC or HPAEC.

Protocol 54. Reversed-Phase (RP) HPLC of PA-O-Glycans

Materials
Column: Cosmosil 5C18-P (150 × 4.6 mm) (Nacalai) or Amino AS-5A (5μm, 250 × 4 mm) (washed with methanol)
Solvent A: 100 mM ammonium acetate, pH 6.0 (0.1 M acetic acid is titrated to pH 6.0 with 4 M aqueous ammonia)
Solvent B: 1% 1-butanol in solvent A
For sialyl-oligosaccharides
A: 0.1 M acetic acid (adjusted to pH 5 with triethylamine)
B: 0.1 M acetic acid with 0.5% 1-butanol

Procedure

1. Equilibrate column with starting eluent [A/B = 99:1 (v/v)] at flow rate of 2 mL/min.
2. Load sample.
3. Elute with gradient from A/B = 99:1 to A/B = 40:60 in 20 min.
4. Monitor eluate at (λ_{ex} 315 nm, λ_{em} 400 nm) (Fig. 8.4).

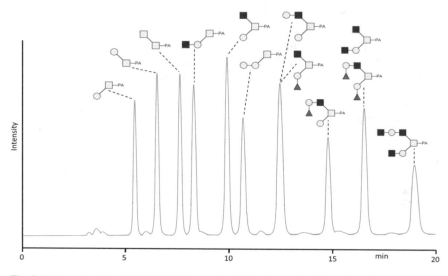

Fig. 8.4 Example of an RP-HPLC elution profile of PA-oligosaccharides (neutral O-glycans)

References

1. Defaus S, Gupta P, Andreu D, Gutiérrez-Gallego R. Mammalian protein glycosylation—structure versus function. Analyst. 2014;139:2944–67.
2. Anumula KR. Single tag for total carbohydrate analysis. Anal Biochem. 2014;457:31–7.
3. Lauber MA, Yu Y-Q, Brousmiche DW, Hua Z, Koza SM, Magnelli P, Guthrie E, Taron CH, Fountain KJ. Rapid preparation of released N-glycans for HILIC analysis using a labeling reagent that facilitates sensitive fluorescence and ESI-MS detection. Anal Chem. 2015;87:5401–9.
4. Zhu Y, Liu X, Zhang Y, Wang Z, Lasanajak Y, Song X. Anthranilic acid as a versatile fluorescent tag and linker for functional glycomics. Bioconjug Chem. 2018;29:3847–55.
5. Ruhaak LR, Zauner G, Huhn C, Bruggink C, Deelder AM, Wuhrer M. Glycan labeling strategies and their use in identification and quantification. Anal Bioanal Chem. 2010;397:3457–81.
6. Royle L, Campbell MP, Radcliffe CM, White DW, Harvey DJ, Abrahams JL, Kim YG, Henry GW, Shadick NA, Weinblatt ME, Lee DM, Rudd PM, Dwek RA. HPLC-based analysis of serum N-glycans on a 96-well plate platform with dedicated database software. Anal Biochem. 2008;376:1–12.
7. Campbell MP, Royle L, Radcliffe CM, Dwek RA, Rudd PM. GlycoBase and autoGU: tools for HPLC-based glycan analysis. Bioinformatics. 2008;24:1214–6.
8. Zhao S, Walsh I, Abrahams JL, Royle L, Nguyen-Khuong T, Spencer D, Fernandes DL, Packer NH, Rudd PM, Campbell MP. GlycoStore: a database of retention properties for glycan analysis. Bioinformatics. 2018;34(18):3231–2.
9. Ruhaak LR, Deelder AM, Wuhrer M. Oligosaccharide analysis by graphitized carbon liquid chromatography-mass spectrometry. Anal Bioanal Chem. 2009;394:163–74.
10. West C, Elfakir C, Lafosse M. Porous graphitic carbon: a versatile stationary phase for liquid chromatography. J Chromatogr A. 2010;1217:3201–16.

11. Ashwood C, Pratt B, MacLean BX, Gundry RL, Packer NH. Standardization of PGC-LC-MS-based glycomics for sample specific glycotyping. Analyst. 2019;144:3601–12.
12. Abrahams JL, Taherzadeh G, Jarvas G, Guttman A, Zhou Y, Campbell MP. Recent advances in glycoinformatics platforms and glycoproteomics. Curr Opin Struct Biol. 2020;62:56–69.
13. Duus JØ, Gotfredsen CH, Bock K. Carbohydrate structural determination by NMR spectroscopy: modern methods and limitations. Chem Rev. 2000;100:4589–614.
14. Gautam S, Peng W, Cho BG, Huang Y, Banazadaeh A, Yu A, Dong X, Mechref Y. Glucose unit index (GUI) of permethylated glycans for effective identification of glycans and glycan isomers. Analyst. 2020;145:6656–67.
15. Morelle W, Michalski JC. Analysis of protein glycosylation by mass spectrometry. Nat Protocol. 2007;2:1585–602.
16. Kobata A. Exo- and endoglycosidases revisited. Proc Jpn Acad Ser B. 2013;89:97–118.
17. Kannicht C, Grunow D, Lucka L. Enzymatic sequence analysis of N-glycans by exoglycosidases cleavage and mass spectrometry: detection of Lewis X structures. In: Kannicht C, editor. Post-translational modification of proteins: tools for functional proteomics, Methods in Molecular Biology, vol. 1934. Springer Science + Business Media; 2019. p. 51.
18. Parekh RB, Deannley J, Ventom A, Edge C, Prime S. Oligosaccharide sequencing based on exo- and endo-glycosidase digestion and liquid chromatography products. J Chromatogr. 1996;720:263–74.
19. Gotz L, Abrahams JL, Mariethoz J, Rudd PM, Karlsson NG, Packer NH, Campbell MP, Lisacek F. GlycoDigest: a tool for the targeted use of exoglycosidase digestions in glycan structure determination. Bioinformatics. 2014;30:3131–3.
20. Taron CH, Walsh I, Shi X, Rudd PM. Recent advances in the use of exoglycosidases to improve structural profiling of N-glycans from biologic drugs. BioPharm Int. 2018;31:16–23.
21. Walsh I, Nguyen-Khuong T, Wongtrakul-Kish K, Tay SJ, Chew D, Tasha J, Taron CH, Rudd PM. GlycanAnalyzer: software for automated interpretation of N-glycan profiles after exoglycosidase digestions. Bioinformatics. 2019;35(4):688–90.
22. Rogers RS, Nightlinger NS, Livingston B, Campbell P, Bailey R, Balland A. Development of a quantitative mass spectrometry multi-attribute method for characterization, quality control testing and disposition of biologics. mAbs. 2015;7(5):881–90.
23. Buettner A, Maier M, Bonnington L, Bulau P, Reusch D. Multi-attribute monitoring of complex erythropoietin beta glycosylation by GluC liquid chromatography–mass spectrometry peptide mapping. Anal Chem. 2020;92:7574–80.
24. Packer NH, Karlsson NG, editors. Glycomics: methods and protocols. Totowa: Humana Press; 2009.
25. Karlsson NG, Jin C, Rojas-Macias MA, Adamczyk B. Next generation O-linked glycomics. Trends Glycosci Glycotechnol. 2017;29:E35–46.
26. Kozak RP, Royle L, Gardner RA, Bondt A, Fernandes DL, Wuhrer M. Improved non-reductive O-glycan release by hydrazinolysis with EDTA. Anal Biochem. 2014;453:29–37.
27. Khoo K-H. Advances toward mapping the full extent of protein site-specific O-GalNAc glycosylation that better reflects underlying glycomic complexity. Curr Opin Struct Biol. 2019;56:146–54.
28. Mao J, You X, Qin H, Wang C, Wang L, Ye M. A new searching strategy for the identification of O-linked glycopeptides. Anal Chem. 2019;91:3852–9.
29. Goso Y. Malonic acid suppresses mucin-type O-glycan degradation during hydrazine treatment of glycoproteins. Anal Biochem. 2016;496:35–42.
30. Kameyama A, Thet Tin WW, Toyoda M, Sakaguchi M. A practical method of liberating O-linked glycans from glycoproteins using hydroxylamine and an organic superbase. Biochem Biophys Res Commun. 2019;513:186–92.
31. Cho BG, Peng W, Mechref Y. Separation of permethylated O-glycans, free oligosaccharides, and glycosphingolipid-glycans using porous graphitized carbon (PGC) column. Metabolites. 2020;10:433–44.

32. Storr SJ, Royle L, Murray A, Dwek RA, Rudd PM. The O-linked glycosylation of secretory/
    shed MUC1 from an advanced breast cancer patient's serum. Glycobiology. 2008;18:456–62.
33. Kozak RP, Urbanowicz PA, Punyadeera C, Reiding KR, Jansen BC, Royle L, Spencer
    DI, Fernandes DL, Wuhrer M. Variation of human salivary O-glycome. PLoS One.
    2016;11:e0162824.
34. Wilkinson H, Saldova R. Current methods for the characterization of O-glycans. J Proteome
    Res. 2020;19:3890–905.

Chapter 9
Analysis of Sialic Acids

Abstract Sialic acids are a family of acidic sugars, in particular, of glycoproteins and glycolipids, where they have multiple functions in determining the overall structure and properties of these biomolecules. Moreover, they are responsible for the exposure of recognition determinants on the cell surface. Sialic acids exhibit great structural diversity (i.e., *O*-acetylation) and play important roles in human health and disease, in relation to the immune system, apoptosis, and in providing receptors for microbes. This chapter contains protocols and examples for the analysis of sialic acids by colorimetric, gas chromatographic, mass spectrometric, and NMR spectroscopic methods.

Keywords Gangliosides · Keto-aldononulosonic acids · KDN · KDO · Neuraminic acid · CMP-sialic acid · Avian influenza virus · Cancer · EPO · Bial's test · GLC-EI/MS · Hydrolysis · Methanolysis · DMB labeling · RP-HPLC · NMR

9.1 General Aspects

The fact that sialic acids are involved in so many biological features during health and diseases makes their characterization of special significance [1, 2]. The roles of these specific sugar residues, which are often present at the outermost end of a glycan chain, are of particular interest. Sialic acid residues protrude most distantly from the protein or lipid backbone. In this way, sialic acids are prominently present on the cell surface and are the first carbohydrate residues met by pathogens that draw near the human cell membrane. Remarkably, extremely high levels of sialic acids are present in human milk and chicken egg yolk. It has been suggested that exogenetic sialic acid can promote brain growth, repair brain damage, and improve study memory behavior. However, it is remarkable that tumor cells of various origins feature increased expression of sialic acids on membrane glycoproteins and glycolipids, which enhances tumor cell growth and progression [3].

The generic term "sialic acids" was designated in 1957 for a specific large family of negatively charged acidic sugars (pK_a 2.6), being α-keto acids (keto-aldononulosonic acids), comprising a six-carbon ring with a three-carbon glycerol

© Springer Nature Switzerland AG 2021
G. J. Gerwig, *The Art of Carbohydrate Analysis*, Techniques in Life Science and Biomedicine for the Non-Expert, https://doi.org/10.1007/978-3-030-77791-3_9

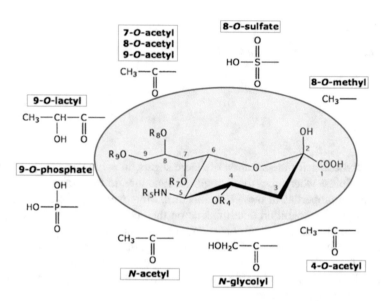

Fig. 9.1 A β-Sialic acid molecule with positions for the different, commonly occurring substituents [4]. Intramolecular lactone, anhydro, and dehydro derivatives are also possible, e.g., Neu2en5Ac where a double bond is formed between C2 and C3. In the case of a hydroxyl group at C5 instead of R$_5$HN, the molecule is called 2-keto-3-deoxy-D-glycero-D-galactononononic acid (KDN), and additionally missing C9, it is called 3-deoxy-D-manno-octulosonic acid (KDO)

chain (C7–C8–C9) attached at ring carbon 6 (C6), next to a carboxylic acid moiety attached at ring carbon 2 (C2).

Natural sialic acids are acyl derivatives of 5-amino-3,5-dideoxy-D-*glycero*-D-*galacto*-non-2-ulosonic acid, called neuraminic acid (Fig. 9.1). The most prevalent forms of sialic acids are N-acetylneuraminic acid (Neu5Ac, with an acetyl group at R5) and N-glycolylneuraminic acid (Neu5Gc, with a glycolyl group at R5). Neu5Ac is common in human cells. Neu5Gc is a nonhuman variant, mainly present in higher animals. Surprisingly, the occurrence of Neu5Gc has been observed in human meconium. Furthermore, in very minor quantities, it is metabolically incorporated into human tissues (in particular epithelial and endothelial cells) from dietary sources (red meat containing animal glycoproteins). Notably, cooked lamb contains a high percentage of Neu5Gc, but it is rare in poultry and fish. How ingested Neu5Gc becomes incorporated in human tissue is still unclear, because, when exposed to Neu5Gc, humoral immunity is activated and initiates the production of anti-Neu5Gc antibodies. The immune response may result in chronic inflammation (allergies). This is one of the reasons that pharmaceutical products, that have animal-based and/ or animal-derived components, must obey strict regulations, because Neu5Gc should be avoided in glycosylated biotherapeutics [5, 6].

More important, it should be noted that Neu5Gc is accumulated in human colon and breast carcinomas [7]. Typically, generally altered sialylation of glycoconjugates is associated with malignant properties such as invasiveness and metastasis

[8]. Consequently, the determination of the type of sialic acids in serum, urine, and tissue cells is important because it can be a valuable marker for certain malignancies. Quantification of total sialic acids or glycolipid-bound sialic acids in serum is helpful to improve the accuracy of clinical diagnosis and therapies. For example, a higher proportion of N-acetyl-9-O-acetylneuraminic acid in gangliosides was found in human melanoma [9, 10] and for T cells in mammary carcinoma.

Depending on cell type, tissue type, developmental stages, and environmental factors, sialic acid is attached to the nonreducing end(s) of the sugar chain of a wide variety of proteins and lipids. At least 20 sialyltransferases exist in mammals, and they catalyze the transfer of different types of sialic acid to the various acceptors to generate mostly ($\alpha2\rightarrow3$)- or ($\alpha2\rightarrow6$)-glycosidic linkage to a galactose (Gal) or N-acetyl galactosamine (GalNAc) residue, using CMP-(β)-sialic acid as a donor. Remarkable, in contrast to other nucleotide sugars (e.g., UDP-GlcNAc, GDP-Man), which are synthesized in the cytosol, CMP-sialic acid is synthesized in the nucleus of the cell. An altered distribution between ($\alpha2\rightarrow3$) and ($\alpha2\rightarrow6$) linkages of sialic acids is associated with cancer, chronic inflammation, and infection diseases [11, 12]. It has been shown that interruption in the biosynthesis of sialic acids is embryonic lethal in mice.

Sialic acids are widely distributed among living creatures, ranging from bacteria, protozoa, fungi, insects, fish, to mammals, but so far, they have not been found in helminths and higher plants [13]. In particular, the sialic acids are abundantly present in higher animal cell membranes, where they are α-glycosidically linked (mostly to galactose) at the exposed, nonreducing terminal positions of the N- and/or O-linked glycans of many glycoproteins and glyco(sphingo)lipids. There, they play pivotal roles in mediating or modulating a variety of physiological processes, including charge, aggregation, agglutination, cell–cell recognition, cell differentiation, Ca^{2+} binding, and receptor–ligand binding. Furthermore, sialic acid residues at the ends of glycan chains of many plasma glycoproteins protect those proteins from uptake and degradation in the liver. Sialic acids might be considered as the most biological important monosaccharide units of glycoconjugates. In the gray matter of the vertebrate brain, sialo-glycolipids (gangliosides) are abundantly present.

Unfortunately, sialic acids are exploited by viruses, bacteria, and toxins that recognize sialylated ligands with high specificity. Sialic acids can be taken up and utilized by bacteria as an energy source or incorporated into capsular or lipopolysaccharide components, protecting the bacteria from host immune recognition, thereby contributing to the pathogenesis of gastrointestinal and systemic infections. Similarly, viruses profit from sialic acids to survive. Isoforms of sialic acid are critical determinants of virus pathogenesis [14, 15].

Additionally, many types of tumor cells have been shown to express aberrantly high levels of sialylation, which protects cancer cells from recognition and eradication by the immune system [16]. Hypersialylation is a prevalent feature of disease progression, also reported in cases of diabetes. On the contrary, a decrease in sialic acids on serum glycoproteins is observed with Alzheimer's disease and is common in older Down's syndrome patients.

The many kinds of modified sialic acids existing in nature (more than 80 already identified) are responsible for a wide spectrum of biological phenomena [1, 4]. For example, sialic acids often function as recognition determinants on the cell surface. The Avian influenza viruses have proteins (haemagglutinin) on their surface that specifically bind to ($\alpha2\rightarrow3$)-linked sialic acid and, in this way, they gain access to the host cells. The disease commonly known as flu is one of the most widely distributed zoonotic infectious diseases in the world, and its pathogen, the influenza virus, is extremely mutable. The human cell membrane receptor for the respirovirus is based on the carbohydrate structures containing Neu5Ac($\alpha2\rightarrow6$)-linked to galactose, which is ($\beta1\rightarrow4$)-linked to N-acetylglucosamine, designated as sialyl lactosamine. O-acetyl groups (in particular at C4 and C9) of sialic acid also play a role. The presence of O-acetyl groups should make the sialic acid molecule partially or completely resistant to release by bacterial and viral neuraminidases. After binding of the virus, a viral sialidase enzyme removes the terminal sialic acid residue from glycans on the outer side of the cell wall, activating the entry of the virus into the host cell. For the same purpose, invading bacteria and parasites also produce sialidases. The cholera toxin also binds to Neu5Ac($\alpha2\rightarrow6$)Gal groups attached to glycolipids on the cell surface of erythrocytes and other cells. In addition to the role of sialic acid in cellular recognition phenomena, their involvement in secretion, immunogenicity, and circulation half-life of glycoproteins, they also form crucial structural elements of some antigenic determinants that have been identified as tumor markers. The most important feature is the altered, mostly elevated, sialylation of the surface of tumor cells (see also Chap. 3). Aberrant cell surface glycosylation is a well-known characteristic of cancer and the increase in cell surface sialylation is one such example, which includes an increase of the tumor antigen sialyl-Tn [Neu5Ac/Gc($\alpha2\rightarrow6$)GalNAc($\alpha1\rightarrow$O-Ser/Thr]. As mentioned, in many tumor types, including colorectal cancer, thyroid cancer, and leukemia, hypersialylation is indicative of poor prognosis. Furthermore, it has to be mentioned that pathological abnormalities and biosynthetic defects in the sialic acid metabolism occur, which will lead to sialic acid storage disease with increased urinary excretion of Neu5Ac (sialuria) due to hypersialylation of glycoprotein O-glycans.

The distribution of sialic acids on therapeutic glycoproteins is extremely important because this modulates the circulating half-life time, safety, and often the biological function of the drug. Nonsialylated glycoproteins are cleared from the blood circulation via lectin receptors (hepatocyte asialo-glycoprotein receptor, AGPR) in the liver. The currently applied regulatory guidelines for the approval of medicinal drugs demand to characterize sialic acid speciation during and after the production process of these biopharmaceuticals. Given their unique diversity and chemical complexity, detailed analysis of sialic acids is a rather difficult task [1, 4, 17–20].

It is obvious that the detection and detailed identification of the sialic acids is of importance to study their biological functions. Due to the lability of O-acetyl groups, which can be present at C4, C7, C8, and/or C9 positions (Fig. 9.1), care has to be taken not to lose them during the qualitative and quantitative analysis of sialic acids. This O-acetylation is one of the major modifications that significantly alter the biological properties of the glycoconjugates of the cellular membrane, affecting physiological and pathological responses [21].

The linkage of sialic acid in human glycoconjugates occur in various forms, such as Neu5Ac($\alpha2\rightarrow3$)Gal, Neu5Ac($\alpha2\rightarrow6$)Gal, Neu5Ac($\alpha2\rightarrow3$)GalNAc, Neu5Ac($\alpha2\rightarrow6$)GalNAc, and Neu5Ac($\alpha2\rightarrow6$)GlcNAc. Polysialic acids are a structurally unique group of linear polymers (DP ~8–200) that consist of Neu5Ac residues joined internally by ($\alpha2\rightarrow8$), ($\alpha2\rightarrow9$), or by alternating ketosidic linkages of these types. This modification can occur at the nonreducing terminal of N-glycans of cell-surface glycoproteins and glycolipids (gangliosides). The expression level is high during the embryonic period and later concentrated to areas of continued neurogenesis. This highly charged and spatially large modification occurs almost specifically on neural cell adhesion molecules (NCAMs). Polysialic acid is involved in neural cell migration and precise neural network formation in children less than 2 years old. Dysregulation of polysialylation and NCAMs have been frequently observed in association with psychiatric disorders, mainly schizophrenia, bipolar disorder, and autism [22]. It must also be noted that polysialic acid is also found at high expression levels on several types of cancer including glioma, neuroblastoma, and lung cancer, which leads to a poor prognosis.

In particular, bacterial polysaccharides can show polysialic acid structures. For instance, colominic acid is a polysaccharide of ($\alpha2\rightarrow8$)-linked N-acetyl neuraminic acids and was isolated from *Escherichia coli* species. The bacterium *Neisseria meningitidis*, responsible for meningitis, contains polymeric ($\alpha2\rightarrow8$)- and ($\alpha2\rightarrow9$)-linked sialic acids, with partially O-Acetyl at C7, C8, and C9 in its capsular polysaccharide. Remarkably, glycoproteins, containing poly-Neu5Gc, are involved in fertilization in fish.

As already mentioned, the type and distribution of sialic acids on therapeutic glycoproteins is of utmost importance because they modulate the biological function, the circulating half-life in the bloodstream, and the safety of these drugs. International regulatory rules require the characterization of sialic acid during and after the bioproduction process of glycoprotein biotherapeutics [23]. Consequently, sialylation, both the abundance and the type of sialylation (including O-acetylation), is a glycosylation critical quality attribute (GCQA). For example, recombinant human erythropoietin (rhEPO) produced in different expression systems can contain glycans differently sialylated than the native human EPO. This feature has been used to detect recombinant EPO abuse during sports [24, 25].

Another reason for analyzing the sialic acid on recombinant glycoprotein is to determine whether there is any N-glycolylneuraminic acid (NeuGc) present. NeuGc is a sialic acid found in many animal cells but is not found in humans because we lack the enzyme required for its synthesis. However, it has now been demonstrated that it is found in normal human tissues, probably picked up from diet. It has been suggested that the increase in carcinoma risk and malignant diseases in humans is due to excessive red meat consumption containing Neu5Gc [7]. There are immunological and chemical methods to detect trace amounts of Neu5Gc in tissues.

9.2   Characterization of Sialic Acid Residues

9.2.1   Colorimetric Determination

At first instance, the presence of sialic acid in a sample can be detected by a TLC spot test (Protocol 14) or by a colorimetric method (see Sect. 4.4.5). Different types of colorimetric methods are available to determine free or conjugated sialic acid. The often-used colorimetric methods for measuring sialic acids are the Warren/thiobarbituric acid (TBA) method for free sialic acids (Protocol 16) and the diphenylamine (DPA) method for free and conjugated sialic acid (Protocol 17). Free sialic acids can also be colorimetrically detected by the Ferric-Orcinol assay, also denoted as Bial's test (Protocol 15). The Bial's test is often used to monitor the presence of sialic acid in fractions during chromatographic separations of biological materials. It is highly recommended to use purified preparations for accurate colorimetric quantifications. Nowadays, these classic wet-chemical methods have been replaced for the most part by modern methods, including HPLC and GLC, aided by MS and NMR [1].

9.2.2   Gas-Liquid Chromatographic Determination

For the quantitative determination of sialic acid in a glycoprotein or glycolipid, the sugar analysis method using methanolysis and gas–liquid chromatography is frequently used (see Sect. 6.2). Methanolysis gives the neuraminic acid methyl ester methyl glycoside, however, with loss of O-acetyl groups and the amide substituents (N-acetyl, N-glycolyl). Thus, discrimination between different species is not possible, although the total amount and general presence of sialic acid can be deduced most accurately by this method. After methanolysis, the sialic acid is re-N-acetylated and derivatized by trimethylsilylation. Gas–liquid chromatography with flame ionization detection and the use of an internal standard provides a quantitative determination [17]. Alternatively, sialic acid released by mild acid hydrolysis and esterified with diazomethane can also be analyzed by GLC(-EIMS) after trimethylsilylation as per TMS methyl ester.

Procedure

1. Lyophilized sialic acid samples are dried in a vacuum desiccator over P_2O_5
2. Dissolve ~50μg sialic acids in 0.5 mL of anhydrous methanol at room temperature.
3. Add 400μL methanol containing Dowex H^+ and mix.
4. Filtrate the mixture through a Pasteur pipet with methanol-prewashed cotton wool.
5. Add diazomethane in diethyl ether to the filtrate until a faint yellow color remains.

6. Evaporate to dryness with a stream of N_2 and dry further over P_2O_5
7. Add 50μL trimethylsilylation reagent (pyridine/hexamethyldisilazane/trimethyl-chlorosilane, 5:1:1 v/v/v).
8. Incubate at room temperature for 2 h with intermittent vortexing.
9. Analyze ~2μL clear sample by GLC-EIMS (EC-1 column, 200–300 °C, 5°/min, injector temperature 230 °C, source temperature 200 °C, electron voltage 70 eV).

The mass spectra give information about the types of sialic acid by characteristic fragment ions (see Sect. 9.3).

9.2.3   Release of Sialic Acids by Mild Acid Hydrolysis

For the structural analysis, the sialic acids must often be detached from the oligosaccharide. Sialic acids are released from glycoconjugates by mildly acidic hydrolysis or enzymatic methods. The amount of liberated sialic acid can be determined by applying colorimetric methods as discussed earlier. Care has to be taken because the substitutions on sialic acids are labile and may be altered or lost during isolations and purifications. In workup procedures, pH values lower than 4 and higher than 6 should be avoided to prevent loss and/or migration of O-acetyl groups. Alkaline conditions cause acetyl migrations and loss of esters. In the case of acidic hydrolysis, the hydrolytic conditions should be carefully chosen because differences in rates of release are influenced by the substitution patterns and the type of glycosidic linkage. Strong acid removes O-acetyl groups and causes de-N-acetylation or de-N-glycolylation from the amino group. Hydrolysis by trifluoroacetic acid (TFA) usually destroys sialic acid completely.

A compromise has to be found for quantitative release of sialic acid with minimal destruction of the O-acyl groups. Note that, when sialic acid is kept for 30 min at 90 °C in 0.01 M HCl, already 20% decomposition occurs. Hydrolysis times can vary for different glycoproteins and glycolipids. It has to be kept in mind that in addition to the loss of O-acetyl groups, migration of acetyl groups can occur, typically from position 7 or position 8 to the thermodynamically more stable position 9 in the sialic acid molecule (see Fig. 9.1). Uncomplete release with preservation of 60–80% of the O-acetyl groups can be achieved by hydrolysis with 0.5 M formic acid, pH 2.1, at 70 °C for 1 h.

Nowadays, hydrolyses with propionic acid (2 M propionic acid for 4 h at 80 °C) and acetic acid (2 M acetic acid for 2 h at 80 °C) are frequently used, causing minimal loss of esters.

After hydrolysis, the sialic acids are isolated by dialysis in the following way:

1. Transfer the reaction mixture into a dialysis tube (MW cutoff 1000 Da)
2. Dialyze, under stirring, in a container containing 20 mL water at 4 °C for 18 h
3. Lyophilize the 20 mL water (containing released sialic acid)
4. Redissolve residue in 500μL water

5. This solution can directly be analyzed by HPLC/HPAEC or derivatized (DMB) for RP-HPLC.

However, hydrolysis with 0.1 M HCl at 70/80 °C for 1 h is still in use to prepare free sialic acids for analysis by HPAEC on a CarboPac PA-1 column. A typical elution program, at a flow rate of 1 mL/min, consists of 100 mM NaOH/50 mM NaOAc for 5 min, 100 mM NaOH/50–180 mM NaOAc in 20 min, 100 mM NaOH/180 mM NaOAc for 5 min [26]. It has to be noted that during HPAEC-PAD of sialic acids O-acetyl groups might get lost due to the alkaline conditions.

9.2.4 Analysis of DMB-Labeled Sialic Acid

A highly sensitive qualitative/quantitative method for sialic acids, in the femtomole range, involves tagging with 1,2-diamino-4,5-methylene-dioxybenzene (DMB) and analysis by RP-HPLC with fluorescence detection and then comparison with a reference panel containing Neu5Ac, Neu5Gc, Neu5,7Ac, Neu5Gc9Ac, Neu5,9Ac$_2$, and Neu5,7,(8),9Ac$_3$Gc [27]. The reaction of sialic acids with DMB is highly specific. RP-HPLC is often performed in combination with ESI-MS. Typically, the method involves:

1. Release of sialic acids using mild acid hydrolysis that preserves the N-acetyl, N-glycolyl, and O-acetyl groups (e.g., 2 M acetic acid for 2 h at 80 °C)
2. Followed by DMB labeling (Protocol 55)
3. Labeled sialic acids are stabilized by sodium hydrosulfite
4. Analysis by RP-HPLC(-EIS-MS) on a Cosmosil 5C18-AR-II (250 × 4.6 mm, Waters) or a C18 Gemini (250 × 4.6 mm, Phenomenex) or on a LiChrosorb RP-18 column (250 × 4 mm, Merck) or on a TSK-Gel ODS-120T column (250 × 4.6 mm) and isocratic elution with acetonitrile/methanol/water (9:7:84, v/v/v). Flow rate 1–1.5 mL/min. Detection of emission at 448 nm, after fluorescence excitation at 373 nm (Fig. 9.2).

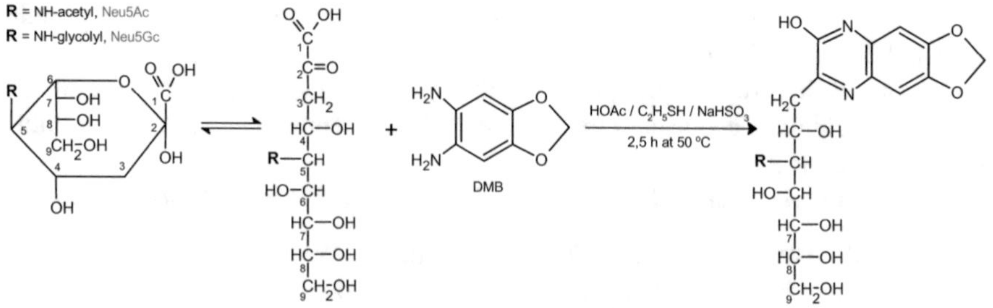

Fig. 9.2 DMB labeling of sialic acid. The carbons C4, C7, C8, and C9 may have O-acetyl groups

Protocol 55. DMB Labeling of Sialic Acids

Materials
Eppendorf tubes with screw cap (2 mL)
Heating block (55 °C)
Glacial acetic acid
1,2-diamino-4,5-methylenedioxybenzene (DMB) dihydrochloride (MW 225.1; Sigma-Aldrich)
β-mercaptoethanol (β-ME, MW 78.13; Sigma)
Sodium hydrosulfite (NaHSO$_3$ MW 174.1; Sigma)
N-acetylneuraminic acid (Neu5Ac; MW 309.3; Sigma)
N-glycolylneuraminic acid (Neu5Gc; MW 325.3; Sigma)
DMB solution: 7 mM DMB dihydrochloride in 1.4 M acetic acid, containing 0.75 M β-mercaptoethanol and 18 mM sodium hydrosulfite

> (Preparation: Take 15.75 mg of DMB, add 804μL of glacial acetic acid and add 528μL of β-ME. Then, add 720μL of a 0.25 M (43.5 mg/mL) solution of sodium hydrosulfite in water. Store in dark at 4 °C). Nowadays, sialic acid release and DMB labeling kits are commercially available from companies like BioLab and Ludger.

Procedure

1. Dissolve ~5μg dry free sialic acids in 100μL of 1 M acetic acid (prepared as 57μL glacial acetic acid in 1 mL water).
2. Add 200μL of DMB solution (wrap tube in aluminum foil).
3. Incubate at 55 °C for 2 h in the dark.
4. Cool on ice, add 200μL water and keep in an ice bath.

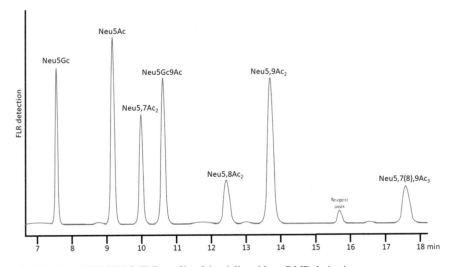

Fig. 9.3 Typical RP-HPLC-FLD profile of the sialic acids as DMB derivatives

5. Take aliquot for RP-HPLC(-EIS-MS) on a Cosmosil 5C18-AR-II (250 × 4.6 mm, Waters) and isocratic elution with acetonitrile/methanol/water (9:7:84, v/v/v). Flow rate 1.5 mL/min. Detection of emission at 448 nm, after fluorescence excitation at 373 nm (Fig. 9.3). DMB-labeled sialic acids are light sensitive and should be analyzed within 24 h.

9.2.5 Release of Sialic Acids by Enzymatic Digestion

Enzymes, called neuraminidases (sialidases), are frequently used to desialylate glycans to establish the number and the linkage type of terminal sialic acid residues. To this end, bacterial sialidases (e.g., from *C. perfringens*, *A. ureafaciens*, or *V. cholerae*) are commercially available. It must be noted that the presence of *O*-acyl groups can make the sialic acid molecule partially or completely resistant to release by the enzymes. In the case of enzymatic hydrolysis with sialidases, linkage specificity, as well as a reduced or complete lack of susceptibility, must be taken into account. In some cases, much lower amounts of sialic acids are released by sialidases than by acid hydrolysis.

Different commercially obtainable sialidases (EC 3.2.1.18) can split the sialyl linkage depending on their substrate specificity and restricted conditions. These enzymes have been isolated from bacteria or viruses and have different specificities. For instance, neuraminidases from bacterial origin are generally able to split ($\alpha2\rightarrow3$) and ($\alpha2\rightarrow6$) linkages, although with different velocity, but almost do not split $\alpha2\rightarrow8$ linkages, while neuraminidases from viruses usually act only on ($\alpha2\rightarrow3$) linkages. The presence of sulfation or phosphorylation and 4-*O*-acetylation on sialic acid resists liberation by neuraminidases. Adequate protocols for the enzymatic release of sialic acids are provided with the commercial enzymes. Usually, an overnight incubation at 37 °C with neuraminidase from *Arthrobacter ureafaciens* (a broad spectrum sialidase) is performed. Free sialic acids predominantly occur in the β-anomeric form.

9.3   Mass Spectrometry of Sialic Acids

In recent years, mass spectrometry (MS) has been widely used in glycan analysis. However, there were various problems related to the sialic acid residues, in that they are labile and thus easily lost. Sialic acids are notoriously unstable under MALDI-MS conditions. MALDI-TOF-MS analysis of sialylated glycopeptides faces the problem that they experience a metastable decay during ionization, resulting in the loss of the sialic acid moiety during the reflector transition [28]. Additionally, the existence of multiple charged forms of individual sialylated glycan also complicates the

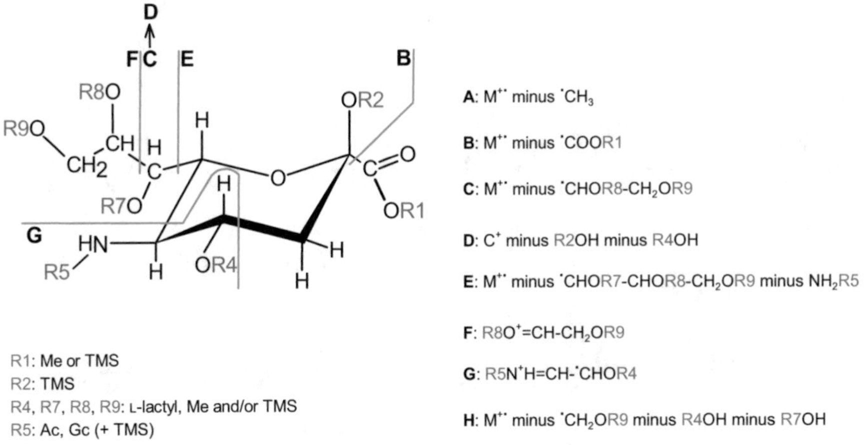

Fig. 9.4 GLC-EI/MS selected fragment ions (**A–H**) for TMS derivatives of *N*-acyl neuraminic acids with *O*-acyl and/or *O*-alkyl substituents. Some typical mass spectra are depicted in Fig. 9.5

interpretation of the mass spectra. Permethylation, which includes methyl esterification, can stabilize the sialic acids (see Chap. 11). A linkage-specific sialic acid permethylation method has been reported to differentiate ($\alpha2\rightarrow3$) and ($\alpha2\rightarrow6$) linkages directly by mass differences due to lactonization, resulting in improved profiling of protein glycosylation by MALDI-TOF-MS [29, 30]. Furthermore, site-specific mapping of sialic acid linkage isomers is possible by ion mobility spectrometry [31]. The reader is referred to an excellent review on sialic acid (linkage-specific) derivatization for mass spectrometry by [32].

For a long time, identification of sialic acid derivatives has been obtained by mass spectrometry after gas-chromatographic separation. Gas–liquid chromatography–electron impact ionization MS (GC-EIMS) provides fragment ions yielding information about the type of sialic acid and possible substitutions within the molecule, according to the fragmentation scheme shown in Fig. 9.4. As examples, in Fig. 9.5, mass spectra of trimethylsilyl derivatized sialic acids are shown. Tables with the characteristic EI/MS fragment ions can be found in "Structural analysis of naturally occurring sialic acids" [4]. In the gas-chromatographic monosaccharide analysis (Chap. 6), the methanolysis conditions cause de-*N*-acetylation and de-*N*-glycolylation and further de-acylation. Nevertheless, quantitatively, all sialic acid types are recovered as TMS *N*-acetyl neuraminic acid methyl ester methyl glycoside, due to the in-built re-*N*-acetylation step, giving the mass spectrum as depicted in Fig. 9.5e. However, it is possible to minimize *N*-deacetylation/*N*-deglycolylation by performing a methanolysis with 0.1 M HCl-methanol for 1 h at 80 °C, but then the yield is poor.

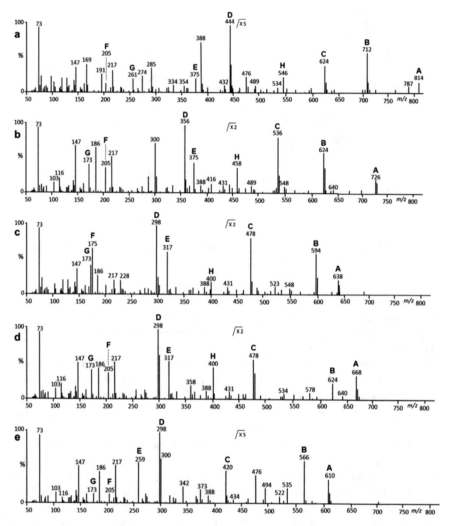

Fig. 9.5 Electron impact mass spectra of pertrimethylsilyl (**a**) Neu5Gc; (**b**) Neu5Ac; (**c**) Neu5,9Ac2; (**d**) Neu5Ac-Me ester; (**e**) Neu5Ac-Me ester Me-glycoside

9.4 ^{1}H NMR Spectroscopy of Sialic Acids

Proton nuclear magnetic resonance (^{1}H NMR) spectroscopy gives detailed information on the structure of *N,O*-acyl sialic acids. Spectra are usually recorded in deuterium oxide (D_2O) solution after the exchange of hydroxyl protons for deuterium. Nowadays, with modern NMR spectrometers with very strong magnetic fields (>1000 MHz for protons), together with advanced computing capabilities, super-high-resolution spectra can be obtained. The complete assignment of all protons is possible in the 1D ¹H NMR spectrum of the different free sialic acids. The protons of C4 to C9 are well separated in the 3.4–4.2 ppm region. The chemical shifts (δ) of the protons of carbon C3 (H-3 axial and H-3 equatorial), around 1.8 and 2.7 ppm, are indicative of the anomeric configuration of sialic acid. Free sialic acid in D₂O has 94% β-configuration. In the case of O-acetyl groups, the number and positions can be determined by well-separated O-acetyl signals in the δ 2.05–2.21 ppm region. The N-acetyl signals are found around 2.03 ppm.

Thus, NMR spectroscopy can give detailed structural information, such as sequence, anomeric configuration, type of linkage, and substituents of sialic acids still attached to glycans. The structural reporter group concept developed for the structural analysis of N- and O-glycans (see Chap. 12) reveals characteristic signals for the linkage position of sialic acids. Here, the protons of carbon C3 (H-3 equatorial and H-3 axial) give the type of linkage with neighboring residues and the further microenvironment of the sialic acid residues (Table 9.1). It is possible to localize substituents on the sialic acid residues.

For complete NMR data of sialic acid-containing structural elements and their further structural-reporter group signals, see [1]. To indicate the effects of chemical shift alterations of structural-reporter group signals by extensions with sialic acids, relevant data of the asialo-glycan is used as a reference. In the case of sialic acid attachments, downfield shifts of the anomeric proton signals of Man and GlcNAc residues in the antennae (branch location) are observed and the anomeric proton signals of Gal shift upfield.

Table 9.1 The ¹H chemical shifts (H-3 equatorial and H-3 axial) of carbon C3 of sialic acids

Sialic acid residue (pD 7)	Chemical shift (δ) (ppm)	
	H-3ax	H-3eq
α-Neu5Gc	1.621–1.644	2.730–2.749
β-Neu5Gc	1.827–1.840	2.208–2.243
Neu5Gc(α2→6)Gal(β1→4)	1.73–1.74	2.69–2.70
Neu5Gc(α2→3)Gal(β1→4)	1.81–1.82	2.77–2.78
Neu5Gc(α2→3)GalNAc-ol	1.721	2.746
α-Neu5Ac	1.621	2.731
β-Neu5Ac	1.827	2.208
β-Neu4,5Ac₂	1.951	2.249
β-Neu5,9Ac₂	1.833	2.221
β-Neu5,8,9Ac₃	1.838	2.189
Neu5Ac(α2→6)Gal(β1→4)...	1.72	2.67
Neu5Ac(α2→3)Gal(β1→4)...	1.80	2.76
Neu4,5Ac2(α2→6)Gal(β1→4)	1.85	2.67–2.68
Neu4,5Ac2(α2→3)Gal(β1→4)	1.926	2.768
Neu5Ac(α2→6)GalNAc-ol	1.700	2.728

References

1. Schauer R, Kamerling JP. Exploration of the sialic acid world. Adv Carb Chem Biochem. 2018;75:1–213.
2. Chen X, Varki A. Advances in the biology and chemistry of sialic acids. ACS Chem Biol. 2010;5:163–76.
3. Büll C, Stoel MA, den Brok MH, Adema GJ. Sialic acids sweeten a tumor's life. Cancer Res. 2014;74:3199–204.
4. Kamerling JP, Gerwig GJ. Structural analysis of naturally occurring sialic acids. In: Brockhausen I, editor. Glycobiology protocols, Methods in molecular biology, vol. 347. Totowa: Humana Press; 2006. p. 69–91.
5. Ghaderi D, Taylor RE, Padler-Karavani V, Diaz S, Varki A. Implications of the presence of N-glycolylneuraminic acid in recombinant therapeutic glycoproteins. Nat Biotechnol. 2010;28:863–7.
6. Anjum C, Chia YC, Kour AK, Adalsteinsson O, Papacharalampous M, Zocchi ML, Kimura I, Sharma R, Macheret L, Arthur B, Chan MKS, Pan SY. Understanding the presence of xeno-derived Neu5Gc in the human body, and its significance: a review. J Stem Cell Res Ther. 2020;6:72–7.
7. Samraj AN, Läubli H, Varki N, Varki A. Involvement of a non-human sialic acid in human cancer. Front Oncol. 2014;4:33
8. Zhang Z, Wuhrer M, Holst S. Serum sialylation changes in cancer. Glycoconj J. 2018;35:139–60.
9. Pearce OMT, Läubli H. Sialic acids in cancer biology and immunity. Glycobiology. 2015;26:111–28.
10. Cavdarli S, Dewald JH, Yamakawa N, Guérardel Y, Terme M, Le Doussal J-M, Delannoy P, Groux-Degroote S. Identification of 9-O-acetyl-N-acetylneuraminic acid (Neu5,9Ac$_2$) as main O-acetylated sialic acid species of GD2 in breast cancer cells. Glycoconj J. 2019;36:79–90.
11. Varki A. Biological roles of glycans. Glycobiology. 2017;27:3–49.
12. Zhou X, Yang G, Guan F. Biological functions and analytical strategies of sialic acids in tumor. Cells. 2020;9:273.
13. Strasser R. Plant protein glycosylation. Glycobiology. 2016;26:926–39.
14. Wasik BR, Barnard KN, Parrish CR. Effects of sialic acid modifications on virus binding and infections. Trends Microbiol. 2016;24:991–1001.
15. Park SS. Post-glycosylation modification of sialic acid and its role in virus pathogenesis. Vaccines. 2019;7:171.
16. Büll C, den Brok MH, Adema GJ. Sweet escape: sialic acid in tumor immune evasion. Biochim Biophys Acta. 2014;1846:238–46.
17. Kamerling JP, Gerwig GJ. Strategies for the structural analysis of carbohydrates. In: Kamerling JP, editor. Comprehensive glycoscience-from chemistry to systems biology. Amsterdam: Elsevier; 2007. p. 1–68.
18. Thaysen-Andersen M, Larsen MR, Packer NH, Palmisano G. Structural analysis of glycoprotein sialylation—part I: pre-LC-MS analytical strategies. RSC Adv. 2013;3:22683–705.
19. Palmisano G, Larsen MR, Packer NH, Thaysen-Andersen M. Structural analysis of glycoprotein sialylation—part II: LC-MS based detection. RSC Adv. 2013;3:22706–26.
20. Medzihradszky KF, Kaasik K, Chalkley RJ. Characterizing sialic acid variants at the glycopeptide level. Anal Chem. 2015;87:3064–71.
21. Shen Y, Kohla G, Lrhorfi AL, Sipos B, Kalthoff H, Gerwig GJ, Kamerling JP, Schauer R, Tiralongo J. O-acetylation and de-O-acetylation of sialic acids in human colorectal carcinoma. Eur J Biochem. 2004;271:281–90.
22. Schnaar RL, Gerardy-Schahn R, Hildebrandt H. Sialic acids in the brain: gangliosides and polysialic acid in nervous system development, stability, disease, and regeneration. Physiol Rev. 2014;94:461–518.
23. Li H, d'Anjou M, Pharmacological significance og glycosylation in therapeutic proteins. Curr Opin Biotechnol. 2009;20:678–84.

24. Llop E, Gutiérrez Gallego R, Belalcazar V, Gerwig GJ, Kamerling JP, Segura J, Pascual JA. Evaluation of protein *N*-glycosylation in 2-DE: erythropoietin as a study case. Proteomics. 2007;7:4278–91.

25. Caval T, Tian W, Yang Z, Clausen H, Heck AJR. Direct quality control of glycoengineered erythropoietin variants. Nat Commun. 2018;9:3342.

26. Rohrer JS, Basumallick L, Hurum D. High-performance anion-exchange chromatography with pulsed amperometric detection for carbohydrate analysis of glycoproteins. Biochemistry. 2013;78:697–709.

27. Chava AK, Chatterjee M, Gerwig GJ, Kamerling JP, Mandal C. Identification of sialic acids on *Leishmania donovani* amastigotes. Biol Chem. 2004;385:59–66.

28. Morelle W, Michalski JC. Analysis of protein glycosylation by mass spectrometry. Nat Protocol. 2007;2:1585–602.

29. De Haan N, Reiding KR, Haberger M, Reusch D, Falck D, Wuhrer M. Linkage-specific sialic acid derivatization for MALDI-TOF-MS profiling of IgG glycopeptides. Anal Chem. 2015;87:8284–91.

30. Jiang K, Zhu H, Li L, Guo Y, Gashash E, Ma C, Sun X, Li J, Zhang L, Wang PG. Sialic acid linkage-specific permethylation for improved profiling of protein glycosylation by MALDI-TOF-MS. Anal Chim Acta. 2017;981:53–61.

31. Guttman M, Lee KK. Site-specific mapping of sialic acid linkage isomers by ion mobility spectrometry. Anal Chem. 2016;88:5212–7.

32. Nishikaze T. Sialic acid derivatization for glycan analysis by mass spectrometry. Proc Jpn Acad Ser B. 2019;95:523–37.

Chapter 10
Carbohydrate Microarray Technology

Abstract Carbohydrate microarrays allow the rapid screening of interactions between glycans and other molecules, such as proteins, other carbohydrates, or even whole bacteria. It is shown that a great variety of proteins with binding properties for specific glycan epitopes are used for lectin microarrays. Nowadays, microarray methods are also used for direct analysis of live mammalian cell-surface glycome. This chapter gives an introduction to the extensive microarray technology.

Keywords Glycobiosensors · Lectins · Antibodies · Glycan-binding protein (GBP) · Fluorescence · GLAD · GlyMDB · CarbArrayART

As has been shown in Chap. 1, carbohydrates play pivotal roles in cell functions. For example, cell–cell recognition, cell signaling, and cell trafficking are mediated through interactions of carbohydrates with other biomolecules, such as (glyco)proteins, (glyco)peptides, and other carbohydrates. A study of the presence or absence of binding between specific molecules can reveal important physiological and structural information for both the binding molecule and the ligand alike.

The carbohydrate microarray technology (also known as glycobiosensors) provides a high-throughput method that enables screening of a multitude of carbohydrates against a multitude of binding partners to determine and evaluate their interactions [1–5]. For a long time, microarray technologies have been used in DNA and RNA research, but it has been only less than two decades ago that this technique was applied in glycoscience. In recent years, because of the increasing development of high-throughput technologies such as lectin arrays and glycan arrays, several databases have been established to store and share the glycan–protein interaction data (see Chap. 13).

In a typical glycan microarray experiment, many different glycan structures are immobilized in a spatially defined and miniaturized fashion on a solid surface, such as a glass slide, for investigation of their interrogation with carbohydrate-binding compounds. The carbohydrates used for immobilization can be either chemically synthesized or isolated from natural sources. Compared to other analytical methods, the amounts of material needed for printing microarrays are very small (femtomole) and use minuscule volumes (picoL). This enables high-density printing of over 1000 spots of different carbohydrates (i.e., potential unique carbohydrate structures)

© Springer Nature Switzerland AG 2021

G. J. Gerwig, *The Art of Carbohydrate Analysis*, Techniques in Life Science and Biomedicine for the Non-Expert, https://doi.org/10.1007/978-3-030-77791-3_10

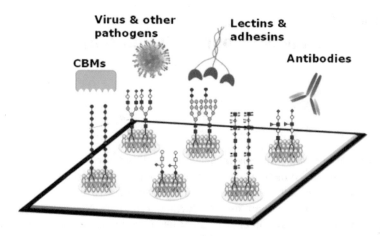

Fig. 10.1 Schematic presentation of a carbohydrate microarray, illustrating possible binding interactions with different ligands (Creative Proteomics, USA)

per microarray the size of a microscope slide. Such a slide can accommodate 100 or more probing experiments at once, making it possible to monitor multiple molecular interactions with biomolecules where glycans are involved (Fig. 10.1) [6, 7]. Additionally, by compartmentalizing the microarray slides with hydrophobic compounds such as Teflon into "multi-well" microarray slides, the volumes of typically precious reagents used for probing are also strongly minimized, enabling efficient, multiplexed, and cost-effective experiments.

The National Consortium for Functional Glycomics (CFG) (https://ncfg.hms. harvard.edu/microarrays) has a microarray resource with the aim to develop carbohydrate (micro)arrays as innovative tools to map highly specific interactions between carbohydrate and protein partners [8–10].

In very general terms, select carbohydrates (mono/oligo/polysaccharides) are bound (printing/blocking) to the microarray surface. This can be obtained by noncovalent interactions through conjugation to hydrophobic molecules and immobilization on nitrocellulose membranes or covalent attachment mediated by functional groups on the oligosaccharides and the microarray surface [9, 11]. The preparation of glycan microarrays requires that the glycans are derivatized with functional groups, such as an amino group. Polysaccharides can be immobilized on a variety of microarray surfaces (e.g., nitrocellulose membranes, and polystyrene) without a need for chemical functionalization because of their size that multiplies the number of nonspecific interactions enabling their adhesion.

Mono- and oligosaccharides do not have this advantage and typically need to be functionalized synthetically to be attached to the solid surface. This can be achieved either by using (a) a functionalized surface and a non-modified carbohydrate moiety or (b) a functionalized surface, and a functionalized carbohydrate.

A general strategy to construct glycan microarrays is the covalent bond formation of glycans containing reactive amine groups (e.g., aminoethyl) at the reducing termini. Using aldehyde-modified glass microscope slides, the primary amines of the modified sugars react with the aldehyde groups to form a Schiff base linkage. Optionally, reducing oligosaccharides may be reductively aminated using an amino lipid, thereby introducing an amphipathic property to the oligosaccharide. The hydrophobic lipid tail facilitates immobilization on silica gel surfaces, plastic microtiter plates, membranes, and microscope slides [12, 13]. Several covalent attachment techniques, for instance using carrier proteins or biotin/streptavidin linkage, can be found in the carbohydrate microarray literature [5].

The immobilized glycans are deposited on slides in a spatially discrete pattern using automated arraying robots, usually at different concentrations in several replicates. Then, the slides are interrogated or incubated with (fluorophore-labeled) compounds that may bind glycans, including cells, (recombinant) proteins, lectins, antibodies, toxins, serum, viruses, and bacteria (Fig. 10.1). Subsequently, binding is detected by direct or indirect fluorescence approaches, which provide an image of fluorescent spots that may be identified to coincide with deposited glycans. The data intensity of fluorescence is measured (scanned) and graphed (quantifying the fluorescence intensity of the bound compound) versus glycan number on the microarray. In this way, glycan microarrays can be used to discover functional glycans, characterize glycan-processing enzymes, and detect pathogens for diagnosis. Binding preferences, inhibition of interactions, enzyme activities, and structure–function relationships can be determined.

Recently, it was found that the fluorescent 2-AA tag of glycans (see Sect. 5.4.2), which is helpful for detection during HPLC, can serve as a linker by using the carboxylic acid functional group for selective amidation with various amines. This opens a method enabling facile immobilization of glycans to solid phases for the preparation of glycan microarrays [14].

In contrast to immobilizing carbohydrates, a multitude of specific lectins (Table 10.1) can be immobilized on the slide to produce an array to assess the specificity of binding properties to carbohydrates [15–18]. Each lectin possesses binding activity toward different glycan epitopes and incubation of the array with a fluorescently labeled (e.g., glycoprotein) sample reveals which of these epitopes are present. These lectin microarrays can be used for rapid analysis of protein glycosylation of therapeutic proteins, although they will not provide a complete structure assignment but a fingerprint pattern [19]. On lectin microarrays, also whole cells and intact extracellular vesicles can be used [20].

From the list of lectins in Table 10.1, there are five that are found in pretty much everybody's lectin toolbox, being Con A, RCA, WGA, PNA, and SNA, because they are specific for binding mainly to one terminal monosaccharide. Most of the other lectins have more complex binding motifs (multivalent), which need careful consideration to avoid wrong conclusions.

Clearly, microarrays are outstanding tools for the study of cellular glycosylation patterns and interactions of biomolecules with carbohydrates. Glycan analogs, nonnatural oligosaccharides, glyco-mimetics, synthetic glycoconjugates, and inhibitors can serve as probes [21]. In particular, the lectin array-based affinity test is becoming an important tool for isomeric glycan analysis [22, 23]. As such, lectin

Table 10.1 Lectins frequently used for carbohydrate identification in microarray and in affinity separation methods (terminal A in abbreviation usually stands for Agglutinin)

Lectin	Source	Carbohydrate specificity
AAA	*Anguilla anguilla*	Fuc(α1→3/6)
AAL	*Aleuria aurantia*	Fuc(α1→6)GlcNAc, Fuc(α1→3)[Gal(β1→4)]GlcNAc
ABA	*Agaricus bisporus*	Gal(β1→3)GalNAc
Con A	*Canavalia ensiformis*	Man(α1→6)[Man(α1→3)]Man (biantennary)
DSL	*Datura stramonium*	GlcNAc(β1→4)GlcNAc, Gal(β1→4)GlcNAc
DBA	*Dolichos biflorus*	GalNAc(α1→3)GalNAc, Terminal α-GalNAc
ECA	*Erythrina cristagalli*	Gal(β1→4)GlcNAc, Terminal GalNAc
GNA	*Galanthus nivalis*	Man(α1→3)Man
GNL	*Galanthus nivalis*	Man(α1→3)Man, High-Man N-glycans
GSL	*Griffonia simplicifolia*	Tri/tetra-antennary N-glycans with terminal GlcNAc, Terminal Gal(α1→3)
HHL	*Hippeastrum* hybrid	High-Man N-glycans, Man(α1→3)Man, Man(α1→6)Man
HAA	*Helix aspersa*	GalNAcα/β
HPA	*Helix pomatia*	GalNAcα/β
LCA	*Lens culinaris*	α-Glc, α-Man in N-glycans with Fuc(α1→6)GlcNAc
LTA	*Lotus tetragonolobus*	Fuc(α1→2)Gal(β1→4)[Fuc(α1→3)]GlcNAc
MAA	*Maackia amurensis*	Neu5Ac(α2→3)Gal(β1→4)GlcNAc
MAL-I	*Maackia amurensis*	Neu5Ac(α2→3)Gal(β1→4)GlcNAc
MAL-II	*Maackia amurensis*	Neu5Ac(α2→3)Gal(β1→3)GlcNAc, Neu5Ac(α2→6)GalNAc
MPA	*Maclura pomifera*	Galα, GalNAcα
NPA	*Narcissus pseudonarcissus*	Terminal Manα
NPL	*Narcissus pseudonarcissus*	High-man N-glycans, Man(α1→6)Man
PHA-E	*Phaseolus vulgaris*	Gal(β1–4)GlcNAc(β1–2)Man with bisecting GlcNAc
PHA-L	*Phaseolus vulgaris*	Tri/tetra-antennary complex N-glycans with terminal Gal
PhoSL	*Pholiota squarrosa*	Fuc(α1→6)GlcNAc
PNA	*Arachis hypogaea*	Gal(β1→3)GalNAc, Gal(β1→4)Glc
PSA	*Pisum sativum*	Manα, Glcα
RCA	*Ricinus communis*	Gal(β1→4)GlcNAc
SBA	Glycine max	GalNAc(α1→3)Gal, Terminal GalNAc
SNA	*Sambucus nigra*	Neu5Ac(α2→6)Gal(NAc), Galβ
TL	Tomato lectin	Poly-*N*-acetyllactosamine
VVA	*Vicia villosa*	Terminal GalNAcα/β
WFA	*Wisturia floribunda*	GalNAc(β1→4)-terminated glycans
WGA	*Triticum vulgaris*	GlcNAc(β1→4)GlcNAc, chitin oligos, Neu5Ac
UEA-I	*Ulex europaeus*	Fuc(α1→2)Gal(β1→4)GlcNAc(β1→

microarrays are a sensitive, rapid, and high-throughput profiling method, which enable the analysis of lectin–glycan interactions in a simultaneous manner. In general, labeled glycoproteins applied on the lectin array side, which are bound, are detected using an array scanner. Lectin arrays are used in the identification of desired glycosylation patterns during clone screening and selection during process development. Furthermore, lectin arrays are excellent tools in biomedical research, useful for pathogen detection, biomarker discovery, and bacterial tropism.

A lectin array-based method can be used for rapid analysis of glycosylation profiles of glycoproteins and mammalian cell surfaces [24]. The method is based on binding an intact glycoprotein to the arrayed lectins, resulting in a characteristic fingerprint that is highly sensitive to changes in glycan composition. A large number of lectins, each with its specific recognition pattern, ensures high sensitivity to changes in the glycan pattern. In fact, the lectin microarray has been recognized as a unique method to analyze glycosylation features of diverse glycoproteins, which include those of crude cell lysates, sera, and bacteria. A set of proprietary algorithms can automatically interpret the fingerprint signals to provide a comprehensive glycan profile output [15]. Sophisticated tools, such as Glycan Array Dashboard (GLAD) and Glycan Microarray Database (GlyMDB), are accessible nowadays [25, 26].

However, there are some conceptual challenges with microarray glycomics because the structural specifics of lectin-epitope binding are not completely understood. There is a strong overlap in carbohydrate binding for many lectins. The detailed characterization of many lectin specificities is still in progress. Most lectins are multimeric and represent avidities rather than one-to-one events, such as neighboring residues. Also, three-dimensional (3D) features of glycans influence the recognition by glycan-binding proteins (GBPs). Thus, the absolute selectivity of a lectin for a class of carbohydrates cannot always be guaranteed. Furthermore, it has to be noted that most lectins seem more effective at the level of intact glycoproteins rather than glycopeptides. The scanning of a microarray, with detection of binding, usually read as fluorescent intensity, the information gathered must be handled with care concerning the quantitative determination. The level of information that a microarray can typically provide is (a) no binding, (b) low binding, (c) medium binding, or (d) high binding. Therefore, treating the data as categorical instead of continuous is generally a more reasonable approach to microarray data analysis.

Nevertheless, a major advantage of lectin arrays is the unbiased glycan analysis over all classes of carbohydrates on intact cells or vesicles, which do not require exhaustive sample preparation. Glycoarrays open the way for a wide variety of biomedical applications, such as the identification of novel sugar markers related to early-stage diseases and the detection and diagnosis of pathogenic infections. Furthermore, the system is a suitable means for quality control of various glycoprotein therapeutics products [24].

A wide array of standard oligosaccharides as pyridylaminated (PA) derivatives is commercially available as a valuable tool for the systematic and detailed analyses of structures and functions of glycans. In recent years, this has led to an explosive growth of data from carbohydrate microarray experiments, coming from multiple research laboratories, each employing their own proprietary technology for spotting the array. However, the often radically different approaches of the spotting of the arrays can change the binding affinities observed and cause problems in comparative analysis and evaluation of the data.

Commercial microarrays and advanced commercial instruments/scanners with dedicated software (e.g., SensoSpot[R] Microarray Analyzers, Array-Pro Analyzer, and GlycoStation) are now becoming available, designed for routine microarray analyses, using fluorescence and colorimetric detection (GP Biosciences, Media Cybernetics, GE Healthcare).

Microarrays can also be prepared from glycopeptides/proteins or glycolipids. Also, mass spectrometry techniques are employed as label-free detection methods in glycan microarrays. The advances in chemical synthesis over the past decade have now made it possible to produce targeted glycans. Since facile methods for producing compound libraries are quickly improving, the use of carbohydrate arrays will rapidly increase.

Despite the several advantages, lectin arrays do not provide detailed information on sialic acid speciation and oligosaccharide sequences and linkages.

The US Consortium for Functional Glycomics (CFG) (http://www.functionalglycomics.org) supported the development of a robot-produced, publicly available microtiter-based glycan array. The currently used printed mammalian glycan microarray format (version 4.1) comprises >450 synthetic and natural glycan sequences representing major glycan structures of glycoproteins and glycolipids [27]. In 2008, a pathogen glycan array was also made available for screening, containing 96 polysaccharides derived from Gram-negative bacteria [28]. Furthermore, there is a role for lectin microarrays in cancer diagnosis [29]. CarbArrayART (Carbohydrate microarray Analysis and Reporting Tool) is a freely available software, based on the GRITS Toolbox [30], for storage, processing, and presentation of carbohydrate microarray data. Since 2020, the GLYCAN GlyMDB covers over 5000 microarray samples information (http://www.glycanstructure.org/glymdb/).

It is evident that for functional studies glycan microarray has become an indispensable technology for screening carbohydrate interactions in biological research and medicine [31–33]. However, implementation of the carbohydrate microarray technique is often hampered by the thought that the method requires an extensive degree of specialized training and a high level of experience. Some methodologies involved in the fabrication of microarrays indeed require moderate expertise in chemistry, starting from modifying carbohydrates through attaching them to solid surfaces, but detailed protocols can make microarray technology accessible even to non-experts through the use of protocols comparable to methodologies as old as Western blot analysis. Studying glycan interactions will be extremely important to find biomarkers for disease.

Here, only a short introduction to glycan microarrays could be provided. The topic is so extensive that it requires a book in its own right to explain all aspects and discuss the many protocols, including array fabrication and elaborate screening procedures. Researchers, interested in starting up microarray experiments, are referred to books, like "Microarrays" [34], "Microarray Technology" [35], and "Carbohydrate Microarrays: Methods and Protocols" [36].

References

1. Horlacher T, Seeberger PH. Carbohydrate arrays as tools for research and diagnostics. Chem Soc Rev. 2008;37:1414–22.
2. Liu Y, Palma AS, Feizi T. Carbohydrate microarrays: key developments in glycobiology. Biol Chem. 2009;390:647–56.
3. Yue T, Haab BB. Microarrays in glycoproteomics research. Clin Lab Med. 2009;29:15–29.
4. Lonardi E, Balog CI, Deelder AM, Wuhrer M. Natural glycan microarrays. Expert Rev Proteomics. 2010;7:761–74.
5. Kračun SK, Fangel JU, Rydahl MG, Pedersen HL, Vidal-Melgosa S, Willats WGT. Chapter 12. Carbohydrate microarray technology applied to high-throughput mapping of plant cell wall glycans using comprehensive microarray polymer profiling (CoMPP). In: Lauc G, Wuhrer M, editors. High throughput glycomics and glycoproteomics, methods and protocols, Methods in molecular biology, vol. 1503. Totowa: Humana Press; 2017. p. 147–65.
6. Hsu KL, Mahal LK. Sweet tasting chips: micro-array-based analysis of glycans. Curr Opin Chem Biol. 2009;13:427–32.
7. Song X, Heimburg-Molinaro J, Smith DF, Cummings RD. Glycan microarrays of fluorescently-tagged natural glycans. Glycoconj J. 2015;32:465–73.
8. Rillahan CD, Paulson JC. Glycan microarrays for decoding the glycome. Annu Rev Biochem. 2011;80:797–823.
9. Park S, Gildersleeve JC, Blixt O, Shin I. Carbohydrate microarrays. Chem Soc Rev. 2013;42:4310–26.
10. Hyun JY, Pai J, Shin I. The glycan microarray story from construction to application. Acc Chem Res. 2017;50:1069–78.
11. De Boer AR, Hokke CH, Deelder AM, Wuhrer M. General microarray technique for immobilization and screening of natural glycans. Anal Chem. 2007;21:8107–13.
12. Beckmann HSG, Niederwieser A, Wiessler M, Wittmann V. Preparation of carbohydrate arrays by using Diels-Alder reactions with inverse electron demand. Chem Eur J. 2012;18:6548–54.
13. Yang S, Li Y, Shah P, Zhang H. Glycomic analysis using glycoprotein immobilization for glycan extraction. Anal Chem. 2013;85:5555–61.
14. Zhu Y, Liu X, Zhang Y, Wang Z, Lasanajak Y, Song X. Anthranilic acid as a versatile fluorescent tag and linker for functional glycomics. Bioconjug Chem. 2018;29:3847–55.
15. Rosenfeld R, Bangio H, Gerwig GJ, Rosenberg R, Aloni R, Cohen Y, Amor Y, Plaschkes I, Kamerling JP, Maya RB. A lectin array-based methodology for the analysis of protein glycosylation. J Biochem Biophys Methods. 2007;70:414–26.
16. Gupta G, Surolia A, Sampathkumar S-G. Lectin microarrays for glycomic analysis. OMICS. 2010;14:419–36.
17. Hirabayashi J, Yamada M, Kuno A, Tateno H. Lectin microarrays: concept, principle and applications. Chem Soc Rev. 2013;42:4443–58.
18. Hirabayashi J, Kuno A, Tateno H. Development and applications of the lectin microarray. Top Curr Chem. 2015;367:105–24.
19. Zhang L, Luo S, Zhang B. The use of lectin microarray for assessing glycosylation of therapeutic proteins. MAbs. 2016;8:524–35.
20. Batista BS, Eng WS, Pilobello KT, et al. Identification of a conserved glycan signature for microvesicles. J Proteome Res. 2011;10:4624–33.
21. Patwa T, Li C, Simeone DM, Lubman DM. Glycoprotein analysis using protein microarrays and mass spectrometry. Mass Spectrom Rev. 2010;29:830–44.
22. Pabst M, Küster SK, Wahl F, Krismer J, Dittrich PS, Zenobi R. A microarray-matrix-assisted laser desorption/ionization-mass spectrometry approach for site-specific protein N-glycosylation analysis, as demonstrated for human serum immunoglobulin M (IgM). Mol Cell Proteomics. 2015;14:1645–56.

23. Gao C, Hanes MS, Byrd-Leotis LA, Wei M, Jia N, Kardish RJ, McKitrick TR, Steinhauer DA, Cummings RD. Unique binding specificities of proteins toward isomeric asparagine-linked glycans. Cell Chem Biol. 2019;26:535–47.
24. Pilobello KT, Mahal LK. Lectin microarrays for glycoprotein analysis. Methods Mol Biol. 2007;385:193–203.
25. Mehta AY, Cummings RD. GLAD: Glycan Array Dashboard, a visual analytics tool for glycan microarrays. Bioinformatics. 2019;35:3536–7.
26. Cao Y, Park SJ, Mehta AY, Cummings RD, Im W. GlyMDB: Glycan Microarray Database and analysis toolset. Bioinformatics. 2020;36:2438–42.
27. Blixt O, Head S, Mondala T, Scanlan C, Huflejt ME, Alvarez R, Bryan MC, et al. Printed covalent glycan array for ligand profiling of diverse glycan binding proteins. Proc Natl Acad Sci U S A. 2004;101:17033–8.
28. Liang PH, Wu CY, Greenberg WA, Wong CH. Glycan arrays: biological and medical applications. Curr Opin Chem Biol. 2008;12:86–92.
29. Syed P, Gidwani K, Kekki H, Leivo J, Pettersson K, Lamminmäki U. Role of lectin microarrays in cancer diagnosis. Proteomics. 2016;16:1257–65.
30. Weatherly DB, Arpinar FS, Porterfield M, Tiemeyer M, York WS, Ranzinger R. GRITS Toolbox—a freely available software for processing, annotating and archiving glycomics mass spectrometry data. Glycobiology. 2019;29:452–60.
31. Huang G, Peng D, Chen X. Using the glycoarray technology in biology and medicine. Curr Pharm Biotechnol. 2013;14:708–12.
32. Geissner A, Seeberger PH. Glycan arrays: from basic biochemical research to bioanalytical and biomedical applications. Annu Rev Anal Chem. 2016;12:223–47.
33. Smith DF, Cummings RD, Song X. History and future of shotgun glycomics. Biochem Soc Transact. 2019;47:1–11.
34. Muller H-J, Roeder T. Microarrays. Oxford: Academic; 2005.
35. Li P, Sedighi A, Wang L. Microarray technology. Totowa: Humana Press; 2016.
36. Chevolot Y. Carbohydrate microarrays: methods and protocols, Methods in molecular biology, vol. 808. Totowa: Humana Press; 2012.

Chapter 11
Analysis of Carbohydrates by Mass Spectrometry

Abstract Mass spectrometry has become the favorite technique for glycomics analysis. This chapter describes the different ionization methods and provides examples of application, with comprehensive literature references. Mass fragmentation of monosaccharide and oligosaccharide derivatives will be discussed. The glycan compositions are usually determined based on accurate masses and tandem mass spectrometry. The coupling of modern separation techniques directly to the mass spectrometer, together with automated tools for interpretation of the MS data using database searching, are common techniques nowadays.

Keywords HPLC-MS · GC-MS · EI-MS · ESI-MS · MALDI-TOF-MS · DHB · ATT · THAP · IgG · N-glycans · O-glycans · Permethylation · CID-MS · MS/MS

11.1 General Aspects

Mass spectrometry (MS) has become an indispensable tool for the primary structural analysis of carbohydrates because of its small sample (picograms/femtomolar range) requirement. Moreover, the adaptability for direct coupling the mass spectrometer to separation techniques (e.g., HPLC) is of enormous value. The combination of gas-liquid chromatography with mass spectrometry (GC-MS) is already in use for a long time (see also Chap. 6).

During the last two decades, the MS field has rapidly evolved with regard to instrument hardware, software, and applications. The development of several new soft ionization techniques, new tandem MS technologies, and the production of sophisticated mass spectrometers for reasonable prices have shown an increased interest in using MS as a highly sensitive and fast analytical instrument in glycomics and glycoproteomics [1–13]. At present, the most widely used approach to characterize glycosylation involves the enzymatic or chemical cleavage of the glycan from the glycoprotein, eventually attaching a chromophore label, followed by purification steps, and subsequent MS analysis. In the case of more than one glycosylation site, MS analysis of intact glycopeptides released by glycoprotein proteolysis might correlate the glycan composition with the different glycosylation sites.

© Springer Nature Switzerland AG 2021

G. J. Gerwig, *The Art of Carbohydrate Analysis*, Techniques in Life Science and Biomedicine for the Non-Expert, https://doi.org/10.1007/978-3-030-77791-3_11

Until recently, barriers to MS glycomics were also related to data analysis, requiring advanced levels of skill. In recent years, several algorithms were developed to assist mass spectra interpretation (see Chap. 13), but the high-resolution MS technique still requires an experienced scientist for instrument operation. Nevertheless, matrix-assisted laser desorption ionization (MALDI) and electrospray ionization (ESI) with quadrupole or time-of-flight (TOF) mass analyzers are now commonly used techniques for glycan analysis in carbohydrate research laboratories. MS has become an ideal analytical technology for high-throughput glycan profiling. In particular, ESI-MS of glycans is most often used in line with HPLC or CE [14, 15]. There is consensus that MS-based strategies provide the most effective means of both identification and quantitation of N- and O-glycans in glycomics studies [16, 17].

In principle, MS of carbohydrates is the production and detection of gas-phase ionized molecular species, separated according to their mass-to-charge (m/z) ratios. The plot of the relative abundance of the ions as a function of their m/z ratio gives a mass spectrum. MS provides two types of structural information, depending on the ionization technique. Using a soft ionization, only the molecular-mass ion and no fragments are formed from the molecule. Using strong ionization, a carbohydrate molecular ion dissociates by forming fragments, which are determined by their mass-to-charge (m/z) ratios [18]. To this end, a mass spectrometer has three components: an ionization source, a mass analyzer to separate ions, and an ion detector.

The structural features of carbohydrates that can be defined by MS methods are (1) type of glycosylation (e.g., N- or O-glycan; high-mannose and hybrid or complex N-glycans), (2) glycosylation sites, (3) glycan branching patterns, (4) the number and lengths of antennae, their composition, and substitution (for instance, with fucose, sialic acid, or sulfate, phosphate, or acetyl esters), and (5) sometimes, the complete sequences of the individual glycan. However, the assignment is according to the masses of pentose, hexose, deoxyhexose, (N-acetyl) hexosamine, and sialic acid without providing the real identity of the sugars. It is important to bear in mind that structural features such as linkage, stereochemistry, and antenna location often require additional experimentation depending on the degree of heterogeneity.

Ionization methods involving electrospray (ESI) and matrix-assisted laser desorption (MALDI), performed with quadrupole, ion trap, time-of-flight, and Fourier transform mass spectrometers have provided detailed information on carbohydrate structures, also by making use of tandem MS/MS techniques. Collision-induced dissociation (CID) fragmentation is an MS/MS technique, whereby selected precursor ions are made to undergo collision with neutral (inert) gas molecules (e.g., argon) to produce controlled fragmentation. The results of fragmentation of glycans strongly depend on collision energy and the type of instrument employed. Orbitrap-hybrid and Q-TOF type mass spectrometers are often used because of high mass/resolution accuracy and higher collisional dissociation (HCD) energy. For a more detailed technical explanation of the different kinds of mass analyzers, such as ion-trap MS analyzers [quadrupole/triple quadrupole (Q), linear trap quadrupole (LTQ), linear quadrupole ion trap (LQIT), Orbi-trap, Q-TOF, TOF-TOF], the reader is kindly referred to more specific mass spectrometry literature [19].

11.2 Electron Impact Mass Spectrometry (EI-MS)

Electron impact mass spectrometry already came up for discussion in Chap. 6. Electron Impact ionization (EI) is one of the oldest strong (hard) ionization modes in MS, and this technique has been used in combination with gas chromatography (GC-MS) for more than four decades for the analysis of derivatized monosaccharides [20]. Directly after leaving the gas chromatograph, via a transfer-line into the mass spectrometer, the sugar molecules are positively ionized under vacuum conditions in the (quadrupole) mass spectrometer by interaction with a beam of energetic electrons (70 eV) emitted from a heated filament. This causes each of the derivatized sugars to ionize and to fragment into smaller ions according to a characteristic and reproducible way. Full-scan mode monitoring (usually 50–600 m/z) shows a high number of fragments of low mass to charge ratio (m/z). In full-scan mode, identifications are based on the comparison of the experimental mass spectrum (abundance of characteristic m/z ions) with standards and theoretical fragment calculations. EI mass fragmentation of TMS derivatives of carbohydrates has been extensively studied already for a long time [21].

Monosaccharide analysis by GC-MS of TMS methyl glycosides is one of the most used methods (see Sect. 6.2). The EI mass spectra are used to identify the peaks by m/z values of characteristic fragments. However, in the EI spectra of TMS-methyl glycosides, stereochemical differences (α/β anomers, enantiomers, diastereomers) do not appear, but it is possible to differentiate between pyranose (p) and furanose(f) ring forms in one type of monosaccharide.

In Sect. 6.2, the EI-MS spectra of some TMS methyl glycosides are shown in Fig. 6.3. Here, in Table 11.1, the m/z fragments are assigned stemming from the molecular mass ion (M) minus typical fragments, such as the CH_3· and CH_3O· radicals, and loss of CH_3SiOH (TMSOH), etc. It has to be noted that the molecular mass ion (M) is not regularly observed, but the $[M - 15]^+$ ion is usually present. The base fragment $(CH_3Si)^+$ at m/z 73 is normally found. The ring form of the TMS methyl glycoside can be deduced from the ratio m/z 204 to m/z 217, as for pyranose m/z 204/217 > 1 and for furanose m/z 204/217 < 1. In the case of HexNAc, this holds for m/z 173/186. Nowadays, every modern GC-MS system has the disposal of an extensive library of spectra of compounds for comparison in its software package.

GC-EIMS is also used in another monosaccharide analysis method, involving alditol acetate derivatives (see Sect. 6.3). In this method, to circumvent a complex multiple peak pattern in the gas chromatogram due to anomers (α and β) and ring configurations (furanose and pyranose) of monosaccharides formed during hydrolysis, an additional step is included after hydrolysis. This step consists of reducing the monosaccharide at the anomeric carbon position with $NaBH_4$, yielding the alditol of each monosaccharide. Subsequently, derivatization by acetylation is performed to obtain volatile compounds for GLC separation. Identification of isomers can be deduced from the difference in retention time. In Sect. 6.3, Fig. 6.4 shows the typical mass fragments which can be found in the EI-mass spectra of peracetylated

Table 11.1 Typical EI-mass fragments for TMS (methyl ester) methyl glycosides

[Fragment ion]+	Hex	Pent	6dHex	HexNH$_2$	HexNAc	HexA
M	482	380	394	409	451	438
M – CH$_3$	467	365	379	394	436	423
M – OCH$_3$	451	349	363	378	420	407
M – CH$_3$ – CH$_3$OH	435	333	347	362	404	391
M – TMSOH	392	290	–	319	361	–
M – CH$_2$OTMS	379	277	–	–	348	–
M – COOCH$_3$	–	–	–	–	–	379
M – CH$_3$ – TMSOH	377	275	289	304	346	333
M – OCH$_3$ – TMSOH	361	259	273	288	330	317
M – CH$_3$ – CH$_3$OH – TMSOH	345	243	257	272	314	301
M – OTMS – TMSOH	303	201	215	230	272	259
M – OTMS – CH$_2$OTMS	290	–	202	217	259	216
M – CH$_2$OTMS – CH$_3$OH – TMSOH	–	–	–	–	226	–
M – CH$_3$ – TMSOH – TMSOH	287	185	199	214	–	243
M – OCH$_3$ – TMSOH – TMSOH	271	169	183	198	–	227
TMSO-CH=C(-OTMS)-CH = +OTMS	305	305	305	–	–	305
TMSO-CH=CH-CH = +OTMS	217	217	217	–	–	217
TMSO-CH-CH = +OTMS	204	204	204	204	–	204
TMSO-CH = +OTMS	191	191	191	–	–	191
TMSO-CH=CH-CH = +NHCOCH$_3$	–	–	–	144	186	–
TMSO-CH-CH = +NHCOCH$_3$	–	–	–	131	173	–
TMSO = +Si(CH$_3$)$_2$	147	147	147	147	147	147

– means "minus", *Hex* hexose (e.g., Man, Gal, Glc), *Pent* pentose (e.g., Ara, Xyl), *6dHex* 6deoxy-hexose (e.g., Rha, Fuc), *HexNH$_2$* hexosamine (e.g., GalNH$_2$, GlcNH$_2$), *HexNAc* N-acetyl-hexosamine (e.g., GalNAc, GlcNAc), *HexA* hexuronic acid (e.g., IdoA, GalA, GlcA)

alditols. As an example, the mass spectra of glucitol hexa-acetate and *N*-acetyl glu-cosaminitol hexa-acetate are shown in Fig. 11.1.

The base peak in the EI mass spectra of alditol acetates is *m/z* 43 [CH$_3$-C=O]+. The primary fragmentation occurs by cleavage of the alditol chain (from the top and bottom directions), followed by fragments produced by elimination of CH$_3$CO$_2$ (*m/z* 59), acetic acid (*m/z* 60), and ketene (*m/z* 42).

As we have seen in Sect. 6.6, GC-EIMS is the perfect technique for the identification of the partially methylated alditol acetates. After an initial permethylation of carbohydrates, followed by hydrolysis, reduction with sodium borodeuteride and acetylation, the mass spectra as shown in Fig. 6.11 in Sect. 6.6.2 reveal the substitution patterns of the monosaccharides. This so-called "methylation analysis" method has become a standard technique for linkage determination in structural polysaccharide analysis.

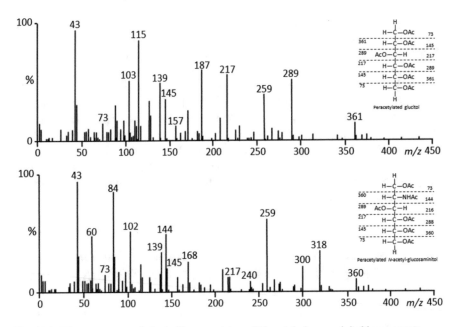

Fig. 11.1 EI-mass spectra of glucitol hexa-acetate and *N*-acetyl glucosaminitol hexa-acetate

11.3   Mass Spectrometry of Glycans

11.3.1   *General*

Mass spectrometry is considered one of the most powerful techniques for glycoproteomic and glycomic study. Several MS techniques are available for the analysis of N- and O-glycans, usually after a tryptic digest or release of the glycans. Mass spectrometry is often used in conjunction with separation methodologies, such as HPLC, affinity chromatography, and capillary electrophoresis (CE). Currently, these are the most popular methods for complex carbohydrate sequencing. However, there are some items to consider. Unless one uses MS in tandem (a rather tedious process), it is nearly impossible to distinguish between isomeric monosaccharides (e.g., glucose, galactose, and mannose are all hexoses with the same mass). This also means that structural isomers due to variations in the sequence and linkages between the monosaccharides cannot be fully differentiated by mass spectrometry.

In general, there are two main strategies for obtaining glycosylation information using MS techniques [22]. The first strategy is for global analysis of the glycans which are released from glycoproteins using endoglycosidases or chemical methods. This is useful for rapid glycan profiling analysis, but the information of the carrier protein is lost. The second strategy is the analysis of intact glycopeptides after proteolytic digestion of the glycoprotein, which provides information on both the glycan composition and the glycosylation site in the protein [23]. Nowadays, high throughput glycomic and glycoproteomic analysis have become possible by using web-side software tools (see Chap. 13). By coupling mass spectrometry with databases containing many carbohydrate structures, automated analysis has almost become possible.

Two MS techniques, being matrix-assisted laser desorption ionization (MALDI-TOF-MS) and electrospray ionization (ESI-MS), are widely used for structural characterization of released, labeled, and unlabeled glycans. The combination of HPLC separation of fluorescent-labeled glycans and mass spectrometry is an often-used technique. Usually, the chemical tag (2-AP or 2-AB) enhances ionization. Furthermore, coupling MS to a separation technique such as capillary electrophoresis (CE-MS) is popular. This technique benefits from short analysis time and excellent peak resolution [24].

11.3.2 Matrix-Assisted Laser Desorption/Ionization-Time of Flight Mass Spectrometry (MALDI-TOF-MS)

At this moment, MALDI-TOF-MS equipped with several software tools plays an important role in carbohydrate research [25, 26]. The analysis of N- and O-glycans from a glycoprotein by MALDI-TOF-MS has become a standard procedure in glycoscience. Compositional information can be derived by high resolution and accurate mass measurement. Although the MALDI technique cannot be directly combined on line with a chromatographic separation technique, MALDI-MS offers several advantages over ESI-MS, like (1) wide mass range, (2) relatively simple spectral interpretation by single charged ions, (3) high-throughput measurements, and (4) repeated measurements of the same sample. In the first instance, MALDI-MS is especially suited for profiling mixtures of glycans. Since fragmentation does not take place, information of the molecular masses present in a glycan mixture is obtained, which provides the degree of heterogeneity of glycans and, for instance, the glycan differences in a native and a recombinant glycoprotein.

The amount of glycan required for mass determination by MALDI-TOF-MS depends on the purity of the sample, i.e., lack of salts and detergents, and on the instrumentation. Typically, samples (oligosaccharide glycans, glycopeptides; in ng/pmol order) are mixed with a low-molecular-weight "matrix" [commonly 2,5-dihydroxybenzoic acid (DHB), 2-azathiothymine (ATT), or 2,4,6-trihydroxyacetophenone (THAP)] that strongly absorbs UV light. DHB (as 10 mg/mL in 50% ACN/50% H_2O) is the most popular matrix for sample preparations of native and derivatized glycans. The sample–matrix combinations are loaded as multiple samples in wells on a small metal MALDI target plate and dried for crystallization. Then, the plate is loaded into the high vacuum chamber of the MS instrument. When a high-energy UV-laser beam of N_2 atoms (337 nm UV) is fired in pulses (3–5 ns) and collides with the light-absorbing matrix, kinetic energy is transferred to the surface analyte molecules, then sputtering out as ions into the vacuum of the ion source of the mass spectrometer. The impact of the laser beam on the different sample spots can be guided and observed by a camera in the instrument. It is a very elegant way of ionization and desorption of carbohydrates to obtain molecular ions in the gas phase by using the positive- or negative-ion mode. Molecular weight information is obtained by the time-of-flight (TOF) from the ion source of the usually single-charged ions $[M + Na]^+$. Also, quadrupole or ion-trap mass analyzers are used. Detailed manuals and protocols are usually provided by the MS instruments supplier.

Thus, MALDI-MS is a very sensitive method for determining the mass of the molecular ion of the intact oligosaccharide. When this technique is applied to glycan mixtures (without preliminary separation), mass information can be obtained, and comparison with databases enables putative compositional assignments. This so-called glycan mass profiling (GMP) is a rapid method for analyzing the complete glycan mixture obtained directly after release from the protein. A compositional profile is obtained with regard to the number of sialic acids, fucoses, hexoses, and N-acetylhexosamines. The principal composition of a glycan can be deduced from the unique molecular mass, often with the help of programs, such as GlycoMod on the ExPASy Server (https://www.expasy.org/glycomod), but the result will mostly be a list of possible alternative compositions.

It also has to be noted that a common phenomenon in reflectron-mode MALDI-TOF-MS is the loss of sialic acids due to in-source decay and metastable decay. Also, sulfate or phosphate groups are lost during the ionization process due to their lability in protonated forms. Preceding MALDI-TOF-MS analysis, glycan samples are often separated into neutral and acidic glycans. Anionic glycans do not respond well in the positive ion mode, whereas neutral glycans do not ionize well in the negative ion mode. Regularly, a permethylation of the hydroxyl and N-acetyl groups is performed, which also includes the esterification of the carboxyl group of sialic acid, minimizing degradation and making the glycans neutral [27]. However, loss of base-labile functional groups occurs during permethylation. Nevertheless, MALDI-TOF-MS of permethylated glycans is intensively used as a profiling method [28, 29]. Permethylation also improves and enhances the ionization efficiency of glycans. In the positive ion mode, molecular masses of glycans are recognized by m/z values of their pseudo-molecular ions, usually single charged as $[M + Na]^+$ because Na^+ is the main adduct often naturally occurring in the samples.

The interpretation of a MALDI spectrum is performed by assigning the composition of a molecular ion signal in terms of the number of monosaccharide constituents (number of hexoses, N-acetylhexosamines, deoxyhexoses, and sialic acids). This is illustrated for three structures in Table 11.2 as an example. An extended list can be found in Appendix B. It contains a list of regularly observed m/z values of pseudomolecular signal ions $[M + Na]^+$ which are related to structures of N-glycans mostly found in glycoproteins. The composition of an oligosaccharide corresponding to a given signal can be deduced by adding the incremental masses of suspected monosaccharide residues, the mass 47 $[OCH_3 + H + CH_3]$ (to account for atoms at the end of the oligosaccharide), and the ionizing species $[Na^+]$ until their sum equals that of the signal m/z value. Most human N-glycans are built from just four

Table 11.2 Pseudomolecular signal ions [M + Na]⁺ in positive ion mode MALDI-TOF-MS of permethylated N-glycans found in glycoproteins (extended list in Appendix B)

Signal (m/z) [M + Na]⁺	Composition				Tentative structure assignment
	Hex ○ ○	HexNAc ■	Fuc ▼	Neu5Ac ◆	
2395.9	9	2	–	–	
3210.5	5	5	1	2	
4586.8	7	6	1	4	

Table 11.3 Incremental residue masses (average) for permethylated N-glycans

Residue	Example	Amu
Hexose	Glc, Gal, Man	204.1
HexNAc	GlcNAc, GalNAc	245.1
deoxy Hex	Fuc, Rha	174.1
Sialic acid	Neu5Ac, Neu5Gc	361.3, 391.4
Pentose	Xyl, Ara, Rib	160.1
HexA	GlcA, GalA, IdoA	218.2
Hexitol (reducing end)	Glc-ol, Gal-ol	251.1

monosaccharide masses, and the combinatorial possibilities are limited due to knowledge of the N-glycan biosynthesis. Incremental masses of some common monosaccharide residues in oligosaccharides are given in Table 11.3. Permethylation has transformed all reactive glycan hydrogens into methyl groups. The carboxyl groups of sialic acid residues have become methyl esters. As an example of MALDI-TOF-MS, Fig. 11.2 shows the spectrum of the permethylated N-glycans (m/z of [M + Na]⁺) released from the immunoglobulin G (IgG) in human serum. In its Fc region, IgG has a single, primarily biantennary N-glycan on each heavy chain.

11.3.3 Electrospray Ionization-Mass Spectrometry (ESI-MS)

ESI is an atmospheric pressure ionization technique. Its principle of operation is to create gas-phase ions, separate them based on their mass-to-charge (m/z) ratio, and quantify them. To this end, a stream of a carbohydrate-containing liquid is injected as a mist (fine spray) of droplets through a metal-coated capillary interface, maintained at a high voltage and high temperature, directly into the ESI source of the mass spectrometer. There, under vacuum, the carbohydrate molecules are stripped from solvent giving charged species, which can be protonated (positive-ion mode)

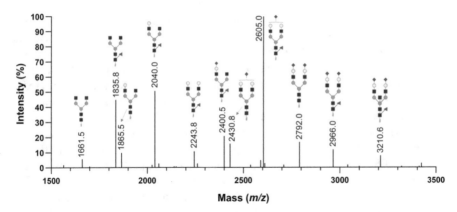

Fig. 11.2 MALDI-TOF-MS of the permethylated N-glycans from human serum IgG. All peaks are single sodium adducts [M + Na]⁺

or deprotonated (negative-ion mode). The *m/z* values of the intact molecules (pseudomolecular ions [M + H]⁺ or [M − H]⁻) are detected, although multiple charged ions are easily generated, which may convolute the mass spectrum. The used "soft ionization" technique is not energetic enough for further breakdown to produce fragment ions. ESI-MS is usually applied for glycopeptides (glycoproteomics) and underivatized oligosaccharides but can also be used for reduced saccharides or glycans labeled with a fluorescent tag or in the permethylated form [18]. ESI-MS of permethylated oligosaccharides remains the method of choice for sequencing of oligosaccharides.

For direct measurement (direct infusion) in the mass spectrometer, carbohydrate samples are preferably dissolved in 5% aqueous acetic acid for determination in the positive ion mode, or in a mild base such as triethylamine for the negative ionization mode. Finally, samples are 50% diluted with an organic solvent (methanol or acetonitrile). Then, ~10µL sample is injected via a syringe pump into the carrier solution flow (~2–4µL/min) [30]. Measuring is performed during the continuous flow of the sample in the carrier solution. ESI-MS is sensitive in the 5–10 pmol range. Care has to be taken to avoid in-source decay (ISD), which means not to generate artificial alterations in glycoforms due to too high capillary temperature (>250 °C) or too high tube lens voltage (>100 V). Furthermore, the pH of the sample and the presence of salts have an influence on the formation of molecular ions and their anionic or cationic adducts. The ESI sensitivity decreases as the mass of the glycan increases.

Nowadays, advanced LC-MS systems are highly sensitive by using extremely low flow rates, in the order of 10–40 nanoliters per minute, applying nano-spray ionization (NSI) [31]. Nano-ESI produces very small ion droplets with a high percentage of ionization, yielding a stronger signal. The use of the nano-flow ESI-MS techniques is growing due to increasing commercial products, such as nano-HPLC Chip systems (Agilent). To overcome in-source fragmentation during ESI, a new sub-ambient pressure ionization with a nano-spray (SPIN) source is in use. In the SPIN source, the ESI emitter is moved from atmospheric pressure to the first vacuum stage of the mass spectrometer and is positioned at the entrance of the electrodynamic ion funnel to allow for the entire electrospray plume to be collected. At this moment, Chip-based RP-HPLC-MS and chip-ESI MS are arising for analysis of glycans [32–34]. Reversed-phase or graphitized carbon (PGC) HPLC directly coupled with electrospray ionization-mass spectrometry are often used systems [35].

A very popular method is PGC-HPLC-ESI-MS for analysis of fluorescent (PA and 2-AB) labeled oligosaccharides. Elution is usually accomplished by aqueous organic solvent mixtures containing low concentrations of acids or volatile buffers; this enables online ESI-MS analysis in positive ion or negative ion mode [36]. The direct LC-ESI-MS coupling provides the possibility of peak integration, precise determination of retention times, and the principal ability to detect all types of glycans. The mild ionization process in ESI-MS yields only (multiple charged) molecular ions, beneficial for the so-called mass fingerprinting of the glycans [37]. More detailed structure analysis can be obtained by tandem mass spectrometry. Discrimination between linkage types and between positional isomers (branching) on the glycan needs advanced MS/MS technologies [38].

11.3.4 Tandem Mass Spectrometry

Glycan structural characterization using MS mainly relies on tandem MS (MS/MS or MS^n) [39, 40]. In this technique, analyte ions of interest are selected and further broken down. The mass of the resulting fragment ions reflects the structure of the precursor. For linkage determination, tandem MS with collision-induced dissociation (CID) is currently a commonly used technique. A tandem mass spectrometer has the capacity to isolate a fragment ion of specific m/z value (called precursor ion) and then degrade it further to smaller fragments ions [8, 35]. A continuous flow of the sample is preferred because then, a single m/z precursor can be analyzed for extended periods of time. Although the sample flow during chromatographic separation is limited to its retention time, tandem MS directly after chromatographic separation (LC-MS/MS) is very well possible.

Hence, the molecular ion of a glycan gives many fragments, which are usually, initially, the result of breakage of the glycosidic bonds. Thereafter, a new precursor ion can be selected for further fragmentation, etc. Multistep tandem mass spectrometry is a method used to repeat the selection and collision of fragment ions. In this way, detailed information on the structure of glycans can be obtained [41–43].

For tandem mass spectrometry (MS^n) operations, where n indicates the number of MS stages, ion traps and triple quadrupoles mass spectrometers are used. Typically, ESI instruments have two analyzers in tandem, for example, a quadrupole Q as the first and an orthogonal TOF (QTOF) as the second. The second analyzer allows the detection of fragment ions produced from molecular ions selected by the first analyzer. They have undergone collisions with an inert gas (argon, nitrogen, helium, or xenon) in a chamber placed between the two analyzers. This is referred to as collision-activated dissociation (CAD) or collision-induced dissociation

(CID). The resulting ions (daughter ions) are separated in the second mass analyzer to yield a spectrum of fragment ions (tandem mass spectra). Using another type of mass spectrometer, the desired precursor ions may be isolated in a quadrupole "ion trap" (QIT) while all other ions are expelled. Then, the precursor ion is subjected to CID. Progress in the improvement of MS/MS techniques is observed by the development of Fourier transform mass spectrometry (FTMS) with extremely high mass accuracy. For detailed information on MS techniques, the reader is referred to [44, 45].

Underivatized glycans can be analyzed by ESI-MS/MS; however, superior data can be obtained if the glycans are first derivatized. There are two possibilities: (1) tagging of the reducing ends, usually with PA or 2-AB and (2) protection of all functional groups, usually by permethylation. Replacement of the hydrogens on hydroxyl groups, amine groups, and carboxyl groups by methyl groups yields a hydrophobic sugar derivative. Permethylation prevents intermolecular hydrogen bonding, thereby increasing the volatility and intensity of the ion signals. This influences its chromatographic behavior as well as enhancing the ion signal strength in MS [18].

Piteously, permethylation is a difficult and time-consuming process that is not easily adapted to large numbers of samples (see Sect. 6.6). However, classic methylation procedures have been supplemented with novel high-throughput, microscale permethylation techniques [27, 46–49]. For instance, permethylation of multiple samples (e.g., released N-glycans) can be performed in a regular polypropylene 96-well plate as follows:

Each N-glycan target is dissolved in 50 μL DMSO in a well. To this, add 75 μL of NaOH/DMSO base reagent (prepared as described in Sect. 6.6) and mix using the pipette. Then, add 25 μL of MeI and mix with the pipette. Incubate for 20 min at room temperature.

Add 100 μL of water and evaporate the excess of MeI (lower layer) by carefully/slowly blowing N_2 through the solution (work in fume hood).

Then, the sample is cleaned by SPE on a commercially available C18 pipette tip:

1. Prewash tip with 2× 200 μL of MeOH and 2× 200 μL of water.
2. Apply sample and wash with 5× 200 μL water.
3. Elute the permethylated glycans with 5× 30 μL of methanol.
4. The eluant from SPE can directly be used for ESI-MS or an aliquot can be mixed with matrix for MALDI-MS.

Nowadays, mass spectrometry of glycans is mostly performed on permethylated samples to increase the overall sensitivity and ionization efficiency, but also to stabilize the glycan structure [50]. Furthermore, the increased hydrophobicity of permethylated glycans allows for efficient separation with reversed-phase liquid chromatography. However, with the sensitivity increase of modern mass spectrometers, underivatized oligosaccharides are also frequently analyzed in their native form. Released native or reduced glycans are typically analyzed in negative ion mode [40, 51]. The pseudomolecular ions [M-H]⁻ and tentative structure assignment of underivatized O-glycans (as alditols) from mucin glycoproteins as found in

Table 11.4 Pseudomolecular signal ions [M − H]⁻ of five O-glycans (as alditols) regularly found in mucin glycoproteins (extended list in Appendix C)

Signal (m/z) [M-H]⁻	Composition				Tentative structure assignment
	Hex ○	HexNAc □ ■	Fuc ▼	Neu5Ac ◆	
716.3	–	2	–	1	
952.4	2	3	–	–	
1024.4	1	2	1	1	
1812.7	3	5	2	–	

ESI mass spectra recorded in the negative ion mode is illustrated for five structures in Table 11.4 as an example. An extended list can be found in Appendix C.

11.3.5   Mass Fragmentation of Oligosaccharides

As described earlier, collision-induced dissociation (CID) is a technique where selected precursor ions are activated to undergo collision with an inert, neutral gas to produce controlled fragmentation. The mass spectra of CID of glycan sodium adducts reveal fragment ions whose m/z values correspond to either glycosidic bond cleavages (using low energy) or cross-ring cleavages (using higher energy). The cleavage of the glycosidic bonds between two neighboring monosaccharide residues occurs on either side of the oxygen. Heteroglycosidic bond cleavage and cross-ring fragmentation of oligosaccharides typically involve the migration of a hydrogen atom. Hence, the fragmentation of oligosaccharides follows a certain pattern, and the widely used nomenclature to designate the different types of glycan fragment ions is shown in Fig. 11.3.

Two major types of product ions are produced by the fragmentation of oligosaccharides: (1) those resulting from glycosidic fragmentation and (2) those from cross-ring fragmentation. For interpretation of the mass spectra, fragment ions are denoted with uppercase letters. Fragment ions started from the nonreducing end are labeled with uppercase letters A, B, and C. Those that started from the reducing end are labeled Z, Y, and X. Cleaved ions are further indicated with subscripts. The A and X ions are produced by the cleavage across the glycosidic ring and are superscript-labeled by assigning each ring bond a number. The superscripts indicate the ring bond number. Ions produced from the cleavage of successive residues are subscript-labeled, starting from the nonreducing end for A, B, etc., and from the reducing end for Z, Y, etc. Y_0 and Z_0 refer to the fragmentation of the bond to the aglycone if present. The fragments produced by double glycosidic cleavage at branching points are labeled with D. The fact that carbohydrates can be branched

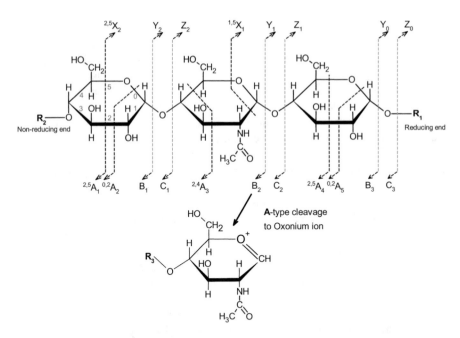

Fig. 11.3 Systematic nomenclature for fragment ions produced by CID-MS of oligosaccharide [52]. Ring-bond numbering indicated in red, cross-ring fragmentation in black, and glycosidic fragmentation in blue. Example of an **A**-type cleavage to an Oxonium ion. R_1 is aglycone; R_2, and R_3 are saccharides. Free hydroxyl groups (and N-acetyl) are usually permethylated before MS

increases the complexity of the proposed nomenclature. The interpretation of a mass spectrum into meaningful information is a time-consuming task and a certain level of expertise is a prerequisite. The use of suitable bioinformatics software is often essential [53].

Low-energy (10–100 eV) CID fragmentation predominantly causes cleavage on either side of the glycosidic bonds yielding nonreducing and reducing end fragments, B/C- and Y/Z-ions, respectively. These ions give information on the monosaccharide composition and the general topology (sequence/branching) of the glycan. In the positive ion mode, cleavage preferably occurs on the nonreducing side to form an oxonium ion. Then, low-molecular-weight ions, such as m/z 163 [Hex+H]$^+$, m/z 204 [HexNAc+H]$^+$, m/z 292 [Neu5Ac+H]$^+$, m/z 366 [HexHexNAc+H]$^+$, m/z 528 [Hex$_2$HexNAc+H]$^+$, and 657 [SiaHexHexNAc+H]$^+$, are specific ions diagnostic for glycan fragmentation. In negative ion mode, neutral losses of hexose (161), N-acetylhexosamine (202), fucose (145), N-acetylneuraminic acid (290), and HexHexNAc (364) can be observed. Higher energy (1–10 keV) CID fragmentation (HCD) generates high abundant low-mass oxonium- and B/C-ions and some Y/Z-ions from glycosidic cleavage, which usually provide some details on the different linkage types [28].

The GlycoMod tool on the ExPASy server and GlycoWorkBench can be used to match MS/MS data with possible glycan compositions (see Chap. 13). The software suite GlycoWorkBench not only provides an interface for the rapid drawing of glycan pictograms (GlycanBuilder) but is also used for the computation and assignments of theoretical MS/MS fragments of a selected structure [54]. The graphical interface of GlycoWorkBench provides an environment in which structure models can be rapidly assembled, automatically matched with MS^n data, and compared to assess the best candidate.

It should be mentioned that frequently all signals cannot be explained solely by fragmentation of one bond. With underivatized samples, sometimes, fragments can be observed that arise from two bond cleavages and loss of water of Y ions. Furthermore, it is known that mass spectrometric rearrangement reactions can occur during CID-tandem MS of glycans and glycopeptides. In particular, this applies to protonated adducts but less to sodium or deprotonated adducts. For instance, internal residue loss (IRL) has been observed for carbohydrates, resulting in misinterpretation of fragments expected from conventional glycosidic bond cleavage, leading to incorrect structural assignments [55]. Monosaccharides, in particular deoxyhexoses, can also migrate in the oligosaccharide molecule, for instance, transfer of terminal antennae fucose to core-fucosylated N-glycans (or vice versa) during positive-mode MS CID-tandem MS fragmentation [56].

Permethylation offers several advantages for MS-based detection, such as leading to prediction of the fragmentation pattern giving interglycosidic linkage information. Applying the positive ion mode, permethylated glycans (oligosaccharides) reduce rearrangements and vastly improve the sensitivity (abundant molecular ions) and facilitate the interpretation of MS/MS data by directing the fragmentation along a few defined, reliable, and predictable pathways. Fragment ions are critical for the sequence of carbohydrates. Permethylated glycans form abundant fragment ions stemming from cleavage at susceptible glycosidic linkages, notably on the reducing side of each HexNAc residue. In contrast, negative ion mode CID analysis results in cross-ring cleavages (A-type) that are more informative than B- and Y-type fragments stemming from cleavage on the nonreducing side of the glycosidic bond to form an oxonium ion, which predominate in positive ion mode analysis. The scar is a nonmethylated hydroxyl group of glycan exposed by tandem MS fragmentation of a glycosidic bond in a permethylated glycan [18, 42, 57].

The combination of mass spectrometry with modern separation methodologies has become a powerful and versatile technique in the structural analysis of glycans. The complexity of acquired data, fragment ion interpretation, and the requirement for reference mass spectra have driven the development of a multitude of glycan and glycoproteomic databases and automation. UniCarbDB, GlycoEpitope, GlycoBase, and GlycoStore are examples of databases that categorize thousands of glycan structures (O- and N-glycans, GSLs, glycosaminoglycans) based on mass (precursor ion and fragment ions/patterns), retention time, and glucose units across a range of derivatization methods (see Chap. 13). Automatic interpretation of MS data will become standard in the future [58]. The automatic interpretation of glycan MS/MS spectra requires high-quality spectra in terms of signal-to-noise ratio, mass precision, and sample purity.

In addition to CID-MS, there are other MS techniques using electron-induced ion activation/fragmentation (in particular for glycopeptides), such as electron capture dissociation (ECD), infrared multiple photon dissociation (IRMPD), electron detachment dissociation (EDD), electronic excitation dissociation (EED or ExD), negative electron transfer dissociation (NETD), and Fourier transform ion cyclotron resonance (FT-ICR-MS), but these techniques are still rarely used in glycomics/glycoproteomics studies due to the cost and complexity of the instruments and the advanced MS expertise needed. But, with the capabilities of modern-day mass spectrometers for hybrid fragmentation, increased applications are expected [59]. While CID and HCD mainly yield B- and Y-ions from the glycans attached to glycopeptides, ETD and ECD primarily fragment the peptide backbone and leave the glycan intact. Interpretation of complex tandem mass spectra generated from glycopeptides usually requires the assistance of sophisticated data analysis software tools and databases [58, 60].

Another promising analytical route seems to be ion mobility-mass spectrometry (IM-MS), in particular in the area of de novo carbohydrate structural identification [61–63]. For specialized information, the reader is referred to the book "Ion Mobility-Mass Spectrometry: Methods and Protocols" [64]. This technique already evolved to $_{PGC}$LC-IM-MS/MS [65]. IM-MS is an ideal analytical technique for isomer/isobaric glycan discrimination by their differing spatial extension. It offers information on the three-dimensional (3D) shape of a molecule from its gas-phase rotationally averaged collision cross-section (CCS values), a function of drift time with a buffer/drift gas [66]. This means that IM-MS can separate and identify ionized molecules (isomeric glycans post-ionization) in the inert gas phase by their size, mass, and shape, illustrated as drift time in the gas phase and charge-to-mass ratio (m/z) in the mass analyzer [65, 67, 68]. This gas-phase separation technique coupled with MS is getting more and more attention in recent years. The separation of the ions is achieved by movement in the presence of a carrier/buffer gas (N_2 or He) through a drift tube, driven by a weak uniform electric field and based on differences in their masses, charges, sizes, and shapes [69, 70]. Different spatial glycan structures permit the possibility of structural isomers being separated. The mobilities may be estimated computationally if the 3D structure of the molecule is known. IM-MS coupled to collision-induced dissociation (CID) mass spectrometry revealed fragment features of glycan isomers which can be used as fingerprints for glycan analysis [71]. New developed IM-MS methods such as gated-trapped IMS (gated-TIMS), drift tube IMS (DTIMS) and traveling wave ion mobility spectrometry (TWIMS) have significantly increased sensitivity and speed. Isomers of $Man_7GlcNAc_2$ and even glucose isomers could be discriminated by ion mobility MS [72]. However, it has to be noted that the application of these techniques in glycobiology demands some expertise and skill in MS. Several aspects have to be considered, for example, charged or neutral glycans, the choice of buffer/drift gas and calibrant, and careful selection of experimental parameters.

Recently, mass spectrometry in dynamic multiple reaction monitoring (dMRM) mode was introduced as a sensitive and selective tandem MS-based quantitative technique, performed on a triple-quadrupole instrument. Although complicated, the method is highly suitable for glycopeptide analysis [73]. In multiple reaction monitoring (MRM), the concentration of an unknown sample is determined by comparing its MS response to that of a known standard. In contrast to the standard LC-MS, which scans all ions within a certain scan range, the instrument is programmed in MRM to specifically look for a select number of predetermined MS and MS^2 ions (transitions). The characteristic oxonium ions, representing hexose (*m/z* 163), HexNAc (*m/z* 204), Neu5Ac (*m/z* 292). HexHexNAc (*m/z* 366), HexHexNAcFuc (*m/z* 512), and HexHexNAcNeu5Ac (*m/z* 657), with or without loss of water, is often used as reporter ions. The first mass analyzer (Q1) is set to only transmit ions of interest, the second mass analyzer (Q2) fragments the ions, and the third mass analyzer (Q3) is set to transmit diagnostic MS fragments only.

Mass spectrometry has now reached sufficient sensitivity to detect low abundant structures when we analyze novel species and more defined cell types [74, 75]. Due to the very limited amounts, de novo characterization of structures is soon getting difficult using NMR technology. Permethylation and MS^n is one pathway that is currently used to identify novel structures in mixtures of sugars of limited supply. Nowadays, sophisticated instruments are commercially available. For example, the SmartMS™-enabled BioAccord System (Waters Corporation) is an easy-to-use LC-MS platform designed for comprehensive analysis of carbohydrate-containing bioproducts with built-in analytical workflows for specific analysis, such as fluorescent labeled N-glycan identification and profiling. It is important to note that when performing MS analysis of released glycans, it is quite common to come across glycans with challenging MS spectra for correct assignments, such as isobaric glycans. Consequently, obtaining complete structural information by MS is hard feasible and still requires a multidisciplinary approach, leading to the integration of LC-MS and NMR techniques [76].

Finally, to fully assign a novel carbohydrate structure, a combination of MS, chemical, and enzymatic treatment, and NMR spectroscopy would be ideal, having the disposal of sufficient material. Unambiguous information regarding the position of glycosidic linkages, the position of branching, monosaccharide composition, and ring sizes can be obtained by gas chromatography-mass spectrometry (GC-MS) by the so-called methylation analysis (see Chap. 6).

References

1. Kailemia MJ, Xu G, Wong M, Li Q, Goonatilleke E, Leon F, Lebrilla CB. Recent advances in the mass spectrometry methods for glycomics and cancer. Anal Chem. 2018;90:208–24.
2. Wada Y, Dell A, Haslam SM, Tissot B, Canis K, et al. Comparison of methods for profiling O-glycosylation. Mol Cell Proteomics. 2010;9:719–27.
3. Pasing Y, Sickman A, Lewandrowski U. N-glycoproteomics: mass spectrometry-based glycosylation site annotation. Biol Chem. 2012;393:249–58.
4. Ly M, Laremore TN, Linhardt RJ. Proteoglycomics: recent progress and future challenges. OMICS. 2010;14:389–99.

5. Tissot B, North SJ, Ceroni A, Pang PC, Panico M, Rosati F, Capone A, Haslam SM, Dell A, Morris HR. Glycoproteomics: past, present and future. FEBS Lett. 2009;583:1728–35.
6. Pan S, Chen R, Aebersold R, Brentnall TA. Mass spectrometry-based glycoproteomics—from a proteomics perspective. Mol Cell Proteomics. 2011;10:R110.003251.
7. Pan S. Quantitative glycoproteomics for N-glycoproteome profiling. Methods Mol Biol. 2014;1156:379–88.
8. Leymarie N, Zaia J. Effective use of mass spectrometry for glycan and glycopeptide structural analysis. Anal Chem. 2012;84:3040–8.
9. Patrie SM, Roth MJ, Kohler JJ. Introduction to glycosylation and mass spectrometry. Methods Mol Biol. 2013;951:1–17.
10. Han L, Costello CE. Mass spectrometry of glycans. Biochemistry. 2013;78:710–20.
11. Klapoetke S. N-glycosylation characterization by liquid chromatography with mass spectrometry. Methods Mol Biol. 2014;1131:513–24.
12. Lazar IM, Deng J, Ikenishi F, Lazar AC. Exploring the glycoproteomics landscape with advanced MS technologies. Electrophoresis. 2015;36:225–37.
13. Xiao K, Han Y, Yang H, Lu H, Tian Z. Mass spectrometry-based qualitative and quantitative N-glycomics: an update of 2017-2018. Anal Chim Acta. 2019;1091:1–22.
14. Zhang Y, Peng Y, Yang L, Lu H. Advances in sample preparation strategies for MS-based qualitative and quantitative N-glycomics. TrAC Trends Anal Chem. 2018;99:34–46.
15. Alley WR, Novotny MV. Structural glycomic analyses at high sensitivity: a decade of progress. Annu Rev Anal Chem. 2013;6:237–65.
16. Ruhaak LR, Zauner G, Huhn C, Bruggink C, Deelder AM, Wuhrer M. Glycan labeling strategies and their use in identification and quantification. Anal Bioanal Chem. 2010;397:3457–81.
17. Brooks SA. Strategies for analysis of the glycosylation of proteins: current status and future perspectives. Mol Biotechnol. 2009;43:76–88.
18. Zaia J. Mass spectrometry and glycomics. OMICS. 2010;14:401–18.
19. El-Aneed A, Cohen A, Banoub J. Mass spectrometry, review of the basics: electrospray, MALDI, and commonly used mass analyzers. Appl Spectroscopy Rev. 2009;44:210–30.
20. Sparkman OD, Penton Z, Kitson F. Gas chromatography and mass spectrometry: a practical guide. 2nd ed. Academic; 2011.
21. Kamerling JP, Gerwig GJ. Strategies for the structural analysis of carbohydrates. In: Kamerling JP, editor. Comprehensive glycoscience-from chemistry to systems biology. Amsterdam: Elsevier; 2007. p. 1–68.
22. Abrahams JL, Taherzadeh G, Jarvas G, Guttman A, Zhou Y, Campbell MP. Recent advances in glycoinformatics platforms and glycoproteomics. Curr Opin Struct Biol. 2020;62:56–69.
23. Levery SB, Steentoft C, Halim A, Narimatsu Y, Clausen H, Vakhrushev SY. Advances in mass spectrometry driven O-glycoproteomics. Biochim Biophys Acta. 2015;1850:33–42.
24. Haselberg R, De Jong GJ, Somsen GW. CE-MS for the analysis of intact proteins 2010–2012. Electrophoresis. 2013;34:99–112.
25. Jeong HJ, Kim YG, Yang YH, Kim BG. High-throughput quantitative analysis of total N-glycans by matrix-assisted laser desorption/ionization time-of-flight mass spectrometry. Anal Chem. 2012;84:3453–60.
26. Cramer R, editor. Advances in MALDI and laser-induced soft ionization mass spectrometry. Springer International Publishing; 2016.
27. Shubhakar A, Pang P-C, Fernandes DL, Dell A, Spencer DI, Haslam SM. Towards automation of glycomic profiling of complex biological materials. Glycoconj J. 2018;35:311–21.
28. North SJ, Hitchen PG, Haslam SM, Dell A. Mass spectrometry in the analysis of N-linked and O-linked glycans. Curr Opin Struct Biol. 2009;19:498–506.
29. Morelle W, Faid V, Chirat F, Michalski JC. Analysis of N- and O-linked glycans from glycoproteins using MALDI-TOF mass spectrometry. Methods Mol Biol. 2009;534:5–21.
30. Azadi P, Heiss C. Mass spectrometry of N-linked glycans. Methods Mol Biol. 2009;534:37–51.
31. Gaunitz S, Nagy G, Pohl NLB, Novotny MV. Recent advances in the analysis of complex glycoproteins. Anal Chem. 2017;89:389–413.

32. Oedit A, Vulto P, Ramautar R, Lindenberg PW, Hankemeier T. Lab-on-a-chip hyphenation with mass spectrometry: strategies for bioanalytical applications. Curr Opin Biotechnol. 2015;31:79–85.

33. Bindila L, Peter-Katalinic J. Chip-mass spectrometry for glycomic studies. Mass Spectrom Rev. 2009;28:223–53.

34. Alley WR, Madera M, Mechref Y, Novotny MV. Chip-based reverse-phase liquid chromatography-mass spectrometry of permethylated N-linked glycans: a potential methodology for cancer-biomarker discovery. Anal Chem. 2010;82:5095–106.

35. Nilsson J. Liquid chromatography-tandem mass spectrometry-based fragmentation analysis of glycopeptides. Glycoconj J. 2016;33:261–72.

36. Ruhaak LR, Deelder AM, Wuhrer M. Oligosaccharide analysis by graphitized carbon liquid chromatography-mass spectrometry. Anal Bioanal Chem. 2009;394:163–74.

37. Dotz V, Haselberg R, Shubhakar A, Kozak RP, Falck D, Rombouts Y, Reusch D, Somsen GW, Fernandes DL, Wuhrer M. Mass spectrometry for glycosylation analysis of biopharmaceuticals. Trends Anal Chem. 2015;73:1–9.

38. Martín-Ortiz A, Carrero-Carralero C, Hernández-Hernández O, Lebrón-Auilar R, Moreno FJ, Sanz MI, Ruiz-Matute AI. Advances in structure elucidation of low molecular weight carbohydrates by liquid chromatography-multiple-stage mass spectrometry analysis. J Chromatogr A. 1612;2020:460644.

39. An HJ, Lebrilla CB. Structure elucidation of native N- and O-linked glycans by tandem mass spectrometry (tutorial). Mass Spectrom Rev. 2011;30:560–78.

40. Wu S, Salcedo J, Tang N, Waddell K, Grimm R, German JB, Lebrilla CB. Employment of tandem MS for the accurate and specific identification of oligosaccharide structures. Anal Chem. 2012;84:7456–62.

41. Thaysen-Andersen M, Packer NH. Advances in LC-MS/MS-based glycoproteomics: getting closer to system-wide site-specific mapping of the N- and O-glycoproteome. Biochim Biophys Acta. 2014;1844:1437–52.

42. Costello CE, Contado-Miller JM, Cipollo JF. A glycomics platform for the analysis of permethylated oligosaccharide alditols. J Am Soc Mass Spectrom. 2007;18:1799–812.

43. Veillon L, Huang Y, Peng W, Dong X, Cho BG, Mechref Y. Characterization of isomeric glycan structures by LC-MS/MS. Electrophoresis. 2017;38:2100–14.

44. Sassaki GL, De Souza LM. Mass spectrometry strategies for structural analysis of carbohydrates and glycoconjugates. Intech; 2013.

45. de Hoffmann E, Stroobant V. Mass spectrometry: principles and applications. 3rd ed. Hoboken: Wiley; 2007.

46. Gao X, Zhang L, Zhang W, Zhao I. Design and application of an open tubular capillary reactor for solid-phase permethylation of glycans in glycoproteins. Analyst. 2015;140:1566–71.

47. Hu Y, Borges CR. A spin column-free approach to sodium hydroxide-based glycan permethylation. Analyst. 2017;142:2748–59.

48. Shajahan A, Supekar NT, Heiss C, Azadi PJ. High-throughput automated micro-permethylation for glycan structure analysis. Anal Chem. 2019;91:1237–40.

49. Kang P, Mechref Y, Novotny MV. High-throughput solid-phase permethylation of glycans prior to mass spectrometry. Rapid Commun Mass Spectrom. 2008;22:721–34.

50. Lin Z, Lubman DM. Permethylated N-glycan analysis with mass spectrometry. Methods Mol Biol. 2013;1007:289–300.

51. Zhang Z, Linhardt RJ. Sequence analysis of native oligosaccharides using negative ESI tandem MS. Curr Anal Chem. 2009;5:225–37.

52. Domon B, Costello CE. Oligosaccharide fragmentation nomenclature. Glycoconj J. 1988;5:397–409.

53. Hu H, Khatri K, Klein J, Leymarie N, Zaia J. A review of methods for interpretation of glycopeptide tandem mass spectral data. Glycoconj J. 2016;33:285–96.

54. Ceroni A, Dell A, Haslam SM. The GlycanBuilder: a fast, intuitive and flexible software tool for building and displaying glycan structures. Source Code Biol Med. 2007;2:3–15.

55. Mucha E, Lettow M, Marianski M, Thomas DA, Struwe WB, Harvey DJ, et al. Fucose migration in intact protonated glycan ions: a universal phenomenon in mass spectrometry. Angew Chem Int Ed. 2018;57:7440–3.
56. Wuhrer M. Glycomics using mass spectrometry (review). Glycoconj J. 2013;30:11–22.
57. Dodds ED. Gas-phase dissociation of glycosylated peptide ions. Mass Spectrom Rev. 2012;31:666–82.
58. Lee LY, Moh ESX, Parker BL, Bern M, Packer NH, Thaysen-Andersen M. Toward automated N-glycopeptide identification in glycoproteomics. J Proteome Res. 2016;15:3904–15.
59. Reiding KR, Bondt A, Franc V, Heck AJR. The benefits of hybrid fragmentation methods for glycoproteomics. Trends Anal Chem. 2018;108:260–8.
60. Scott NE, Cordwell SJ. Enrichment and identification of bacterial glycopeptides by mass spectrometry. Methods Mol Biol. 2015;1295:355–68.
61. Yang H, Shi L, Zhuang X, Su R, Wan D, Song F, Li J, Liu S. Identification of structurally closely related monosaccharide and disaccharide isomers by PMP labeling in conjunction with IM-MS/MS. Sci Rep. 2016;6:28079.
62. Hofmann J, Pagel K. Glycan analysis by ion mobility-mass spectrometry. Angew Chem Int Ed. 2017;56:8342–9.
63. Manz C, Pagel K. Glycan analysis by ion mobility-mass spectrometry and gas-phase spectroscopy. Curr Opin Chem Biol. 2018;42:16–24.
64. Paglia G, Astarita G. Ion mobility-mass spectrometry: methods and protocols. vol. 2084, Methods in molecular biology. Springer; 2020.
65. Jin C, Harvey DJ, Struwe WB, Karlsson NG. Separation of isomeric O-glycans by ion mobility and liquid chromatography-mass spectrometry. Anal Chem. 2019;91:10604–13.
66. Chen Z, Glover MS, Li L. Recent advances in ion mobility-mass spectrometry for improved structural characterization of glycans and glycoconjugates. Curr Opin Chem Biol. 2018;42:1–8.
67. Pagel K, Harvey DJ. Ion mobility-mass spectrometry of complex carbohydrates: collision cross sections of sodiated N-linked glycans. Anal Chem. 2013;85:5138–45.
68. Wei J, Wu J, Tang Y, Ridgeway ME, Park MA, Costello CE, Zaia J, Lin C. Characterization and quantification of highly sulfated glycosaminoglycan isomers by gated-trapped ion mobility spectrometry negative electron transfer dissociation MS/MS. Anal Chem. 2019;91:2994–3001.
69. Hofmann J, Hahm HS, Seeberger PH, Pagel K. Identification of carbohydrate anomers using ion mobility-mass spectrometry. Nature. 2015;526:241–4.
70. Gabelica V, Marklund E. Fundamentals of ion mobility spectroscopy. Curr Opin Chem Biol. 2018;42:51–9.
71. Gray CJ, Thomas B, Upton R, Migas LG, Eyers CE, Barran PE, et al. Application of ion mobility mass spectrometry for high throughput, high resolution glycan analysis. Biochim Biophys Acta. 2016;1860:1688–709.
72. Gaye MM, Nagy G, Pohl NLB, Clemmer DE. Multidimensional analysis of 16 glucose isomers by ion mobility spectrometry. Anal Chem. 2016;88:2344–55.
73. Xu G, Amicucci MJ, Cheng Z, Galermo AG, Lebrilla CB. Revisiting monosaccharide analysis—Quantitation of a comprehensive set of monosaccharides using dynamic multiple reaction monitoring. Analyst. 2017;143:200–7.
74. Ruhaak LR, Xu G, Li Q, Goonatilleke E, Lebrilla CB. Mass spectrometry approaches to glycomic and glycoproteomic analyses. Chem Rev. 2018;118:7886–930.
75. Dong X, Huang Y, Cho BG, Zhong J, Gautam S, Peng W, Williamson SD, Banazadeh A, Torres-Ulloa KY, Mechref Y. Advances in mass spectrometry-based glycomics. Electrophoresis. 2018;39:3063–81.
76. Fellenberg M, Behnken HN, Nagel T, Wiegandt A, Baerenfaenger M, Meyer B. Glycan analysis: Scope and limitations of different techniques—a case for integrated use of LC-MS(/MS) and NMR techniques. Anal Bioanal Chem. 2013;405:7291–305.

Chapter 12
Analysis of Carbohydrates by Nuclear Magnetic Resonance Spectroscopy

Abstract NMR spectroscopy has shown to be extremely useful for the characterization of glycan structures. Carbohydrates have two natural NMR-active nuclei: ^{1}H and ^{13}C. This chapter describes the identification of monosaccharides and glycans by the assignment of their protons and carbons, using the structural-reporter group concept and 2D NMR techniques, which additionally provide linkage and sequence information. All this is illustrated by some NMR spectra of glycans and polysaccharides and the NMR measuring procedure is summarized.

Keywords Cryoprobe · Fourier transformation · CASPER · Free induction decay · Chemical shift · Vicinal coupling constant · Deuterium · Bulk region · COSY · TOCSY · HMBC · NOESY · ROESY · HSQC · MAS-NMR · N-glycan · Polysaccharide · α-glucan

12.1 Introduction

Nuclear magnetic resonance (NMR) spectroscopy was occasionally used in carbohydrate research since 1958, but at the beginning of the 1980s, the application of NMR in glycobiology enormously accelerated the primary structural analysis of carbohydrates, in particular for the glycans belonging to glycoproteins [1–3]. The information that can be deduced from an NMR spectrum of an underivatized complex oligosaccharide is wide. It comprises the number and types of constituting monosaccharides, including ring size and anomeric configuration, the position of glycosidic linkages as well as the type and position of noncarbohydrate substituents. A complete standardized experimental setup allows a fully reliable comparison of data measured on interchanging laboratories and spectrometers. In addition to primary structural data, NMR can also provide information on the conformation and molecular dynamics (three-dimensional structure) of oligosaccharide molecules in a solution state. NMR has the advantage of being nondestructive, allowing the recovery of the intact sample after NMR analysis so that it can later be used for other analytical methods and/or biological activity tests. Despite the high costs of an NMR spectrometer and the level of expertise required for recording and interpreting NMR spectra,

© Springer Nature Switzerland AG 2021
G. J. Gerwig, *The Art of Carbohydrate Analysis*, Techniques in Life Science and Biomedicine for the Non-Expert, https://doi.org/10.1007/978-3-030-77791-3_12

the technique will remain one of the most powerful methods for detailed structural analysis in carbohydrate research for the time being, next to mass spectrometry.

The development of novel helium-cooled superconducting materials to make ultra-strong magnets has made high-resolution NMR spectrometers beyond 1000 MHz [1.2 GHz, magnetic field strengths >25 Tesla (T)] possible and commercially available. The disadvantage of the early days needing substantial amounts of material to record a reasonable spectrum have been minimized nowadays due to the applications of newly designed nanoprobes, RF-microcoils, and cryogenically cooled NMR probes (cryoprobes), showing high resolution and sensitivity for detailed structural analysis of low nano/picomolar amounts of oligo/polysaccharides and glycoconjugates [4].

Techniques such as dynamic nuclear polarization provide several orders of magnitude improvement in sensitivity as well as the use of optimized pulse sequences. Furthermore, sophisticated fast computer programs allow a series of different multidimensional measuring experiments and extended handlings of spectra. Although automatic sample changers and system automations are of common use nowadays, NMR experiments can still be time-consuming. Additionally, NMR spectroscopy often requires an experienced scientist for instrument operation, data processing, and interpretation of spectra. A detailed discussion of the principle and theory of NMR spectroscopy, in terms of sophisticated multiple-pulse sequences in modern NMR techniques, is beyond the scope of this book. The interested reader is referred to specialized articles and recent textbooks [5–9].

12.2   Principle of NMR Spectroscopy

Briefly, NMR spectroscopy is based on the behavior of some atomic nuclei (^1H, ^{13}C, ^{31}P, and ^{15}N) in a strong homogeneous magnetic field when they absorb energy, caused by a pulsed radio frequency (RF) irradiation. For carbohydrates, mainly ^1H and ^{13}C nuclei (intrinsic spin quantum number $I = 1/2$) are used. They can absorb energy at characteristic radio frequencies (e.g., at 11.74 T: ^1H 500.0 MHz and ^{13}C 125.7 MHz). A very short (10–60µs), high-power (several kilowatts), electromagnetic radiation pulse, (short enough that the uncertainty band in frequency will excite all nuclei into resonance) is used to monitor carbohydrate compounds. This is possible because of differences in the magnetic states of the ^1H and ^{13}C nuclei, involving very small transitions in energy levels. These atomic nuclei are magnetic because they are charged and have a spin. Spinning an electrical charge generates a magnetic field. Due to interaction with a uniform external strong magnetic field, they align themselves in a direction either parallel (mostly) or antiparallel to the magnetic field. The two orientations have different energies; parallel is lower than antiparallel. It is possible to change the orientation from parallel to antiparallel (and vice versa), causing the nuclear spins to resonate. The energy difference (ΔE)

corresponds to a precise electromagnetic frequency (ν). When exposed to an appropriate radiofrequency, (depending on the strength of the external magnetic field) transitions between the energy levels of the magnetic nuclei will occur when the energy gap and the applied frequency are in resonance (exactly matched). When the radio frequency is removed, the absorbed energy will be released at characteristic frequencies as the spins try to restore their thermodynamic equilibrium orientation (relaxation process). Measurement of this change in energy state for each of the nuclei in the sample provides the basic information in NMR spectroscopy. This process is repeated in consecutive scans, and time-domain signals are accumulated into computer memory for analysis by Fourier transformation. Several tools that aim at decreasing the time needed for peak assignment and spectra interpretation have been developed, like ProspectND for NMR spectra processing and CASPER, for structure determination of carbohydrates. However, basic knowledge about NMR and signal assignment is a prerequisite for working with these tools.

12.3 NMR Spectroscopy for Carbohydrates

Luckily, carbohydrates contain a lot of hydrogen atoms (protons), which are the most easily detected magnetic nuclei for NMR spectroscopy. Applying proton (^{1}H) NMR spectroscopy, different chemical environments of the hydrogen atoms in a carbohydrate molecule result in different ^{1}H resonances, which can be detected and collected as a (time-domain signal) free induction decay (FID). Using Fourier transformation (FT), the added-up FIDs are converted and presented in an NMR spectrum showing the frequencies at which the nuclei responded.

The displacement of such a proton frequency from that of an internal reference, tetramethylsilane (TMS) [$Si(CH_3)_4$], is referred to as its chemical shift (δ) and is quoted in parts per million (ppm).

$$\text{Chemical shift} (\delta)(\text{ppm}) = \frac{\text{Separation between Signal}(Hz)\text{ and TMS}(Hz) \times 10^6}{\text{Spectrometer radiofrequency}(Hz)}$$

It is evident that a number of effects in the neighborhood of the nucleus determine its chemical shift. It can be said that the chemical shift represents the chemical environment of the nucleus. For protons, the shielding effect of the surrounding electron cloud generally has the greatest influence; this effect can be related to the electronegativity of a nearby atom. The NMR signal of a proton usually shows more than one peak in the spectrum. This is called "splitting," resulting from interactions with nearby nuclei sharing bonding electrons, and is expressed as ^{1}H spin-spin or scalar coupling (J) (intramolecular connectivity). The distance between the peaks reflects the vicinal coupling constants ($^nJ_{Hx,Hy}$), quoted in Hz, which characterize scalar (through-bond) interactions between neighboring protons (n is the number of bonds separating the coupled protons, and the subscript defines the protons involved). For example, the anomeric ^{1}H signal at C1 of a glycosyl residue usually has a doublet fine structure (two equal-height peaks) caused by its neighbor proton (H-2) at C2. Likewise, the H-2 shows up as a pair of doublets due to coupling with H-1 and H-3. The size of the coupling constant (Hz) depends on the orientation of the protons in the molecule.

In short, chemical shifts and coupling constants are often diagnostic of both the nature of the monosaccharide and the type of linkage but do not directly determine the sequence of the residues. Computer manipulations (zero filling and phase correction) are used to enhance sensitivity and resolution. Spectral integration may determine relative intensities (i.e., peak areas) of signals, which are proportional to the number of contributing protons, making NMR a primary ratio-quantitative method. However, despite the relatively small size of oligosaccharides, their proton NMR spectra are remarkably complex due to the poor dispersion of ring protons showing severe resonance overlap and the increased probability of strong coupling effects.

12.3.1   Sample Preparation and Measuring

For ^{1}H NMR spectroscopy of carbohydrates, usually, the protons of the hydroxyl and amino groups (-OH and -NH) are exchanged for deuterium (D or ^{2}H), making only the protons linked to carbons detectable, which reduces the complexity of the spectrum. The exchange is done by at least two D_2O-dissolution/lyophilization cycles. Residual water wil be observed in the ^{1}H spectrum as an HOD signal. Although nonaqueous solvents, such as deuterated dimethyl sulfoxide (DMSO-d_6), deuterochloroform (CDCl$_3$), and pyridine-d_5, are used, NMR spectra of underivatized carbohydrates are mostly recorded in deuterium oxide (D$_2$O) at ambient temperature, 27 °C (300 K), and at pD 7. The chemical shifts of basic or acidic carbohydrates are strongly pH-dependent.

Typically, NMR analysis of carbohydrate samples is carried out in high-quality glass 5-mm NMR tubes, using a volume of 600μL. Nowadays, dedicated low-volume, special NMR tubes (e.g., Shigemi-tubes, 50–100μL; <5 nmol sugar) are available for use in probes with narrow coil diameters. Furthermore, notwithstanding the requirement of complex setups, the coupling of separation techniques, like high-performance liquid chromatography (HPLC) or capillary electrophoresis (CE), directly to NMR instruments is making progress.

Careful preparation of the NMR sample is of vital importance for successful NMR spectroscopy. The sample must contain a sufficient amount of material (preferably >5 nmol oligosaccharide for 1D ^{1}H spectrum, but >50μmol for 2D spectra) and must be free of paramagnetic impurities (metal ions). It is important to remove any proton- and non-proton-containing contaminants, like chelators, salts, and/or buffers [acetate, lactate, sodium dodecyl sulfate (SDS) and N,N,N′,N′-ethylenediamine-tetra-acetic acid (EDTA)]. High salt concentrations can affect the NMR spectrometer's tuning, matching, and shimming process. Salts, such as NaCl and phosphates, and pH differences cause line broadening and shift of the HOD peak. Furthermore, the viscosity of the sample (in particular with polysaccharides) can lead to signals broadening but can be minimized by measuring at higher temperatures. Higher temperatures also increase the resolution due to a decrease in the correlation time, but the influence on the chemical shifts (δ) of protons is minimal.

The position of the HOD signal in the spectrum significantly varies with temperature (0.005 ppm/°C), so its chemical shift can be adjusted to avoid possible overlap with sugar resonances. By raising the temperature, the HOD signal moves upfield (to lower ppm at the right in the spectrum). A ^{1}H NMR spectrum of a carbohydrate shows an assembly of peaks (signals) that can be thought of as a distribution curve of protons as a function of their structural environment (electron density) in the molecule. Thus, the total spectrum is characteristic of the structure of the carbohydrate. The chemical shift defines the location of an NMR signal along the horizontal RF axis. Chemical shifts (δ) are expressed in ppm downfield from the internal standard tetramethylsilane (TMS) ($\delta = 0$ ppm for ^{1}H spectra), but often, due to the insolubility of TMS in D_2O, they are measured compared to the (internal standard) acetone CH_3 signal at 2.225 ppm for ^{1}H and 31.08 ppm for ^{13}C at 300 K.

In practice, usually, suppression of the residual solvent (HOD) signal is applied. The simplest way to suppress the solvent signal is by fast pulsing because carbohydrate protons have much shorter relaxation times than the HOD proton. Furthermore, presaturation is one of the widely used ways to eliminate a strong solvent signal, in which a long low-power pulse is applied on the HOD resonance as part of the pulse. After saturation of this resonance at the transmitter frequency, a nonselective pulse with a wide excitation profile is then applied to place all remaining spins in the transverse plane for detection. Pulse sequence protocols for solvent signal suppression and further explanation can usually be found in the NMR spectrometer's manual.

In Fig. 12.1A, the ^{1}H NMR spectrum of glucose is shown after deuterium exchange of the hydroxyl groups. The residual HOD signal at δ 4.76 has been removed in this picture. In an aqueous solution, there is an equilibrium between α-anomeric (34%) and β-anomeric (66%) forms. Figure 12.1B shows the ^{13}C spectrum of glucose (see Sect. 12.3.3). The ^{1}H and ^{13}C NMR chemical shift data of some common monosaccharides are presented in Tables 12.1, 12.2, and 12.3, respectively. Small deviations in the chemical shifts can occur when these monosaccharides are constituents of glycans and complex carbohydrates due to steric hindrance, hydrogen bonding, and electronic effects (proximity effects).

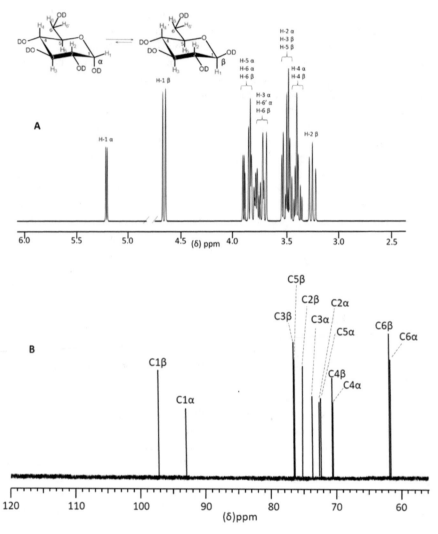

Fig. 12.1 ¹H (**A**) and ¹³C (**B**) NMR spectrum of glucose in D₂O at 300 K. Note the anomeric equilibrium

12.3.2   ¹H NMR Structural-Reporter-Group Concept

For the structural analysis of glycans, in particular N-glycans, spectral interpretation methods have been developed based on sets of "structural-reporter groups" [10,
11]. The essence of the structural-reporter group (SRG) concept is that it is sufficient to inspect only certain regions in the ¹H NMR spectrum to ascertain the primary structure of the glycan. This means that the chemical shifts (δ) of protons

Table 12.1 ¹H chemical shifts of some common free monosaccharides in D₂O at 300 K

Monosaccharide	H-1	H-2	H-3	H-4	H-5	H-6s	NAc
α-D-Glcp	5.22	3.54	3.73	3.42	3.84	3.83, 3.76	
β-D-Glcp	4.64	3.25	3.49	3.42	3.47	3.89, 3.72	
α-D-Galp	5.26	3.80	3.85	3.98	4.05	3.80, 3.75	
β-D-Galp	4.59	3.50	3.63	3.90	3.74	3.81, 3.67	
α-D-Manp	5.18	3.93	3.85	3.66	3.83	3.87, 3.76	
β-D-Manp	4.90	3.96	3.66	3.58	3.38	3.87, 3.74	
α-L-Fucp	5.20	3.77	3.84	3.80	4.10	1.21 (CH₃)	
β-L-Fucp	4.55	3.46	3.63	3.74	3.79	1.26 (CH₃)	
α-L-Rhap	5.12	3.92	3.78	3.45	3.83	1.29 (CH₃)	
β-L-Rhap	4.85	3.94	3.59	3.37	3.39	1.31 (CH₃)	
α-D-Xylp	5.19	3.52	3.58	3.62	3.68, 3.67	–	
β-D-Xylp	4.57	3.24	3.43	3.61	3.92, 3.32	–	
α-L-Arap	5.23	3.80	3.88	3.99	4.01, 3.64	–	
β-L-Arap	4.52	3.51	3.65	3.94	3.89, 3.68	–	
α-D-Ribp	4.85	3.81	3.93	3.87	3.92, 3.60	–	
β-D-Ribp	4.91	3.51	4.04	3.87	3.82, 3.67	–	
α-D-GlcpNAc	5.21	3.88	3.75	3.49	3.86	3.85, 3.77	2.06
β-D-GlcpNAc	4.72	3.65	3.56	3.46	3.46	3.91, 3.75	2.06
α-D-GalpNAc	5.28	4.19	3.95	4.05	4.13	3.80, 3.79	2.06
β-D-GalpNAc	4.68	3.90	3.77	3.98	3.72	3.84, 3.82	2.06
α-D-ManpNAc	5.14	4.32	4.07	3.65	3.68	3.84, 3.84	2.09
β-D-ManpNAc	5.02	4.43	3.84	3.53	3.45	3.81, 3.90	2.05
α-D-GlcpA	5.24	3.59	3.74	3.54	4.09	–	
β-D-GlcpA	4.65	3.31	3.52	3.55	3.73	–	
α-D-GalpA	5.31	3.83	3.92	4.30	4.40	–	
β-D-GalpA	4.55	3.51	3.69	4.24	4.04	–	
α-D-ManpA	5.22	3.90	3.87	3.84	4.05	–	
β-D-ManpA	4.89	3.93	3.65	3.73	3.64	–	

Table 12.2 ¹H chemical shifts of D-fructose equilibrated in D₂O at 300 K

Monosaccharide	H-1	H-1'	H-3	H-4	H-5	H-6	H-6'
α-D-Frup	3.69	3.65	4.03	3.95	3.88	3.87	3.70
β-D-Frup	3.71	3.58	3.85	3.80	4.00	4.03	3.72
α-D-Fruf	3.67	3.64	4.11	4.00	4.06	3.82	3.71
β-D-Fruf	3.59	3.55	4.12	4.12	3.85	3.81	3.68

resonating at clearly distinguishable positions in the spectrum, together with their splitting patterns (coupling constants J) and the line widths of their signals, usually bear the information essential to permit assigning of the primary structure. Hence, the structural reporter group signals are found outside the so-called "bulk region" (δ 3.1–4.0 ppm in ¹H spectrum). The bulk region comprises a severe overlap of the nonanomeric ring-skeleton protons.

Table 12.3 ^{13}C NMR chemical shifts of some common free monosaccharides in D^{2}O at 300 K

Monosaccharide	C1	C2	C3	C4	C5	C6
α-D-Glc*p*	93.0	72.5	73.7	70.5	72.3	61.6
β-D-Glc*p*	97.0	75.1	76.7	70.6	76.7	61.8
α-D-Glc*f*	104.0	77.7	76.6	78.8	70.7	64.2
β-D-Glc*f*	103.9	81.6	75.8	82.3	70.7	64.7
α-D-Gal*p*	93.7	69.3	70.2	70.3	71.3	62.1
β-D-Gal*p*	97.4	72.7	73.8	69.7	75.9	61.9
α-D-Gal*f*	96.0	77.5	75.6	81.9	73.5	63.8
β-D-Gal*f*	102.0	82.0	76.9	83.4	71.8	63.7
α-D-Man*p*	94.9	71.6	71.3	67.9	73.5	62.1
β-D-Man*p*	94.5	72.2	73.8	67.6	76.7	61.9
α-D-Man*f*	109.7	77.9	72.5	80.5	70.6	64.5
β-D-Man*f*	103.6	73.1	71.2	80.7	71.4	64.4
α-L-Fuc*p*	93.2	69.1	70.3	72.8	67.1	16.3
β-L-Fuc*p*	97.2	72.7	74.0	72.4	71.6	16.3
α-L-Rha*p*	94.9	71.9	71.1	73.2	69.2	17.7
β-L-Rha*p*	94.4	72.3	73.8	72.8	72.8	17.6
α-D-Xyl*p*	93.2	72.7	74.1	70.5	62.1	–
β-D-Xyl*p*	97.6	75.2	77.0	70.3	66.3	–
α-D-Xyl*f*	103.5	77.8	76.2	79.3	61.6	–
β-D-Xyl*f*	109.7	81.0	76.1	83.6	62.2	–
α-L-Ara*p*	93.5	69.5	69.8	70.0	63.7	–
β-L-Ara*p*	97.7	72.8	73.6	69.6	67.4	–
α-D-Ara*f*	102.1	81.9	77.0	84.5	62.3	–
β-D-Ara*f*	96.2	77.4	75.5	82.6	63.1	–
α-D-Rib*p*	94.4	71.0	70.3	68.2	64.8	–
β-D-Rib*p*	94.8	72.1	69.8	68.4	64.0	–
α-D-Rib*f*	97.2	71.8	70.8	84.4	62.1	–
β-D-Rib*f*	101.8	75.1	71.4	83.5	63.4	–
α-D-Fru*p*	63.0	97.7	71.0	71.4	69.0	63.2
β-D-Fru*p*	64.3	99.2	68.2	70.2	70.0	64.2
α-D-Fru*f*	63.5	105.2	82.1	76.5	82.0	63.0
β-D-Fru*f*	63.4	102.3	75.4	74.7	80.8	63.5
α-D-Glc*p*NAc	91.8	55.3	71.7	71.3	72.5	61.8
β-D-Glc*p*NAc	95.9	57.9	74.8	71.1	76.8	61.9

(continued)

Table 12.3 (continued)

Monosaccharide	C1	C2	C3	C4	C5	C6
α-D-GalpNAc	92.0	51.3	68.4	69.6	71.4	62.1
β-D-GalpNAc	96.3	54.8	72.0	68.9	76.0	61.9
α-D-ManpNAc	94.1	54.4	70.1	68.1	73.4	61.7
β-D-ManpNAc	94.0	55.2	73.2	67.8	77.5	61.8
α-D-GlcpA	93.0	72.3	73.5	72.9	72.5	177.5
β-D-GlcpA	96.8	75.0	76.5	72.7	76.9	176.5
α-D-GalpA	93.1	69.0	70.3	71.6	72.3	176.5
β-D-GalpA	96.9	72.6	73.8	71.2	76.5	175.6
α-D-ManpA	94.8	71.5	71.1	70.0	73.7	177.6
β-D-ManpA	94.5	72.0	73.9	69.7	77.0	176.8

The ^{1}H NMR structural-reporter group signals for a spectrum of glycoprotein-derived N- and O-glycans, can be summarized as:

- The anomeric proton signals (H-1 atoms, δ 4.3–5.8 ppm) are easy to spot because they have shifted to a downfield (higher ppm) position due to their relative unshielding by the ring oxygen atom (having electron-withdrawing properties). They give information on the kind of monosaccharide residue as well as on the type and configuration (α/β) of its glycosidic linkage. The number of monosaccharide residues may be determined by integrating the anomeric proton resonance peaks.
- The H-2 (and H-3, in some peculiar cases) atoms of mannose residues (δ 4.0–4.3 ppm) indicate the type of substitution of the N-glycan mannotriose branching core.
- The N-acetyl-CH$_3$ protons (δ 2.0–2.2 ppm) indicate N-acetylhexosamines and sialic acids.
- The sialic acid H-3 protons give the type and configuration of the glycosidic linkage and location of sialic acid in the chain.
- The 6-deoxysugar (fucose) H-5 and H-6 (CH$_3$) protons (δ 1.1–1.3 ppm), together with H-1 give the type and configuration of its glycosidic linkage.
- The galactose H-3 and H-4 (H-5 of GalNAc-ol) atoms characterize the type and configuration of the glycosidic linkage between galactose and its substituent.
- Protons that can be discerned outside the bulk region, as a result of glycosylation shifts, or under influence of substituents such as sulfate, phosphate, alkyl, and acyl groups.
- Alkyl and acyl substituents, like methyl (δ ~1.2 ppm) and acetyl (δ ~2.0–2.1 ppm), glycolyl, pyruvate, and similar groups.
- The relative intensities of the structural-reporter group signals, obtained by spectral integration, provide molar ratios of the monosaccharide residues.

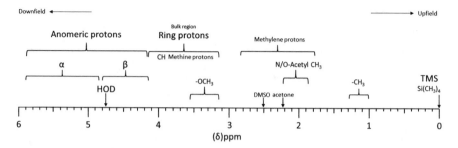

Fig. 12.2 Predictable regions of chemical shifts of carbohydrate protons in a ¹H NMR spectrum

Typical chemical shifts of notable resonances observed in ¹H spectra of carbohydrates are illustrated in Fig. 12.2.

12.3.3   ¹³C NMR Spectroscopy

¹³C NMR spectroscopy is also of great value for the study of carbohydrates, despite being inherently less sensitive compared to ¹H NMR spectroscopy due to the low natural abundance (~1%) of carbon-13. But, in contrast to ¹H spectra, ¹³C resonances are much more dispersed, having a wider spectral range (~200 ppm). These spectra, usually recorded under ¹H decoupling conditions, consist of singlets only. The much sharper singlet signals also show less overlap. Furthermore, ¹³C chemical shifts values of a monosaccharide unit in an oligosaccharide usually do not differ much (Δ <0.2 ppm) from the corresponding values in the free monosaccharide, with the exception of the carbon at a linked position, shifting at least 5–8 ppm downfield, with a small upfield displacement of the signals of the adjacent carbon atoms. Also, O-alkylation (Δ 7–10 ppm) and O-acylation (Δ 2–3 ppm) move the ¹³C chemical shifts downfield. In addition, the next neighboring carbon in the ring slightly moves upfield (Δ 0.5–1.5 ppm). The chemical shift of carbon in an α-linkage (δ ~97–101 ppm) is, in general, less than that of an anomeric carbon in a β-linkage (δ ~103–105 ppm).

In principle, it can be said: (1) anomeric carbon atoms in pyranoses and furanoses resonate at lowest field (90–110 ppm), (2) carbon atoms bearing primary hydroxyl groups appear at highest field (60–64 ppm), (3) carbon atoms having secondary hydroxyl groups give signals at 65–85 ppm, and (4) signals of alkoxylate carbon atoms (including C5 in penta-pyranoses and C4 in furanoses) are shifted 5–10 ppm downfield when compared to the corresponding hydroxy-substituted carbon atoms. Also, for ¹³C, carbons of furanose sugars often resonate at lower field compared to those of the pyranose forms. For 6-deoxysugars, the -CH₃ signals appear at 15–20 ppm high-field. Typical chemical shifts of notable resonances observed in ¹³C NMR spectra of carbohydrates are illustrated in Fig. 12.3.

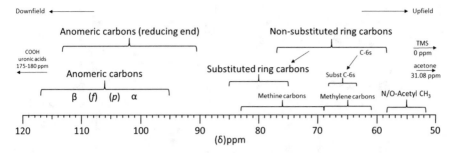

Fig. 12.3 Predictable regions of chemical shifts of carbohydrate carbons in a ^{13}C NMR spectrum

12.4 NMR Spectroscopy for Glycans

For studies of the primary structure of N- and O-linked glycans by NMR, intact glycoproteins are not prudent because of molecular mass, (micro)heterogeneity and the possible occurrence of multiple glycosylation sites. So, partial compounds like glycopeptides, oligosaccharides, or oligosaccharide-alditols, obtained via enzymatic and/or chemical degradation protocols, followed by rigorous purification, are more suitable. Fluorescent- or UV-active labels do not substantially interfere in the NMR spectra of the glycans.

Nowadays, single-pulse, one-dimensional (1D) ^{1}H, and ^{13}C NMR spectroscopy are often used as a "fingerprinting" technique because there is the possibility of comparing the spectra with an overwhelming library of reference compounds. Proton spectra recorded with NMR instruments operating at a minimal 600 MHz contain sufficient details to be used as "identity cards" of a carbohydrate. Given the high sensitivity of proton chemical shifts to their local environment, the ^{1}H NMR spectrum of an oligosaccharide with a particular sequence is essentially unique. Already, a first glance at the ^{1}H spectrum gives a good estimate of how many residues there are (one anomeric proton per residue) and the anomeric type (α/β) of each residue. The profile of ^{1}H-signals in the 1D NMR spectrum also offers a degree of purity.

As an example, a simplified ^{1}H spectrum (the bulk region of ring protons between 3 and 4 ppm is not shown) of an N-glycan is depicted in Fig. 12.4.

The typical monosaccharides in N-glycans have characteristic chemical shift values. These chemical shifts (resonance positions) and coupling patterns of the structural reporter groups are translated into structural information based on a comparison to patterns in a library of relevant reference compounds. This means that linkage and sequence information may be derived from ^{1}H and ^{13}C chemical shifts in cases where these values have been reported for specific linkages. These chemical shift data of many N- and O-glycans, compiled in tables, are found in the early literature [11–13], but is, nowadays, also easily accessible through online NMR databases. See the www.glycosciences.de web portal for instance (see also Chap. 13). Nevertheless, a certain level of expertise is required for interpreting NMR spectra of different types of carbohydrate structures. For excellent introductions to the

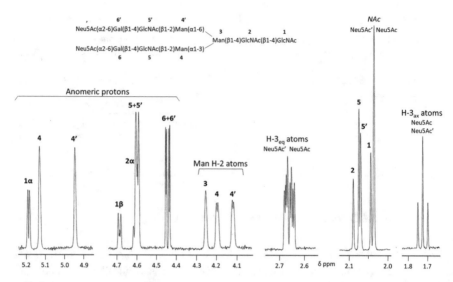

Fig. 12.4 ¹H NMR spectrum showing the structural-reporter group regions of a di-antennary-*N*-acetyllactosamine-type N-glycan. The bulk region has been omitted. The scale of the NAc region differs from that of the rest of the spectrum. For the spectral interpretation and assignments of resonances, see Sect. 12.4.2

structural determination of carbohydrate by NMR spectroscopy, the reader is advised to consult [1, 14–16].

When the 1D ¹H spectrum does not resemble that of a known oligosaccharide structure, the combination of multiple-pulse ¹H NMR techniques may be applied for de novo sequencing of the carbohydrate. Two-dimensional (2D) spectra contain information, spread into two dimensions displaying signal intensity as a function of two independent frequencies, effecting a dramatic improvement in resolution to assign overlapping signals. It is outside the scope of this book to describe the technical details of the different pulse programs of the various methods available for recording these 2D spectra. At present, 2D NMR belongs to the standard package for the NMR spectroscopist. For a detailed explanation of these NMR techniques for carbohydrates, the reader is referred to specialized books and appropriate articles [17, 18].

12.4.1   Two-Dimensional NMR Spectroscopy

In brief, homonuclear correlation spectroscopy (¹H-¹H COSY and TOCSY) is a way to assign the ¹H signals within an individual monosaccharide by providing correlations (cross-peaks) between spins in a coupled system (i.e., proton/proton signals) to identify the monosaccharide residues.

The 2D spectrum has a common layout that the diagonal corresponds to the conventional 1D spectrum, while the off-diagonal cross-peaks contain information about the connections between resonances on the diagonal. Usually, proton resonance assignments begin with the well-resolved anomeric proton. Magnetization is transferred around the glycosidic ring through scalar couplings of protons until it is disrupted by adjacent C-H with small coupling constants (two vicinal protons), for example, the $^3J_{H4,H5}$ (~2 Hz) in a D-galactose residue. By tuning the duration of the TOCSY spin lock mixing time (20–200 ms) in subsequent experiments, the complete homonuclear scalar coupling pathway (H-1 to H-6s) can be achieved. In other words, the determination of the inter-ring connectivities by their cross-peaks gives the chemical shift values of the protons H-1, H-2, H-3, H-4, H-5, and H-6s. So, TOCSY permits sub-spectral editing for each constituting glycosyl residue and, consequently, the complete assignment of nearly all the multiplet patterns in the region between 3 and 4 ppm (bulk region).

Once the individual resonances have been assigned to the specific monosaccharide residues, then protons, in close proximity (i.e., typically less than 4 Å), which are not physically linked, can be detected by 2D nuclear Overhauser effect spectroscopy (NOESY) via dipolar through space interactions (called long-range correlations, cross-relaxation). For instance, the correlation between the anomeric proton of one residue and the proton of the ring carbon of another residue (inter-residue NOE) may indicate the involvement of a glycosidic linkage. By this, the sequence of the glycosyl residues can be inferred, including the positions of the glycosidic linkages. Notably, the inter-residue NOE connectivities depend upon the solution conformation of the oligosaccharide. It should be noted that also intraresidue NOEs are observed (for instance, cross-peaks H-1 to H-3 and H-1 to H-5 in a β-glycopyranosyl residue or H-1 to H-2 in an α-glycopyranosyl residue). Equally, the same data can be obtained by Rotating frame Overhauser effect spectroscopy (ROESY), preferably used for small oligosaccharides (DP <10, 1–3 kDa). Depending on the size of the molecule, mixing times from 100 to 800 ms are used for these experiments.

When ^1H and ^{13}C NMR are used together, NMR spectroscopy demonstrates its real power in the de novo delineation of carbohydrate structures from the first principles. These techniques are the most informative NMR experiments for carbohydrate molecules. A heteronuclear ^1H-^{13}C single quantum coherence (^1H-^{13}C HSQC) experiment shows signals as single-bond correlations between ^{13}C carbon and directly attached protons [$^1J_{H1-C1}$, $^1J_{H2-C2}$, etc.], enabling the assignments of the carbon resonances. Ring carbons that are substituted by glycosylation are recognized by downfield shifts (Δ ~5–10 ppm) (Fig. 12.3). Heteronuclear multiple-bond correlation (HMBC) spectroscopy detects long-range heteronuclear connectivities (^{13}C-^1H through-bond correlations), for example, between the anomeric proton and the carbon atom on the opposite side of the glycosidic linkage as well as the anomeric carbon and the proton of the adjacent sugar residue. Hence, both intra- and inter-residue correlations can be seen. In the case that a glycan is phosphorylated, ^1H-^{31}P heteronuclear experiments are performed for the assignment of ^{31}P resonances.

With the advanced 2D NMR techniques available today, it is possible to decipher the structures of very complex carbohydrates [19]. Moreover, 2D homo- or heteronuclear experiments, such as COSY, NOESY, and HSQC, can be combined into 3D or higher experiments for carbohydrates to allow the resolution of the many overlapping and/or closely resonating signals. Furthermore, high-resolution magic-angle (54°44/ relative to the external magnetic field) spinning NMR (HR-MAS NMR) spectroscopy is a special technique to reduce line width, which can be used to examine protein gly-cosylation from intact, whole cells (i.e., cell glycans with sufficient slow relaxation). TOCSY, NOESY, and HSQC can also be performed by MAS-NMR. For instance, capsular polysaccharides are often well suited to be studied by HR-MAS NMR because they are usually long-chain rigid carbohydrates. However, since the mobility of cell-associated glycans vary and is difficult to predict, here, the MAS method can require trial and error of each system in the initial stages. Additional to the above-dis-cussed liquid-state NMR techniques, which provide fine chemical and structural infor-mation on soluble carbohydrates, there exist high-resolution solid state NMR spectroscopic techniques applicable for large polymers as well as whole cells [20]. For a detailed explanation, readers are suggested to read the earlier referred NMR books.

12.4.2   Spectral Interpretation and Assignments

As mentioned, the NMR structural-reporter group concept has the potential for structural characterization of a carbohydrate chain without assigning all ^{1}H-NMR signals, but the complete structural elucidation of an unknown glycan requires the full assignment of both the ^{1}H and the ^{13}C resonances of the constituent monosac-charide residues of the oligosaccharide in the NMR spectra. Total assignment can be accomplished by a combination of 2D NMR techniques as discussed earlier.

In an aqueous solution (D$_2$O), there is a tautomeric equilibrium between the α- and the β-anomer (mutarotation) of a free monosaccharide. Also, the monosaccha-ride at the reducing end of the oligosaccharide shows this effect. The two anomeric protons exhibit downfield resonances in the spectrum, corresponding to the α and β anomers, well separated from the group of other (ring) protons. The anomeric configurations may be determined using the vicinal coupling constant between H-1 and H-2. The anomeric proton signals (doublets) of most monosaccharide residues (having an axial H-2) resonate in the low-field region of the ^{1}H NMR spectrum, typically at 4.5–4.9 ppm for β-anomeric protons, having coupling constants $^3J_{1,2}$ of 6–8 Hz because of a *trans*-diaxial coupling (consult NMR textbook for detailed explanation). Under the same conditions, α-anomeric protons are found at 4.7–5.6 ppm, having coupling constants $^3J_{1,2}$ of 1–4 Hz (due to axial-equatorial coupling). However, for mannose (having an equatorial H-2), the β- and α-anomeric protons resonate close together at δ ~4.7 ($^3J_{1,2}$ < 1 Hz) and at δ ~5.1 ($^3J_{1,2}$ ~ 1.2 Hz), respectively. Small couplings ($^3J_{1,2}$ 3–4 Hz) are also found for equatorial/equatorial protons. It is of interest to note that the anomeric protons of furanose sugars often resonate at a lower field compared to those of the pyranose forms, and equatorial ring protons generally resonate at a lower field than their axial counterparts.

Usually, α-anomeric proton signals resonate more downfield than β-ones in the majority of the D-sugar units, having $^4C_1$ conformation, and α-anomers usually exhibit smaller splitting of the characteristic doublet signal (coupling constants) than β-anomers. Hence, the anomeric configurations of the constituent residues are determined using the vicinal coupling constants between H-1 and H-2, but also by one bond ^{13}C-^{1}H coupling constants, in particular for pyranoses. Reducing anomeric α-carbons usually resonate in the region 91–96 ppm, while β-anomeric carbons resonate at 94–98 ppm. The number of monosaccharide residues may be determined by integrating the anomeric resonance peaks in the spectrum.

As remarked before, the proton spectra of carbohydrates are often complicated by the overlap of several proton resonances, particularly in the case of oligosaccharides. The use of heteronuclear 2D NMR techniques, for example, ^{13}C coupled to ^{1}H, becomes valuable and complementary to assign all chemical shifts. As already mentioned, the proton and carbon chemical shifts are sensitive to the attachment of a noncarbohydrate group (methyl, acetyl, sulfate, or phosphate), giving a downfield shift of ~0.2–0.5 ppm for protons and even higher shifts for carbons, where the group is located. It is increasingly necessary to use good software programs to handle complex NMR data.

Substitution of ring carbons (e.g., with another sugar unit) is usually clearly indicated by the downfield shift of the ^{13}C chemical shift of the concerning carbon to values of δ 77–82 ppm. Furthermore, the position of substituent groups (acetate, methyl, sulfate, or phosphate) affects the resonances of neighboring proton and carbon chemical shifts. The absence of ^{13}C signals of nonanomeric ring carbons at a lower field than δ 82 ppm indicates that the monosaccharides are in pyranose forms. For furanosides, the ring carbons can have ^{13}C chemical shifts >82 ppm (Table 12.2).

The assignments of the ^{1}H and ^{13}C chemical shifts are usually listed in a table, as shown in Tables 12.1 and 12.2. As an example, Table 12.4 lists the significant ^{1}H chemical shift values for a complex-type N-glycan, an oligomannose-type N-glycan, and a typical O-glycan, having the following structures:

Complex-type N-glycan:

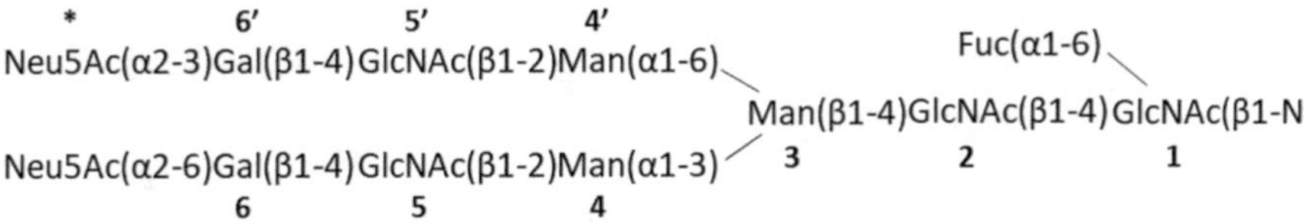

```
   *          6'          5'          4'
Neu5Ac(α2-3)Gal(β1-4)GlcNAc(β1-2)Man(α1-6)              Fuc(α1-6)
                                            \                      \
                                             Man(β1-4)GlcNAc(β1-4)GlcNAc(β1-N
Neu5Ac(α2-6)Gal(β1-4)GlcNAc(β1-2)Man(α1-3)/    3          2          1
              6          5          4
```

Oligomannose-type N-glycan:

```
D3         B
Man(α1-2)Man(α1-6)     4'
                  \
D2         A       Man(α1-6)
Man(α1-2)Man(α1-3)/          \
                              Man(β1-4)GlcNAc(β1-4)GlcNAc(β1-N
                             /   3          2          1
Man(α1-2)Man(α1-2)Man(α1-3)/
   D1         C         4
```

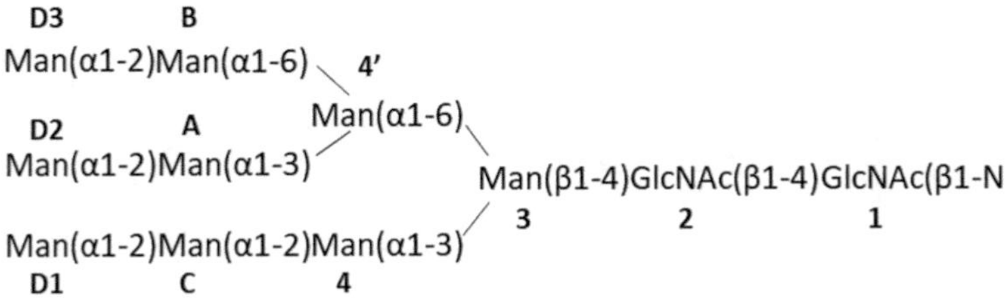

Table 12.4 ^{1}H NMR chemical shifts of structural-reporter groups of the constituent monosaccharides of two N-glycans and an O-glycan

Reporter group	N-glycan Complex-type		N-glycan Oligomannose-type		O-glycan	
	Residue	δ (ppm)	Residue	δ (ppm)	Residue	δ (ppm)
H-1 of	GlcNAc-1	5.07	GlcNAc-1	5.09	GalNAc-ol	
	GlcNAc-2	4.68	GlcNAc-2	4.61	H-2	4.38
	Man-3	4.77	Man-3	4.77	H-3	4.07
	Man-4	5.13	Man-4	5.34	H-4	3.53
	Man-4′	4.92	Man-4′	4.87	H-5	4.24
	GlcNAc-5	4.61	Man-A	5.40	H-6	3.48
	GlcNAc-5′	4.58	Man-B	5.14	NAc	2.04
	Gal-6	4.55	Man-C	5.31		
	Gal-6′	4.46	Man-D1	5.05	Gal	
			Man-D2	5.06	H-1	4.54
			Man-D3	5.04	H-2	3.60
					H-3	4.12
H-2 of	Man-3	4.25	Man-3	4.23	H-4	3.85
	Man-4	4.20	Man-4	4.11	H-5	3.74
	Man-4′	4.11	Man-4'	4.15	H-6'	3.80
			Man-A	4.10		
			Man-B	4.02	Neu5Ac^3	
			Man-C	4.11	H-3ax	1.80
			Man-D1	4.07	H-3eq	2.77
			Man-D2	4.07	NAc	2.03
			Man-D3	4.07		
					Neu5Ac^6	
H-3 of	Gal-6	<4			H-3ax	1.70
	Gal-6′	<4			H-3eq	2.73
					NAc	2.03
H-3ax	Neu5Ac*	1.80				
	Neu5Ac	1.72				
H-3eq	Neu5Ac*	2.76				
	Neu5Ac	2.67				
H-1 of	Fuc	4.88				
H-5 of	Fuc	4.12				
CH$_3$ of	Fuc	1.21				
NAc of	GlcNAc-1	2.014	GlcNAc-1	2.011		
	GlcNAc-2	2.094	GlcNAc-2	2.065		
	GlcNAc-5	2.068				
	GlcNAc-5′	2.045				
	Neu5Ac*	2.032				
	Neu5Ac	2.030				

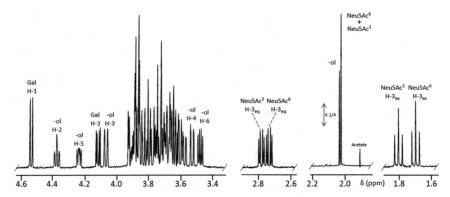

Fig. 12.5 ¹H NMR spectrum of an O-glycan as alditol. The scale of the NAc proton region (1.9–2.2 ppm) differs from that of the rest of the spectrum

Remember, as the result of D_2O exchange, only the CH protons are detected. It should be mentioned that the chemical shift values are not absolute and can differ in the last decimal for other similar structures, such as in tri- and tetra-antennary N-glycans containing substituents, like fucose influencing the neighboring residues. This also holds for similar O-glycans differently elongated. The tri-, tri'-, and tetra-antennary structures will show new anomeric signals and characteristic patterns. The chemical shift data for the primary structure characterization of the glycan part of small glycopeptides is basically identical to that of free glycans.

The oligomannose-type structures, which can contain three to nine mannose residues (or more), are identified by the patterns and relative intensities of the Man H-1 and H-2 signals. For the complex-type N-glycan, the Man H-1 and in particular the H-2 signals reflect the number of antennae (di- to penta-) and the attachment positions. The CH_3 signals from the GlcNAc *N*-acetyl groups are reporters of the type and position of substituents on the individual GlcNAc residues. For sialic acids, having no real anomeric proton, the H-3$_{axial}$ and H-3$_{equatorial}$ chemical shifts, together with the NAc signal, are typical for the type and position of sialic acid residues [21]. The assignment of glycan structures based on NMR data has been facilitated by the characteristic NMR chemical shifts and coupling constants of the various glycans present in modern online databases (see Chap. 13).

Based on both ¹H and ¹³C chemical shift values for most monosaccharides, an assignment of the individual residues in a glycan can be made. It is often sufficient to inspect only certain regions of a spectrum to ascertain the primary structure of a common carbohydrate structure. The structural-reporter group concept has proven its usefulness for the identification of numerous carbohydrates, in particular, the glycoconjugate glycans that form an ensemble of closely related compounds. Accurate calibration of the experimental conditions, like sample concentrations, measuring temperature, solvent pH, is a prerequisite. Obviously, monosaccharide composition and branching information established through chemical means (GC-MS) will contribute to the interpretation of the NMR spectra, particularly when dealing with novel carbohydrate structures.

NMR spectroscopy of glycolipids (>100µg) is usually done with 2% D_2O in DMSO-d_6 or pyridine-d_5. During high-resolution 500 or 600 MHz ^{1}H NMR analysis of native glycolipid, all the ^{1}H signals from the ceramide component can be assigned while the proton signals from the oligosaccharide component do not overlap with the ceramide signals. The anomeric proton of the glucose residue in Glc(β1–1)Cer is found at 4.10–4.17 ppm, while those of Gal(α1–4) and Gal(β1–4) are at δ 4.84 and δ 4.27 ppm, respectively. Likewise, GalNAc(α1–3) and GalNAc(β1–3) at δ 4.74 and δ 4.56 ppm, respectively.

12.5   NMR Spectroscopy of Polysaccharides

NMR spectroscopy is one of the most useful techniques for the determination of the primary structure of polysaccharides. Polysaccharides often consist of mixtures of complex compounds differing in molecular mass and built up from irregular elements that are unevenly distributed over the chain (see Sect. 2.3). Nevertheless, NMR spectroscopy can provide detailed structural information, including identification of the monosaccharide composition, determination of α- and/or β-anomeric configurations, and the establishment of linkage patterns, and sequences of the sugar residues [10]. However, a preliminary monosaccharide and linkage analysis (Sects. 6.2 and 6.6) helps a lot in interpreting the NMR spectra.

For NMR spectroscopic analysis, the polysaccharide is preferably dissolved in deuterium oxide (D_2O) after a full exchange of the hydroxyl groups (–OH) into – OD groups. Since polysaccharides often have high molecular mass, solubility in D_2O is sometimes problematic. In that case, a preliminary very mild hydrolysis (0.5 M TFA for 5 min at 100 °C), which slightly lowers the molecular mass, will help. This is also beneficial when NMR resonances are broad. Furthermore, higher measuring temperatures are helpful in producing better-resolved spectra. Computer mathematical manipulations are also used that artificially narrow the resonances and optimize the appearance of a spectrum.

Every polysaccharide gives a unique ^{1}H and ^{13}C NMR spectrum, but the broad signals are frequently crowded in a narrow region, especially for the ^{1}H spectrum mostly between 3 and 5 ppm. For the total assignment of the chemical shifts of the constituting monosaccharide residues, 2D NMR experiments (COSY, TOCSY, HMBC, HSQC, NOESY, ROESY) are necessary. Moreover, for the complete elucidation of the structure of a complex polysaccharide, NMR and MS studies on oligosaccharide fragments obtained from partial hydrolysis and/or specific degradation methods, together with chemical technologies (monosaccharide and methylation analysis by GLC-MS), are unavoidable to arrive at the insight of a reliable polymeric structure. Fortunately, polysaccharides can generally be isolated in sufficient amounts for NMR and additional experiments.

Most bacterial polysaccharides are heteropolysaccharides built up from regular repeating units. Sometimes, a large polysaccharide can give a relatively simple NMR spectrum, which appears like the spectrum of the repeating oligosaccharide unit. But as said, typically, the structural determination of a polysaccharide by NMR must be combined with chemical methods and mass spectrometry techniques. By applying partial hydrolysis, the repeating unit can often be isolated as an oligosaccharide, which structure can "easily" be elucidated [22]. As examples, the ^{1}H NMR spectra of two heteropolysaccharides with repeating units are shown in Figs. 12.6 and 12.7. The spectrum of the extracellular polysaccharide (EPS) from a

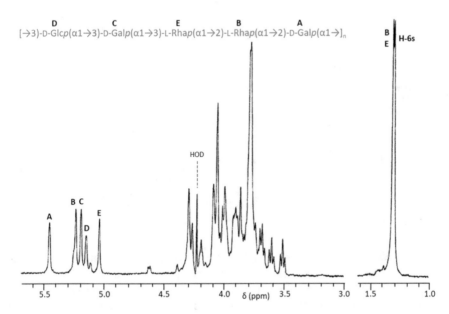

Fig. 12.6 ^{1}H NMR spectrum of EPS produced by *Lactobacillus delbreuckii* ssp. *Bulgaricus* LBB. B332, recorded in D$_2$O at 350 K [23]

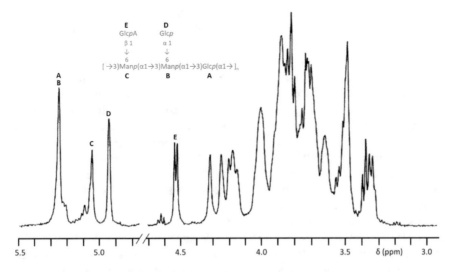

Fig. 12.7 ^{1}H NMR spectrum of EPS produced by *Propionibacterium freudenreichii* 109 recorded in D$_2$O at 300 K. HOD signal at δ 4.76 has been omitted [24]

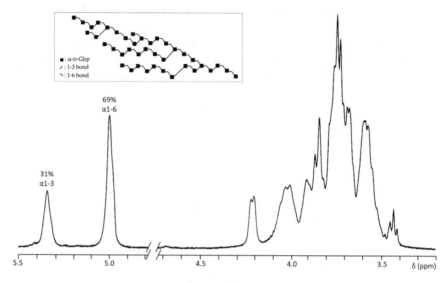

Fig. 12.8 ¹H NMR spectrum of an α-glucan EPS produced by *Lactobacillus reuteri* 180 recorded in D₂O at 300 K. HOD signal at δ 4.76 has been omitted [25]

Lactobacillus species demonstrates that an elevated measuring temperature (350 K instead of 300 K) can produce relatively sharp peaks. Notably, the HOD signal shifted from δ 4.76 to δ 4.23 ppm. The EPS from a *Propionibacterium* species shows the typical broad resonance signals. However, 2D homonuclear correlation NMR experiments (TOCSY) could assign the chemical shifts of the overlapping proton signals, while the complete assignment of ¹³C-resonances was achieved by ¹H-¹³C heteronuclear correlation (HSQC) spectroscopy. Long-range correlation techniques, such as NOESY and HMBC, provided sequence information of the polysaccharides. Together with monosaccharide and methylation analyses, the depicted structures of the polysaccharides were elucidated.

Polysaccharides without repeating units are more difficult to structurally analyze since the intact structures can be built up in various ways, ranging from a random distribution of the monomers to organized domains. In this case, the preparation of small fragments through degradation is usually necessary. Then, the structure determination of isolated fragments yields information that might allow deduction of the structure of the polymer. Figure 12.8 shows the ¹H NMR spectrum of an α-glucan (a polysaccharide built up only from α-glucose residues) produced by glucansucrase enzymes from a lactic acid bacterium. NMR spectroscopic investigation of its oligosaccharide fragments, ranging from di- to hepta-saccharides, led to the representation of the α-glucan polymer structure.

A structural-reporter group concept for the primary structural characterization of α-glucans was developed, featuring the constituting monosaccharides, their conformations, and their substituting patterns, after an NMR study of the proton chemical shift patterns of known α-D-glucopyranose di- and tri-saccharide structures [26].

In the ¹H NMR spectrum (Fig. 12.8), the αH-1 protons show downfield signals, well separated from the group of other protons. The signals for the anomeric protons of (α1–3)- and (α1–6)-linked glucosyl units (δ 5.35 and δ 4.97, respectively) are clearly separated, but the overlap of (α1–3)- with (α1–4)-linked glucosyl units [δ 5.39] is possible. In that case, the more powerful 2D NMR experiments contribute to the structural elucidation. Table 12.5 shows the ¹H and ¹³C chemical shifts of all possible substitution patterns of glucose residues in α-glucans. Very small deviations in the chemical shifts can occur due to spatial organization, causing steric hindrance, hydrogen bonding, and proximity effects. In an aqueous solution, for free glucose, there is a tautomeric equilibrium between α-Glc and β-Glc (see Fig. 12.1). A doublet (³J 4Hz), due to axial-equatorial coupling, is observed at δ 5.22 ppm for α-ᴅ-glucopyranose. When the glucose residue is in a terminal position and linked via different linkages, this results in a variation of δ H-1 and some ring protons as indicated (yellow) in Table 12.5. Furthermore, substitution at different

Table 12.5 ¹H and ¹³C chemical shifts of (substituted) glucose residues in α-glucans

	Glcp α		Glcp β	
H-1/C-1	5.22	93.0	4.64	96.7
H-2/C-2	3.54	72.5	3.24	75.1
H-3/C-3	3.72	73.8	3.49	76.7
H-4/C-4	3.42	70.6	3.41	70.6
H-5/C-5	3.84	72.3	3.46	76.8
H-6a/C-1	3.84	61.7	3.89	61.7
H-6b/C-1	3.75		3.72	

	Glcp α		Glc(α1→2)		Glc(α1→3)		Glc(α1→4)		Glc(α1→6)	
H-1/C-1	5.22	93.0	5.10	97.3	5.33	100.0	5.40	100.3	4.96	99.3
H-2/C-2	3.54	72.5	3.56	72.3	3.58	72.6	3.59	72.6	3.56	72.3
H-3/C-3	3.72	73.8	3.78	73.7	3.75	73.9	3.68	73.8	3.73	73.9
H-4/C-4	3.42	70.6	3.45	70.4	3.45	70.6	3.42	70.3	3.43	70.5
H-5/C-5	3.84	72.3	3.95	73.0	4.00	72.8	3.73	73.7	3.74	72.8
H-6a/C-6	3.84	61.7	3.84	61.4	3.86	61.5	3.85	61.5	3.85	61.6
H-6b	3.75		3.78		3.79		3.76		3.77	

	Glc(α1→2)		→2)Glc(α1→2)		→3)Glc(α1→2)		→4)Glc(α1→2)		→6)Glc(α1→2)	
H-1/C-1	5.10	97.3	5.27	94.8	5.08	97.7	5.08	97.4	5.09	97.6
H-2/C-2	3.56	72.3	3.66	76.9	3.66	70.8	3.61	72.1	3.58	72.2
H-3/C-3	3.78	73.7	3.88	72.1	3.93	80.6	4.03	74.0	3.78	73.9
H-4/C-4	3.45	70.4	3.51	70.5	3.69	70.6	3.66	78.3	3.55	70.5
H-5/C-5	3.95	73.0	3.94	72.7	3.93	72.6	4.03	71.5	4.09	71.4
H-6a/C-6	3.84	61.4	3.86	61.5	3.83	61.3	3.88	61.4	3.98	66.8
H-6b	3.78		3.78		3.78		3.81		3.76	

	Glc(α1→3)		→2)Glc(α1→3)		→3)Glc(α1→3)		→4)Glc(α1→3)		→6)Glc(α1→3)		→3,6)Glc(α1→3)	
H-1/C-1	5.33	100.0	5.51	97.3	5.35	100.1	5.33	99.8	5.30	101.2	5.34	100.1
H-2/C-2	3.58	72.2	3.69	76.8	3.69	71.4	3.63	72.5	3.56	73.5	3.70	71.2
H-3/C-3	3.75	73.9	3.86	72.2	3.91	80.6	4.01	74.2	3.72	75.0	3.90	81.5
H-4/C-4	3.45	70.6	3.50	70.5	3.68	71.1	3.65	78.4	3.51	71.5	3.79	70.7
H-5/C-5	4.00	72.8	4.05	72.4	4.02	72.9	4.08	71.4	4.14	72.0	4.16	71.7
H-6a/C-6	3.86	61.5	3.85	61.5	3.86	61.7	3.90	61.4	3.95	67.0	3.99	66.8
H-6b	3.79		3.77		3.76		3.82		3.78		3.76	

	Glc(α1→4)		→2)Glc(α1→4)		→3)Glc(α1→4)		→4)Glc(α1→4)		→6)Glc(α1→4)		→3,6)Glc(α1→4)		→4,6)Glc(α1→4)	
H-1/C-1	5.40	100.3	5.55	97.9	5.36	100.8	5.39	100.7	5.38	100.6	5.37	100.8	5.36	100.5
H-2/C-2	3.59	72.6	3.69	77.4	3.69	71.3	3.63	72.5	3.63	72.5	3.71	71.3	3.65	72.6
H-3/C-3	3.68	73.8	3.80	72.3	3.84	81.3	3.96	74.3	3.70	74.1	3.84	81.5	4.00	74.4
H-4/C-4	3.42	70.3	3.49	70.4	3.68	70.6	3.65	77.9	3.49	70.4	3.78	70.6	3.65	78.2
H-5/C-5	3.73	73.7	3.75	73.5	3.75	73.4	3.85	72.2	3.94	71.8	3.92	71.9	4.02	72.1
H-6a/C-6	3.85	61.5	3.87	61.6	3.84	61.5	3.88	61.5	3.97	66.7	3.98	66.7	3.99	67.0
H-6b	3.76		3.78		3.77		3.82		3.74		3.75		3.85	

	Glc(α1→6)		→2)Glc(α1→6)		→3)Glc(α1→6)		→4)Glc(α1→6)		→6)Glc(α1→6)		→3,4)Glc(α1→6)		→3,6)Glc(α1→6)	
H-1/C-1	4.96	99.3	5.16	95.0	4.97	98.9	4.97	98.9	4.97	98.8	4.97	98.3	4.98	99.0
H-2/C-2	3.56	72.3	3.67	79.0	3.67	71.5	3.61	72.5	3.58	72.5	3.77	71.5	3.68	71.0
H-3/C-3	3.73	73.9	3.79	72.8	3.84	81.5	4.02	74.3	3.73	74.4	4.08	82.0	3.87	81.7
H-4/C-4	3.43	70.5	3.50	70.6	3.68	70.8	3.65	78.4	3.51	70.6	3.89	76.3	3.78	70.8
H-5/C-5	3.74	72.8	3.75	72.3	3.75	72.5	3.85	71.3	3.92	71.3	3.87	71.3	3.91	71.2
H-6a/C-6	3.85	61.6	3.87	61.5	3.84	61.5	3.88	61.5	3.98	66.7	3.92	61.7	3.98	66.7
H-6b	3.77		3.78		3.77		3.84		3.76		3.88		3.75	

positions influences the chemical shifts of protons, as indicated. Moreover, the position of substitution on a glucose residue is clearly indicated by the downfield ^{13}C shift of the involved carbon (grey in Table 12.5).

12.5.1 Summarized NMR Measuring Procedure

In summary, for de novo structure determination of carbohydrates, NMR is still of inestimable value. Sample preparation is a crucial step for NMR spectroscopy. Metal ions are conveniently removed from the sample by passing it through a small column of Calix (Bio-Rad). Buffers and salts should be removed in advance as much as possible. Oligosaccharides are usually desalted on a carbon column. Subsequently, samples are treated twice with deuterium oxide (D_2O) with intermediate lyophilization and then dissolved in D_2O (high quality, >99.96% D). At least a 500-MHz pulse-FT NMR spectrometer [using 500.13 MHz for ^{1}H and 125.75 MHz for ^{13}C]) is appropriate. Spectrometer setup procedures and pulse programs are usually described in manuals and information booklets from spectrometer manufacturers.

   The first step in the structural elucidation of an oligosaccharide by NMR is the recognition of resonance signals belonging to closed spin systems of the individual residues, and this can best be achieved with the aid of correlation spectroscopy (COSY, TOCSY). The protocols for different 2D NMR experiments are given in the manuals provided with the NMR instrument. Starting with a given anomeric proton, one can assign the other protons that belong to the same monosaccharide unit using cross-peaks showing magnetization transfer from H-1 to H-2, H-2 to H3, etc. Consequently, from their chemical shifts and coupling data, the structure of the constituent monosaccharides can be elucidated. The success of this method depends on the fact that protons of each monosaccharide constitute a separate spin system (J-network), characteristic of the type of monosaccharide and its anomeric configurations. Intraresidual NOE effects are useful in determining anomeric configurations, that is, for α-anomeric configuration (NOE between H-1 and H-2) and for β-anomeric configurations NOEs between H-1 and H-3 and between H-1 and H-5. Inter-residue contracts establish interglycosidic linkages and the position of appended groups. Glycosidic linkages are deduced by NOE cross-peaks. For instance, a cross-peak from H-1 of residue **A** to H-4 of residue **B** reveals a (1→4) linkage between **A** and **B**.

   Assignments of ^{13}C resonances can be accomplished via ^{1}H detected [^{1}H, ^{13}C] one-bond correlation experiments. ^{1}H resonance assignments achieved by homonuclear techniques can readily be correlated with ^{13}C frequency in HSQC experiments. Once the individual resonances have been assigned to specific sugar residues, then, sites of glycosidic linkage can be established by employing glycosylation-induced shifts for ^{1}H and ^{13}C resonances. These can be confirmed by means of ^{1}H-detected multiple-bond correlation spectra optimized for long-range coupling ($^2J_{CH}$ and $^3J_{CH}$) or by searching for NOEs between each of the anomeric protons and the relevant protons of the adjacent glycosidically-linked sugar residue in a two-dimensional NOE (NOESY, ROESY) spectrum, as mentioned earlier.

NMR chemical shifts are well suited for computational purposes. A chemical shift can often be assigned to a single atom in a given structure, depending on the local chemical environment of the atom. This is mainly determined by the type of bonds with directly adjacent atoms. Since chemical shifts in oligosaccharides only slightly differ from those in free monosaccharides in a predictable manner, the experimentally obtained shift list can be used to computationally estimate structures with similar chemical shifts. Nowadays, several web-based tools and databases are available, allowing the prediction of possible structures from 1D and 2D data sets by correlating chemical shifts with structural features (see Chap. 13). The prediction of proton and carbon chemical shifts can be obtained by a computer database tool CASPER (Computer-Assisted SPectrum Evaluation of Regular polysaccharides) (http://www.casper.organ.su.se/casper/) [27–29]. The program is able to perform a structural analysis of both linear and branched oligo- and polysaccharides using ^{1}H and ^{13}C chemical shift data and $^1J_{CH}$ or $^3J_{HH}$ scalar coupling constants. Simulation of spectra is also possible. From the input of the structure of an oligosaccharide, including composition, linkages, conformations, and branching, the expected ^{1}H and ^{13}C chemical shifts of the monosaccharides are provided. Another approach to estimating ^{1}H and ^{13}C NMR chemical shifts of glycans is GlyNest, available through a web interface [30].

References

1. Jiménez-Barbero J, Peters T, editors. NMR spectroscopy of glycoconjugates. Weinheim: Wiley-VCH; 2003.
2. Brisson J-R, Vinogradov E, McNally DJ, Khieu NH, Schoenhofen IC, Logan SM, et al. The application of NMR spectroscopy to functional glycomics. Methods Mol Biol. 2010;600:155–73.
3. Schubert M, Walczak MJ, Aebi M, Wider G. Posttranslational modifications of intact proteins detected by NMR spectroscopy: application to glycosylation. Angew Chem Int Ed. 2015;54:7096–100.
4. Fellenberg M, Coksezen A, Meyer B. Characterization of picomole amounts of oligosaccharides from glycoproteins by ^{1}H NMR spectroscopy. Angew Chem Int Ed. 2010;49:2630–3.
5. Kwan AH, Mobil M, Gooley PR, King GF, MacKay JP. Macromolecular NMR spectroscopy for the non-spectroscopist. FEBS J. 2011;278:687–703.
6. Friebolin H. Basic one- and two-dimensional NMR spectroscopy. Weinheim: Wiley-VCH; 2011.
7. Günther H. NMR spectroscopy: basic principles, concepts and applications in chemistry. Weinheim: Wiley-VCH; 2013.
8. Zerbe O, Jurt S. Applied NMR spectroscopy for chemists and life scientists. Weinheim: Wiley-VCH; 2013.
9. Hore PJ. Nuclear magnetic resonance. Oxford: Oxford University Press; 2015.
10. Vliegenthart JFG. Introduction to NMR spectroscopy of carbohydrates. ACS Symp Ser. 2006;930:1–19.
11. Vliegenthart JFG, Kamerling JP. ^{1}H NMR structural-reporter-group concepts in carbohydrate analysis. In: Kamerling JP, editor. Comprehensive glycoscience-from chemistry to systems biology. Amsterdam: Elsevier; 2007. p. 133–91.
12. Vliegenthart JFG, Dorland L, Van Halbeek H. High-resolution, ^{1}H-nuclear magnetic resonance spectroscopy as a tool in the structural analysis of carbohydrates related to glycoproteins. Adv Carbohydr Chem Biochem. 1983;41:209–374.

13. Kamerling JP, Vliegenthart JFG. High-resolution ¹H-nuclear magnetic resonance spectroscopy of oligosaccharide-alditols released from mucin-type O-glycoproteins. Biol Magn Res. 1992;10:1–194.
14. Duus JØ, Gotfredsen CH, Bock K. Carbohydrate structural determination by NMR spectroscopy: modern methods and limitations. Chem Rev. 2000;100:4589–614.
15. Pomin VH. Unraveling glycobiology by NMR spectroscopy. Intech; 2012.
16. Battistel MD, Azurmendi HF, Yu B, Freedberg DI. NMR of glycans: shedding new light on old problems. Prog Nucl Magn Reson Spectrosc. 2014;79:48–68.
17. Leeflang BR, Vliegenthart JFG. Glycoprotein analysis: using nuclear magnetic resonance. In: Meyers RA, editor. Encyclopedia of analytical chemistry. Hoboken: Wiley; 2000. p. 821–34.
18. Leeflang BR, Vliegenthart JFG. Glycoprotein analysis: using nuclear magnetic resonance. In: Encyclopedia of analytical chemistry. Wiley Online Library; 2012.
19. Kato K, Peters T, editors. NMR in glycoscience and glycotechnology. London: Royal Society of Chemistry; 2017.
20. Kang X, Zhao W, Dickwella Widanage MC, Kirui A, Ozdenvar U, Wang T. CCMRD: a solid-state NMR database for complex carbohydrates. J Biomol NMR. 2020;74:239–45.
21. Kamerling JP, Gerwig GJ. Structural analysis of naturally occurring sialic acids. In: Brockhausen I, editor. Glycobiology protocols, Methods in molecular biology, vol. 347. Totowa: Humana Press; 2006. p. 69–91.
22. Gerwig GJ, Dobruchowska JM, Shi T, Urashima T, Fukuda K, Kamerling JP. Structure determination of the exopolysaccharide of *Lactobacillus fermentum* TDS030603—a revision. Carbohydr Res. 2013;378:84–90.
23. Sánchez-Medina I, Gerwig GJ, Urshev ZL, Kamerling JP. Structural determination of a neutral exopolysaccharide produced by *Lactobacillus delbrueckii* ssp. *bulgaricus* LBB.B332. Carbohydr Res. 2007;342:2735–44.
24. Dobruchowska JM, Gerwig GJ, Babuchowski A, Kamerling JP. Structural studies on exopolysaccharides produced by three different propionibacteria strains. Carbohydr Res. 2008;343:726–45.
25. Van Leeuwen SS, Kralj S, van Geel-Schutten IH, Gerwig GJ, Dijkhuizen L, Kamerling JP. Structural analysis of the α-D-glucan (EPS180) produced by the *Lactobacillus reuteri* strain 180 glucansucrase GTF180 enzyme. Carbohydr Res. 2008;343:1237–50.
26. Van Leeuwen SS, Leeflang BR, Gerwig GJ, Kamerling JP. Development of a ¹H NMR structural-reporter-group concept for the primary structural characterization of α-D-glucans. Carbohydr Res. 2008;343:1114–9.
27. Jansson P-E, Stenutz R, Widmalm G. Sequence determination of oligosaccharides and regular polysaccharides using NMR spectroscopy and a novel web-based version of the computer program CASPER. Carbohydr Res. 2006;431:1003–10.
28. Lundborg M, Widmalm G. Structural analysis of glycans by NMR chemical shift prediction. Anal Chem. 2011;83:1514–7.
29. Fontana C, Li S, Yang Z, Widmalm G. Structural studies of the exopolysaccharide from *Lactobacillus plantarum* C88 using NMR spectroscopy and the program CASPER. Carbohydr Res. 2015;404:87–94.
30. Loss A, Stenutz R, Schwarzer E, Von der Lieth C-W. GlyNest and CASPER: two independent approaches to estimate ¹H and ¹³C NMR shifts of glycans available through a common web-interface. Nucleic Acids Res. 2006;34:W733–7.

Chapter 13
Glycobioinformatics

Abstract Carbohydrates, with their diverse structures, are involved in a myriad of biological processes. The enormous size of complex data obtained by glycomics studies has grown beyond manual interpretation. Advanced bioinformatic programs are required to capture the overwhelming information. This resulted in the development of several databases, collecting all discovered carbohydrate structures. In addition, programs and algorithms that assign glycan structures based on MS and NMR spectra have been developed. Nowadays, desktop tools are available to address standardized HPLC separation data to predict glycan structures. During the last three decades, different Research Institutes have initiated several glycobioinformatics systems. This chapter provides an historical overview of these initiatives, including their capacities and the URL access addresses at the moment of writing. Existing databases and tools are critically needed but their sustainability is far from being ensured.

Keywords Database · Genomics · Proteomics · CyberSpace · HPLC-MS · NMR · Glycoforum · GlycoPedia · MIRAGE · ExPASy

Modern scientific research has come to rely more and more on databases and computer-assisted tools. The data from studies concerning the involvement of carbohydrates in life sciences and biomedicine have been accumulated enormously in the last two decades and will keep growing steadily. A current challenge is to keep up and adapt to the increasing levels of data. This means that there is an urgent need for a comprehensive, multidisciplinary glycomics database. This must be a freely available, well-organized, regularly updated, database allowing carbohydrate structures to be defined, archived, searched and annotated, and equipped with bioinformatic tools for computational purposes. This bioinformatics platform must be cross-linked to other databases with related genomic and proteomic information [1, 2].

However, in comparison with genes (genomics) and proteins (proteomics), the development of glycan-related databases and glycobioinformatics sources is considerably lagging behind [3–6]. Significant information can be found in [7–9]. Furthermore, an excellent review "A Traveler's Guide to Complex Carbohydrates in the CyberSpace" is online at http://www.glycopedia.eu/IMG/pdf/.

© Springer Nature Switzerland AG 2021

G. J. Gerwig, *The Art of Carbohydrate Analysis*, Techniques in Life Science and Biomedicine for the Non-Expert, https://doi.org/10.1007/978-3-030-77791-3_13

Source: Shutterstock

In a wide sense, glycobioinformatics should include the field of computational and statistics related to diverse glycobiological phenomena and glycochemistry. The translation of the extensive carbohydrate structures, detected so far, into their biological functions in life science is not straightforward due to their biosynthetic intricacy and enormous structural complexity. At this moment, it is still not possible to accurately predict the structures of glycans that an organism can produce, notwithstanding the knowledge of that organism's genome or proteome. The integration of glycan structural data repositories and databases could be helpful to get more insight. One strategy in functional glycomics would be the comparison of the glycan repertoire found in normal tissue with that of diseased or treated tissue. Knowledge of the glycan synthetic pathways and the type of tissue/glycoprotein/cell from which the glycans originate can help to decipher many of the unknown features of these carbohydrate structures.

As we now begin to realize that the glycome of specific biological systems is relatively small and finite, a functional database for the identification of oligosaccharides would have considerable value. It is obvious that a repository will have to be established for glycomics research [10]. At this moment much glyco-related information can be found at Glycoforum (http://www.glycoforum.gr.jp) and at GlycoPedia (http://www.glycopedia.eu), which collects e-chapters on basic, intermediary, and advanced topics of glycoscience. Excellent material for educational purposes. Both sites compile various sources and links to carbohydrate databases and relevant readings to the scientific community are suggested. Additionally, the GlycoNAVI Database (http://www.glyconavi.org) is a support system for carbohydrate research and glycoscience. It provides glycoproteomics information, including glycan-related genes and three-dimensional (3D) structures of carbohydrates.

The problem with constructing a perfect carbohydrate database is how to deal with the structural complexity of the glycan structures. For genes and proteins, the relatively small constituent residues (4 nucleotides in DNA/RNA and 20 amino acids in proteins) can be summarized in a simple, linear, one-letter code, easy to digitize. However, the number of naturally occurring sugar residues for glycans is much larger, and each pair of monosaccharide residues can be linked in several ways. Moreover, one residue can be connected to three or four other residues (branching). It is evident that glycomics requires sophisticated algorithmic approaches.

In the scientific literature, several structural descriptions of carbohydrates are still in use, depending on whether you are dealing with medically or biologically oriented scientists or (bio)chemists. A symbolic representation of glycans by a series of colored geometric shapes (see Table 2.1 in Chap. 2) is widely used in publications, because they are easily recognized and allowed rapid comparison of structures, but this is not an ideal computer-readable format [11, 12]. A simple notation that covers both the variety of monosaccharides and the handling of linkage positions and branching is required.

For glyco/bioinformaticians, it is a major challenge to integrate the conceptually diverse biological and chemical information, stored in several different data repositories, and represented by various distinct formats into a universal digital format. The MIRAGE Commission (Minimum Information Required for A Glycomics Experiment) (http://www.beilstein-institut.de/en/projects/mirage), established in 2011, has developed guidelines and reporting standards for sample preparation, glycomic HPLC and CE, and glycan array data that need to be implemented in databases. Databases that provide authoritative information about glycan and glycoconjugate structures, including insight into the biological functions and consequences of glycosylation, need many different types of digital algorithms tools, ranging from basic visualization software to software that assists in the interpretation and structural annotation of glyco-analytical data (e.g., chromatography and arrays, NMR and mass spectra) and that identify correlations between glycosylation and other biological phenomena (e.g., diseases, gene expression, and cell differentiation) [9].

During the last three decades, several initiatives have been launched to develop bioinformatics systems, using mathematical and statistical algorithms and computational methodologies to analyze the data and to build glycobiology databases, in a way of libraries of carbohydrate structures, HPLC profiling and mapping methods, enzymatic reactions, glycan biomarker predictions, and mass and NMR spectra assignments. Fast success was hampered because no standard procedures existed to deposit glycan structures found in the various species, organs, tissues, or cells. This created that carbohydrate sequence information was spread in incompatible formats over several unconnected databases. Moreover, several databases, promising-started, have not been updated since the late 1990s, and only a few of them are curated.

This chapter of "The Art of Carbohydrate Analysis" gives a synopsis of the most important databases established during the last three decades. Researchers can access these databases with their Web Browser (i.e., Internet Explorer, FireFox, Safari, etc.). It is impossible to give all the protocols to use the different databases. In general, you have to sit down and find out the procedures of the system, which could take some time of mouse-clicking.

The CAZy Database (Carbohydrate-Active enZYmes) (http://www.cazy.org) is the oldest classification system, correlating the structure and molecular mechanism of carbohydrate-active enzymes (glycosidases, glycosyltransferases, polysaccharide lyases, and carbohydrate esterases) of various living organisms and provides the EC number and accession number for each gene. It describes structurally related catalytic and carbohydrate-binding modules (or functional domains) of enzymes that degrade, modify, or create glycosidic bonds. At the moment, it is the major public database related to carbohydrates and glycogenes (genes of glycosyltransferase, sugar-nucleotide synthase, sugar-nucleotide transporters, etc.). A cross-link to access 3D structures of the enzymes from the Protein Data Bank (PDB) (http://www.rcsb.org) is available [13, 14].

To store information on human glycogenes, the dedicated GlycoGene Database (GGDB) (http://www.acgg.asia/ggdb2/) was constructed at the Research Center for Medical Glycoscience in Ibaraki, Japan, in association with the Asian Community of Glycoscience and Glycotechnology. The properties, such as gene name, enzyme name, DNA sequence, tissue expression, substrate specificity, homologous gene, and EC number of over 150 glycogenes, are reported with links to other databases.

In contrast to proteins, there is no database containing an inclusive and closed set of sequences representing all possible carbohydrate structures. The Complex Carbohydrate Structures Database (CCSD), better known as CarbBank, was created in 1989 at the Complex Carbohydrate Research Center (CCRC), University of Georgia, USA. It was one of the first databases to collect published carbohydrate structures and distribute the structures on a CD-ROM. It contained a large collection of carbohydrate primary structures, including data about sources, analytical methods, and literature references. GLYDE-II (GLYcan Data Exchange), an XML-based format for glycans and glycoconjugates, developed by the CCRC was first agreed upon as the standard format for exchanging carbohydrate structure data [15].

When funding for CCSD was stopped in 1996, the data, more than 15,000 annotated natural glycans and glycoconjugates, were adopted by other databases, for instance, the database of KEGG (Kyoto Encyclopedia of Genes and Genomes) in Japan.

The KEGG Database (http://www.genome.jp/kegg/) is a web-based resource for understanding functions and utilities of the biological system, including pathways of biosynthesis and metabolism of complex carbohydrates [16]. It provides, among others, the resources: KEGG GLYCAN (http://www.genome.jp/kegg/glycan), a database containing >10,000 glycan structures, some of which are methylated, phosphorylated and sulfated, with a GIF image file for display and KEGG REACTION, the glycosyltransferase and the glycosidase reaction library. The KEGG PATHWAY database represents molecular interaction networks, including metabolic and regulatory biosynthetic pathways of carbohydrates, with links to around 100 glycan biosynthesis enzymes. Glyco-related genes are organized in KEGG BRITE. The KEGG Carbohydrate Matcher (KCaM) is a software tool for analyzing the structures of carbohydrate sugar chains [17].

Another Japanese database is RINGS (Resource for INformation of Glycomics at Soka) (http://www.rings.t.soka.ac.jp) which provides a tool to detect, characterize, and perform relative quantitation of N-glycopeptides based on LC-MS runs (glycan profiling MS data), including GlycoMiner [18]. RINGS software also has analytical tools for predicting glycan-binding patterns from glycan array data (GlycomeAtlas) and a drawing tool DrawRINGS (http://www.rings.t.soka.ac.jp/drawRINGS-js/) [19].

Also, at the same time, when funding for CarbBank was discontinued, in Europe, the SWEET-DB project was initiated to annotate and cross-reference existing carbohydrate-related data collections and make them available using modern web-based techniques. This finally led to the development of the GLYCOSCIENCES.de website (http://www.glycosciences.de/), which was hosted at the German Cancer Research Center (DKFZ). It addresses 3D structures of carbohydrates and provides a collection of databases and bioinformatic tools for glycobiology and glycomics in an excellent way. The structures and reference data from CarbBank were combined with NMR data, including several computational tools, using a specially adjusted notation for carbohydrate structures LINUCS (Linear Notation for Unique description of Carbohydrate Sequences). The database contains over 25,000 glycan structure entries of N-glycans, O-glycans, and glycolipids, over 3400 NMR spectra and mass spectrometry information, as well as >20,000 bibliographic references. Furthermore, it contains LiGraph (http://www.glycosciences.de/tools/LiGraph/), a tool for drawing oligosaccharides, however, with a limited number of monosaccharides. The main glycan structure database of the Glycosciences.de portal, now named "Glycosciences.DB" (http://www.glycosciences.de/database/), is one of the best at this moment [20]. For the analysis of glycoprotein sequences, there is the GlySeq web-interface (http://www.glycosciences.de/tools/glyseq/) providing a detailed statistical analysis of the frequency of amino acids found around N- and O-glycosylation sites based on protein sequences from SWISS-PROT and the Protein DataBank (PDB), and there is Glypeps (http://www.dkfz.de/spec/glypeps/) for glycoprotein detection.

To get insight into carbohydrate function–structure relationships, it is important to obtain, next to chemical composition, 3D information (spatial or tertiary structures) of the glycan molecules [21]. Within the conformational analysis of carbohydrates, computational methods play an important role. The results are often presented as the structure of the lowest energy conformation or as an adiabatic potential energy surface with the glycosidic torsion angles, φ and ψ, as axes. The extensive database GlycoMapsDB (http://www.glycosciences.de/modeling/glycomapsdb/) contains φ/ψ-maps of di- to pentasaccharides obtained by molecular dynamic (MD) simulations using the MM3 force field [22].

With SWEET-II (http://www.glycosciences.de/modeling/sweet2/doc/index.php), it is possible to perform Molecular Modelling (MM) calculations (MM3 force field) after manual input of a carbohydrate structure by its alphanumeric IUPAC nomenclature, allowing the generation of reliable 3D spatial structures [5]. An extended ensemble of glycan conformations is present in the Glydict database (http://www.glycosciences.de/modeling/glydict/), including a web tool for predicting N-glycan structures based on MD-simulations with the MM3 force-field.

GlyProt (http://www.glycosciences.de/modeling/glyprot/php/main.php) is a web-based tool that evaluates whether a potential N-glycosylation site is spatially accessible. It enables meaningful N-glycan conformations to be attached to potential N-glycosylation sites of a known three-dimensional (3D) protein structure (in-silico glycosylation of proteins). The aim is to provide rapid access to reliable 3D models of glycoproteins. GlyTorsion (http://www.glycosciences.de/tools/glytorsion/) detects, by input of a disaccharide, the torsion angles of the glycosidic linkages between the two sugar rings, which dominantly determine the 3D shape of the carbohydrate structure. To describe 3D structures of carbohydrates is not straightforward. Due to their inherent flexibility, oligosaccharides typically exist in solution or on proteins as an ensemble of configurations.

The 3D structures of polysaccharides established from various experimental methods (X-ray, neutron and electron diffraction, molecular modeling, and high-resolution NMR spectroscopy) have been incorporated into an annotated database, Polysac3DB [23]. Correspondingly, since 2017, there is EPS-DB (http://www.eps-database.com) giving 3D structure models.

The GLYCAM-Web (http://glycam.org) provides an online search interface, GlyFinder, for searching relevant carbohydrate-containing structures and tools for modeling 3D structures of oligosaccharides, glycans, glycosaminoglycans, and glycoproteins. It provides all files necessary for the user to perform molecular dynamics simulations of these systems with the AMBER software package. Additional tools facilitate the prediction of the specificity of glycan-binding proteins, such as lectins and antibodies, and provide theoretical 3D structures for these complexes. Glyco3D (http://glyco3d.cermav.cnrs.fr/home.php) is a single-entry portal to access three-dimensional features of (1) carbohydrate and carbohydrate polymers in different physical and biological conditions and (2) carbohydrate-binding proteins crystallized with their carbohydrate ligands (lectins, monoclonal antibodies against carbohydrates, and glycosyltransferases). Most data are structural data arising from diffraction experiments, but 3D models resulting from theoretical calculations are also included [24].

The US Consortium for Functional Glycomics (CFG) (http://www.functionalglycomics.org/), established in 2001, started to develop a glycan database, taking structures from CarbBank and the commercial GlycoMinds database. Furthermore, diverse data sets were derived from (1) gene expression of glycosyltransferases (enzyme database) and glycan-binding proteins (GBPs), (2) phenotypic analysis of transgenic mice, (3) mass spectrometric profiling of glycan structures isolated from cells and tissues, and (4) screening glycan affinity of proteins using glycan arrays. Thus, MS spectra and microarray profiling of glycans expressed in cells and tissues can be seen in this database. At this moment, it is the major public source for glycan array data. It is one of the largest databases and source for glycomic services now.

In 2005, EUROCarbDB (http://www.eurocarbdb.org) initiated an infrastructure for carbohydrate-related data (structure and function) [25]. The aim was to develop a collection of tools and workflows to assist the interpretation of HPLC, MS, and NMR experimental data. Tools were integrated for streamlining European glycomics research through the development of databases, bioinformatics standards, efficient analysis and search algorithms, and Web-based software components. It launched GlycoWorkBench (http://www.code.google.com/archive/p/glycoworkbench/downloads) or (http://download.glycoworkbench.org/) as a freely downloadable software tool for a rapid drawing of glycan structures and a tool to assist the interpretation of MS/MS analysis data by matching a theoretical list of fragment masses against the experimental peak list derived from a spectrum (Glyco-Peakfinder) [26, 27]. The GlycanBuilder tool, as a part of GlycoWorkBench (EUROCarbDB), is a software library and contains a set of tools to allow for the rapid drawing of glycan structures with support for all of the most common symbolic notation formats [27, 28]. Unfortunately, EUROCarbDB ceased operation in 2011. GlycoWorkBench is still active. Today, there is a newer version of GlycanBuilder, named GlycanBuilder2, which can be downloaded at http://www. rings.t.soka.ac.jp/downloads.html [29].

GlycanBuilder is also part of UniCarbKB (http://www.unicarbkb.org/builder) (see later). The tool uses the SNFG notation to display glycan structures and provides an excellent interface for glycan building. Additionally, SugarSketcher (http:// www.glycoproteome.expasy.org/sugarsketcher/) provides users with a simple and efficient interface for drawing glycan molecules with the logos defined in the "Symbol Nomenclature for Glycans" (SNFG) [11, 30]. For a recent overview of currently active drawing tools for glycans, the reader is referred to [31].

In 2008, a new carbohydrate metadatabase, called GlycomeDB (http://www. glycome-db.org/), was initiated to overcome the isolation of the carbohydrate structure databases and to create a comprehensive index of all available structures with references back to the original databases. To achieve this goal, most structures of the freely available databases were translated to the GlycoCT sequence format, which was developed as part of the EUROCarbDB project [32]. A freshly designed JAVA software application, called GlycoUpdateDB, downloads data from seven major public databases, including CarbBank, CFG, CSDB, GlycoBase (Dublin), GlycoBase (Lille), GLYCOSCIENCES.de, and KEGG. The encoded sequences and corresponding IDs from the source databases were stored in this new database GlycomeDB, which was weekly updated. All of the taxonomic annotations available from the various databases are gathered and harmonized. In this way, it was possible to get an overview of all glycan structures in the different databases using a single web interface and to cross-link common structures [33, 34]. In 2016, GlycomeDB was integrated into GlyTouCan, a new international glycan structure repository.

GlyTouCan (http://glytoucan.org/) is an uncurated registry for glycan structures that assigns globally unique accession numbers to any glycan independent of the level of information provided by the experimental method used to identify the structure(s). GlyTouCan records next to glycan structures also experimental data, and biological sources [35]. On the website "Glycobiodiversity" available from the University of Lille (USTL, France), there is a database GlycoBase (Lille) (http:// www.glycobase.univ-lille1.fr/base/), a glycan sequence database which includes taxonomy and NMR data and providing a link to GlyTouCan, containing GlycomeDB. The latter, as mentioned earlier, is a freely available glycan structure registry with associated accession numbers that can be used by diverse glycobiology data systems [36]. GlycanBuilder, an earlier mentioned tool that allows for the drawing of glycan structures (N- and O-glycans, glycosaminoglycans, glycosphingolipids, milk oligosaccharides), is a part of GlyTouCan.

Besides GlycoBase (Lille), there is the database GlycoBase 3.2 (Dublin) (http://
www.nibrt.ie/glycobase/) containing >700 N- and O-linked carbohydrate structures
and corresponding references. Furthermore, it contains U/HPLC (HILIC) chromatographic retention data (expressed in GU values) for >700 2-AB-labeled N- and
O-linked glycan structures of mammalian glycoproteins and provides exoglycosidase sequencing and mass spectrometry data [37]. GlycoExtractor is a web-based
tool for processing HPLC-glycan data [38]. GlycoStore (http://glycostore.org/) is a
curated chromatographic, electrophoretic, and mass-spectrometry database of N-,
O-, glycosphingolipid (GSL)-glycans and free oligosaccharides, associated with a
range of glycoproteins, glycolipids and biotherapeutics [39]. N-glycan structures
can be assigned based on their HILIC-UPLC retention time converted to glucose
units (GUs). The database is built on publicly available experimental data sets from
GlycoBase (NIBRT, Ireland).

In 2009, the UniCarbDB (http://unicarb-db.expasy.org) started as a platform for
presenting glycomics information, including glycan structures and mass fragment
data characterized by HPLC-MS/MS strategies [10, 40–42]. It contains structural
and fragmentation data of *N*- and *O*-glycans is currently one of the largest experimental glycomics MS databases, integrated in ExPASy portal.

From 2010, another database UniCarbKB (http://www.unicarbkb.org), where
KB stands for Knowledge Base, includes the data (literature-based curated glycan
structures, glycoprotein site/global information) previously available from
GlycoSuiteDB [41]. UniCarbKB has a link to the proteomics knowledgebase
UniProtKB (http://www.uniprotkb.org), which is a protein sequence database with
information on glycosylated proteins and is also cross-referenced to SWISS-PROT,
which is a curated protein sequence database, which contains nearly 1000 annotated
and verified glycosylated amino acid sequons. The previously mentioned
GlycanBuilder tool, that allows for the drawing of glycan structures, is part of
UniCarbKB. The UniCarbKB database (rebuilt as UniCarbKB 2.0 since 2019) also
provides a series of new user interfaces and functionalities and has been made
accessible and linked to protein data on the ExPASy proteomics server. The database Unipep (http://www.unipep.org) contains N-glycoproteins and glycopeptides.

A structure database, called UniCorn (http://www.unicarbkb.org), containing
theoretical structures of N-glycan sites, was built from a library of 50 human glycosyltransferases. The enzyme specificities were sourced from major online databases,
including Kyoto Encyclopedia of Genes and Genomes (KEGG) Glycan, Consortium
for Functional Glycomics (CFG), Carbohydrate-Active enZYmes (CAZy),
GlycoGene DataBase (GGDB), UniProtKB, and BRENDA (http://www.brenda-
enzymes.org/). Based on the known activities, a library of theoretically possible
N-glycan structures, composed of 15 or less monosaccharide residues, was generated and compared to those contained in UniCarbKB, a mentioned database that
stores experimentally described glycan structures (>3200) reported in the literature.
It demonstrates that UniCorn can be used to aid in the assignment of ambiguous
structures whilst also serving as a discovery database [43]. Later, the data repository
UniCarb-DR (http://www.unicarb-dr.biomedicine.gu.se/) was created, which serves

as temporary storage of experimental MS fragment data and structures before transition into the UniCarb-DB database [44].

At this moment, an important bioinformatics resource portal is ExPASy of the Swiss Institute of Bioinformatics (http://www.expasy.org/glycomics), having links to many databases and several links to online glycomics software tools [45]. GlyConnect (http://glyconnect.expasy.org) is a platform that provides a resource for the study of the relationships between glycans, the proteins that carry them (mainly glycoproteins with N- and O-linked glycans), the enzymes that synthesize or degrade them, and proteins that bind them [46]. The user can navigate from glycan structures to proteins and back while exploring possible associations between tissue expression and site-specific glycosylation. GlyConnect also integrates tools otherwise accessible individually in the glycomics collection of ExPASy. GlycoDigest, an integrated feature of UniCarbKB (http://unicarb.org/glycodigest/), gives exoglycosidase digest prediction. GlycoDigest is a tool that simulates the action of these exoglycosidases on released oligosaccharides. It assists glycobiologists in designing enzyme mixtures that can be used to guide the precise determination of glycan structures.

The Sugar Bind Database (http://www.sugarbind.expasy.org) contains literature references on lectin adhesins of viral and bacterial pathogens and biotoxins and their known human carbohydrate ligands (carbohydrate sequences).

MzJava (http://www.mzjava.expasy.org/) is an open-source Java library for the analysis of mass spectrometry data. It provides algorithms and data structures for processing mass spectra and their associated biological molecules, such as small molecules, glycans, proteins, and peptides with posttranslational modifications. MzJava includes methods to perform mass calculation, protein digestion, peptide and glycan fragmentation, MS/MS signal processing, and scoring for spectra-spectra and peptide/glycan-spectra matches.

GlycoMod (http://www.expasy.org/glycomod/) was the first online software tool to predict possible oligosaccharide (glycan) structures of glycopeptides/proteins from experimentally determined molecular-ion MS fragments, taking into account several derivatizations and reducing-end modifications. With another web-based program GlycanMass (http://www.expasy.org/glycanmass/), you can calculate the mass of an oligosaccharide structure from its composition.

Since mass spectrometry in combination with modern separation methodologies has become a powerful and versatile technique for structural analysis of glycans, many attempts have been made to develop algorithms for automatic interpretation of MS spectra of glycans [47]. New software tools to analyze glycoproteomics data, in a way of automated interpretation of mass spectrometry data, keep appearing, such as tools for data-mining in MS-based glycoproteomics [48, 49]. Web tools to support the interpretation of mass spectra of complex carbohydrates and glycopeptides are GlycoFragment, which, by input of a glycan, calculates theoretically possible mass fragments and GlycoSearchMS which finds, by input of a list of *m/z* values, glycans whose fragmentation pattern match best. Most of the software tools use databases of known glycan structures and compare the experimental tandem MS data to find a fit. Success is depending on the completeness of the database. GLYCH (http://www.omictools.com/glych-tool) was developed to predict any type of glycan structure from mass spectra, using a dynamic programming method and a listing of all possible fragments of glycans [50].

Another automated tool for the interpretation of MS spectra is the commercial SimGlycan™ (http://www.premierbiosoft.com). It predicts the structure of glycan/glycopeptides from LC-MS/MS data, matching them with a database of theoretical fragmentation of over 9000 glycans. Other information, such as glycan class, reaction, pathway, and enzymes, are also made available [51]. GlyPID (http://www.mendel.informatics.indiana.edu/) is another tool for the characterization of protein glycosylation from CID and HCD MS/MS data. Recently, GlycoDeNovo, an efficient algorithm for accurate de novo glycan topology reconstruction from tandem mass spectra, was developed [52].

The Japanese Consortium for Glycobiology and Glycotechnology (JCGG) explores a carbohydrate metadatabase (JCGGDB) (http://www.jcggdb.jp/). It contains search systems as GlycoChemExplorer and Keyword/Structure Search, tools for drawing carbohydrate/glycan structures, but also to find glycogene information, glycomics-related protocols, and cross-references. JCGGDB assembles GlycoEpitope, GlycoProtDB, CabosDB, L*f*DB (Lectin Frontier DataBase), SGCAL (containing 1500 unique carbohydrate structures), and GlycoPOD (GlycoScience Protocol Online Database).

GlycoEpitope (http://www.glycoepitope.jp/epitopes) is an integrated database of carbohydrate epitopes (antigens) and antibodies and is useful for glycobiologists and life science researchers. In this database, information has been assembled on a large number of poly- and monoclonal antibodies (>600) that have been used for analyzing the expression of various carbohydrate chains and their functions. GlycoEpitope provides information including lists of glycoproteins that express carbohydrate antigens, glycolipids of which part of the structure is a carbohydrate antigen, enzymes that take part in the synthesis and degradation of epitopes, the times and sites of expression of carbohydrate antigens, diseases to which carbohydrate epitopes are related, and suppliers from which carbohydrate-recognizing antibodies can be obtained.

The GlycoProtein DataBase (GlycoProtDB) (http://www.jcggdb.jp/rcmg/gpdb/) provides information on N-glycosylated proteins and their glycosylated site(s), in particular, from *C. elegans* tissues and mouse liver. The database is searchable using gene ID and gene name as query. In GlycoProtDB, glycoproteins and glycopeptides can be searched by their name, amino acid length, and molecular weight, giving information on glycosylation sites. The database can also be found at the National Institute of Advanced Industrial Science and Technology (AIST).

CabosDB (CArBOhydrate Sequencing DataBase) consists of an oligosaccharide database, a mass spectra database, a lectin affinity database (LADB), and a glycoprotein database. As said, it belongs to The Japanese Consortium for Glycobiology and Glycotechnology (JCGGDB) (http://www.jcggdb.jp/). The mass spectra database stores mass spectra and experimental conditions. The oligosaccharide database contains oligosaccharide structures and provides a web service for the rapid identification of oligosaccharide structures from mass spectra data. LADB contains >200 lectin molecules and provides comprehensive affinity information about lectins and glycans, a lectin map. The lectin molecules are organized by family, carbohydrate recognition domain (CRD), formal nomenclature, amino acid sequence, function, tertiary structures, and binding sites. LADB allows researchers to compare the characteristics of lectins of interest with structural features of their target glycans in a simultaneous manner, thereby giving insights into these molecular interactions.

The Lectin Frontier Database (L/DB) (http://www.acgg.asia/lfdb2/) is a resource of carbohydrate–protein interaction data. It provides detailed information on lectins and quantitative data of their interactions with glycans having specific structures, obtained by automated frontal affinity chromatography with fluorescence detection (FAC-FD). GlycoPOD stands for GlycoScience Protocol Online Database and contains a collection of protocols for performing experiments on glycans. It is a very useful database, providing step-by-step protocols for isolating and analyzing glycans. Relevant lectin information can also be obtained from the UniLectin3D database (http://www.unilectin.eu/unilectin3D/).

In JCGGDB Report, you will find a collection of mini-reviews about the latest information for research and applications in glycoscience. At this moment, the JCGG conducts 15 original databases in AIST (Advanced Industrial Science and Technology) and 6 cooperative databases in alliance. There among is GMDB (Glycan Mass DataBase) (http://www.jcggdb.jp/rcmg/glycodb/Ms_ResultSearch), which is a database of glycan mass spectral data providing a tool for use in glycomics research. It stores MS^n spectra of N- and O-linked glycans and glycolipid glycans. Glycan structures can be estimated through a comparison of MS(-MS) spectra (m/z values).

Tools for automated mass interpretation (LC-MS/MS) of glycopeptides are still scarce [53]. Software, such as Sweet-Heart, utilizes machine-learning algorithms to predict N-glycopeptides from HCD fragments [54]. GlycoFragWork [55] and the commercial Byonic™ (ProteinMetrics/Proteome Discoverer v2.0 Thermo Scientific) allow the tentative identification of glycopeptides [56]. In addition, there is GP-Quest for the assignment of tandem mass spectra [57].

The GRITS Toolbox (http://www.grits-toolbox.org/) is a freely available modular software suite, which includes modules for collecting, annotating, and comparing mass spectral data (MALDI or LC-MS/MS), manipulating the corresponding metadata, and generating reports. It can process different types of open file format MS data of released glycans with different chemical derivatives. Annotation is performed using the integration annotation module Glycomics Elucidation and Annotation Tool (GELATO) using Qrator software [58]. DANGO (http://www.ms-dango.org) is an extension of the GRITS Toolbox software that allows the (semi-)automated annotation of intact glycosphingolipid (GSL) MS data and protocols for glycolipid MS analysis. As a carbohydrate structure builder, the GRITS software uses GlycanBuilder (GlycoWorkBench).

Another important database is the Carbohydrate Structure Database (CSDB) (http://csdb.glycoscience.ru/database) of the Russia Zelinsky Institute. CSDB is composed of two parts: Bacterial & Archaeal (BCSDB) and Plant & Fungal (PFCSDB). It is a database for bacterial, archaea, fungi, and plant glycan structures, and contains more than 20,000 compounds from >10,000 organisms with references to >9000 publications. The search criteria can be, for instance, structures (fragments) and NMR (chemical shifts) [59–61]. CSDB claims to be fully curated. Due to the fact that various established carbohydrate databases use different sequence formats to encode carbohydrate structures, cross-linking between these databases was not possible. The lack of cross-links between databases means that users had to visit different database web portals to retrieve all the available information on a specific carbohydrate structure.

During glycobiology studies, in many cases, nuclear magnetic resonance (NMR) spectroscopy techniques can lead to a full structural characterization of oligosaccharides, including the monosaccharide stereochemistry, the anomeric configuration, the linkage type, and the complete sugar sequence. However, for a long time, the process of assigning NMR resonances of biomolecules was time-consuming and laborious. With the increase in advanced instruments with robotic sample exchange, together with automated start-up equilibration procedures and spectral databases, measuring time and interpretation of spectra have been shortened significantly. NMR is now also frequently used for quality control during drug development.

The GLYCOSCIENCES.de database (http://www.glycosciences.de/) contains more than 3400 NMR spectra, stored as lists of chemical shifts, which can be searched online by atom and residue names and by chemical shift values. It contains a good NMR tool for finding glycans whose spectrum matches best with the input spectrum according to NMR chemical shifts, both ^{1}H and ^{13}C [62]. It is possible to perform computational prediction of chemical shifts of structures for which no experimental data are present ("shift estimation"). The NMR chemical shifts are well suited for computational purposes. Another, long-existing, very valuable, excellent online NMR tool is CASPER (Computer Assisted SPectrum Evaluation of Regular polysaccharides) (http://www.casper.organ.su.se/casper/). The CASPER program can facilitate rapid structural determination of glycans. The ^{1}H and ^{13}C chemical shifts of a user-defined structure can be predicted, simulating NMR spectra. The Complex Carbohydrate Magnetic Resonance Database (CCMRD) (http://www.ccmrd.org) contains a database for solid-state NMR chemical shifts of over 400 complex carbohydrates from various organisms [63].

Recently, at the Complex Carbohydrate Research Center (CCRC), University of Georgia, USA, a new initiative supported by the NIH Common Fund, called GlyGen, has started, providing computational and informatics resources for Glycosciences (http://glygen.org). It comprises a comprehensive data repository that integrates diverse types of data, including glycan structures, glycan biosynthesis enzymes, glycoproteins, and 3D glycoprotein structures along with genomic and proteomic knowledge. Additionally, in April 2017, the GlyCosmos Glycoscience Web Portal (http://glycosmos.org) was released by the Japanese Society for Carbohydrate Research (JSCR) to provide up-to-date information about data related to glycans, including glycogens, glycoproteins, glycolipids, pathways, and diseases. A user-friendly web interface allows non-specialists to browse the available information. There are links to UniProtKB and GlycoProtDB. Glycosylation sites on proteins can be visualized and compared. It is an excellent location from which glycoscience-related web resources can be accessed. It includes access to the GlyTouCan repository (http://www.glytoucan.org/). GlyCosmos forms together with GlyGen in the United States and Glycomics@ExPASy in Europe, the GlySpace Alliance [64], which is committed to sharing all data among its members and ensuring that the quality of the data is maintained.

From 2020, there is the ProCarbDB (http://www.procarbdb.science/procarb/), a database covering protein–carbohydrate complexes and containing >5000 3D structures. Protein–carbohydrate complexes are also subject matter in the GLYCAN GBSDB database (http://www.glycanstructure.org/gbs-db/pdb/). For information on Glycoscience research in Europe, visit www.carbomet.eu, "Metrology of Carbohydrates for Enabling European BioIndustries", providing also relevant databases and tools.

To further advance our understanding of the roles that glycans play in development and disease, bioinformatics tools will become more and more important. Since glycobioinformatics would profit from consolidating initiatives, the international Glycome Informatics Consortium (GLIC) was founded in 2015 to provide and maintain a centralized software resource, enabling the development of cooperative databases and tools. The importance of bioinformatics resources for glycoscience is also recognized by the National Institutes of Health. As part of the Common Fund Glycoscience Program, new methodologies and resources to study glycans, including the development of data integration and analysis tools, are being developed. A standard for the depiction of monosaccharides and complex glycans using various colored, geometric shapes, called Symbol Nomenclature for Glycans (SNFG), has now mostly been adopted for use in publications. For high-quality databases, much attention has to be paid to data quality and annotation accuracies during the recording of the information with an emphasis on data management. The present bioinformatics tools appear to aid and facilitate many aspects of glycomics and glycoproteomics experiments. Further enlargement of database contents will assist in the automated interpretation of structures of glycans. New bioinformatics tools are necessary also for 3D structural analysis of glycans. Serious efforts in the development and design of new bioinformatics tools, algorithms, and data collections are required to manage and analyze successfully the large amount of data that will be produced by new glycomics projects in the near future.

References

1. Pérez S, Mulloy B. Prospects for glycoinformatics. Curr Opin Struct Biol. 2005;15:517–24.
2. Liu G, Neelameghan S. Integration of systems glycobiology with bioinformatics toolboxes, glycoinformatics resources, and glycoproteomics data. Wiley Interdiscip Rev Syst Biol Med. 2015;7:163–81.
3. Von der Lieth C-W. Databases and informatics for glycobiology and glycomics. In: Kamerling JP, editor. Comprehensive glycoscience—from chemistry to systems biology, vol. 2. Amsterdam: Elsevier; 2007. p. 329–46.
4. Lütteke T. Web resources for the glycoscientist. ChemBioChem. 2008;9:2155–60.

5. Frank M, Schloissnig S. Bioinformatics and molecular modeling in glycobiology. Cell Mol Life Sci. 2010;67:2749–72.
6. Egorova KS, Toukach PV. Glycoinformatics: bridging isolated islands in the sea of data. Angew Chem Int Ed. 2018;57:14986–90.
7. Von der Lieth C-W, Lütteke T, Frank M, editors. Bioinformatics for glycobiology and glycomics: an introduction. Hoboken: Wiley-Blackwell; 2009.
8. Lütteke T, Frank M, editors. Glyco-informatics. Totowa: Humana Press; 2015.
9. Aoki-Kinoshita K, editor. A practical guide to using glycomics databases. Springer; 2017.
10. Lisacek F, Mariethoz J, Alocci D, Rudd PM, Abrahams JL, et al. Databases and associated tools for glycomics and glycoproteomics. Methods Mol Biol. 2017;1503:235–64.
11. Harvey DJ, Merry AH, Royle L, et al. Proposal for a standard system for drawing structural diagrams of N- and O-linked carbohydrates and related compounds. Proteomics. 2009;9:3796–801.
12. Neelamegham S, Aoki-Kinoshita K, Bolton E, Frank M, Lisacek F, Lütteke T, O'Boyle N, Packer NH, Stanley P, Toukach P, Varki A, Woods RJ, The SNFG Discussion Group. Updates to the symbol nomenclature for glycan guidelines. Glycobiology. 2019;29:620–4.
13. Cantarel BL, Coutinho PM, Rancurel C, Bernard T, Lombard V, Henrissat B. The Carbohydrate-Active EnZymes database (CAZy): an expert resource for glycogenomics. Nucleic Acids Res. 2009;37:D233–8.
14. Lombard V, Golaconda Ramulu H, Drula E, Coutinho PM, Henrissat B. The carbohydrate-active enzymes database (CAZy) in 2013. Nucleic Acids Res. 2014;42:D490–5.
15. Sahoo SS, Thomas C, Sheth A, Henson C, York WS. GLYDE—an expressive XML standard for the presentation of glycan structure. Carbohydr Res. 2005;340:2802–7.
16. Hashimoto K, Goto S, Kawano S, Aoki-Kinoshita KF, Ueda N, Hamajima M, Kawasaki T, Kanehisa M. KEGG: a glycome informatics resource. Glycobiology. 2006;16:63R–70R.
17. Aoki-Kinoshita K. Using databases and web resources for glycomics research. Mol Cell Proteomics. 2013;12:1036–45.
18. Ozohanics O, Krenyacz J, Ludanyi K, Pollreisz F, Vekey K, Drahos L. GlycoMiner: a new software tool to elucidate glycopeptide composition. Rapid Commun Mass Spectrom. 2008;22:3245–54.
19. Akune Y, Hosoda M, Kaiya S, Shinmachi D, Aoki-Kinoshita KF. The RINGS resource for glycome informatics analysis and data mining on the web. OMICS. 2010;14:475–86.
20. Böhm M, Bohne-Lang A, Frank M, Loss A, Rojas-Macias MA, Lütteke T. Glycosciences DB: an annotated data collection linking glycomics and proteomics data (2018 update). Nucleic Acids Res. 2019;47:D1195–201.
21. Scherbinina SI, Toukach PV. Three-dimensional structures of carbohydrates and where to find them. Int. J. Mol. Sci. 2020;21:7702.
22. Frank M, Lütteke T, Von der Lieth CW. GlycoMapsDB: a database of the accessible conformational space of glycosidic linkages. Nucleic Acids Res. 2007;35:287–90.
23. Sarkar A, Pérez S. PolySac3DB: an annotated data base of 3 dimensional structures of polysaccharides. BMC Bioinformatics. 2012;13:302.
24. Pérez S, Sarkar A, River A, Breton C, Imberty A. Glyco3D: a portal for structural glycosciences. Methods Mol Biol. 2015;1273:241–58.
25. Von der Lieth C-W, Freire AA, Blank D, Campbell MP, Ceroni A, et al. EUROCarbDB: an open-access platform for glycoinformatics. Glycobiology. 2011;21:493–502.
26. Ceroni A, Maass K, Geyer H, Geyer R, Dell A, Haslam SM. GlycoWorkBench: a tool for computer-assisted annotation of mass spectra of glycans. J Proteome Res. 2008;7:1650–9.
27. Damerell D, Ceroni A, Maass K, Ranzinger R, Dell A, Haslam SM. The GlycanBuilder and GlycoWorkbench glycoinformatics tools: updates and new developments. Biol Chem. 2012;393:1357–62.
28. Ceroni A, Dell A, Haslam SM. The GlycanBuilder: a fast, intuitive and flexible software tool for building and displaying glycan structures. Source Code Biol Med. 2007;2:3–15.

29. Tsuchiya S, Aoki NP, Shinmachi D, Matsubara M, Yamada I, Aoki-Kinoshita KF, Narimatsu H. Implementation of GlycanBuilder to draw a wide variety of ambiguous glycans. Carbohydr Res. 2017;445:104–16.

30. Alocci D, Suchánková P, Costa R, Hory N, Mariethoz J, Varekova RS, Toukach P, Lisacek F. SugarSketcher: quick and intuitive online glycan drawing. Molecules. 2018;23:3206.

31. Lal K, Bermeo R, Perez S. Computational tools for drawing, building and displaying carbohydrates: a visual guide. Beilstein J Org Chem. 2020;16:2448–68.

32. Herget S, Ranzinger R, Maass K, Von der Lieth CW. GlycoCT—a unifying sequence format for carbohydrates. Carbohydr Res. 2008;343:2162–71.

33. Ranzinger R, Herget S, Wetter T, Von der Lieth C-W. GlycomeDB—integration of open-access carbohydrate structure databases. BMC Bioinformatics. 2008;9:384–97.

34. Ranzinger R, Frank M, von der Lieth CW, Herget S. Glycome-DB.org: a portal for querying across the digital world of carbohydrate sequences. Glycobiology. 2009;19:1563–7.

35. Tiemeyer M, Aoki K, Paulson J, Cummings RD, York WS, Karlsson NG, Lisacek F, Packer NH, Campbell MP, Aoki NP, et al. GlyTouCan: an accessible glycan structure repository. Glycobiology. 2017;27:915–9.

36. Aoki-Kinoshita K, Agravat S, Aoki NP, Arpinar S, Cummings RD, Fujita A, Fujita N, Hart GM, Haslam SM, Kawasaki T, et al. GlyTouCan 1.0—the international glycan structure repository. Nucleic Acids Res. 2017;44:D1237–42.

37. Campbell MP, Royle L, Radcliffe CM, Dwek RA, Rudd PM. GlycoBase and autoGU: tools for HPLC-based glycan analysis. Bioinformatics. 2008;24:1214–6.

38. Artemenko NV, Campbell MP, Rudd PM. GlycoExtractor—a web-based interface for high throughput processing of HPLC-glycan data. J Proteome Res. 2010;9:2037–41.

39. Zhao S, Walsh I, Abrahams JL, Royle L, Nguyen-Khuong T, Spencer D, Fernandes DL, Packer NH, Rudd PM, Campbell MP. GlycoStore: a database of retention properties for glycan analysis. Bioinformatics. 2018;34(18):3231–2.

40. Hayes CA, Karlsson NG, Struwe WB, Lisacek F, Rudd PM, Packer NH, Campbell MP. UniCarb-DB: a database resource for glycomic discovery. Bioinformatics. 2011;27:1343–4.

41. Campbell MP, Peterson R, Mariethoz J, Gasteiger E, Akune Y, Aoki-Kinoshita KF, Lisacek F, Packer NH. UniCarbKB: building a knowledge platform for glycoproteomics. Nucleic Acids Res. 2014;42:D215–21.

42. Campbell MP, Nguyen-Khuong T, Hayes CA, et al. Validation of the curation pipeline of UniCarb-DB: building a global glycan reference MS/MS repository. Biochim Biophys Acta. 2014;1844:108–16.

43. Akune Y, Lin CH, Abrahams JL, Zhang J, Packer NH, Aoki-Kinoshita KF, Campbell MP. Comprehensive analysis of the N-glycan biosynthetic pathway using bioinformatics to generate UniCorn: a theoretical N-glycan structure database. Carbohydr Res. 2016;431:56–63.

44. Rojas-Macias MA, Mariethoz J, Andersson P, Jin C, Venkatakrishnan V, Aoki NP, et al. Towards a standardized bioinformatics infrastructure for N- and O-glycomics. Nat Commun. 2019;10:3275.

45. Mariethoz J, Alocci D, Gastaldello A, Horlacher O, Gasteiger E, Rojas-Macias M, Karlsson NG, Packer NH, Lisacek F. Glycomics@ExPASy: bridging the gap. Mol Cell Proteomics. 2018;17:2164–76.

46. Alocci D, Mariethoz J, Gastaldello A, Gasteiger E, Karlsson NG, Kolarich D, Packer NH, Lisacek F. GlyConnect: glycoproteomics goes visual, interactive and analytical software tools and data resources. J Proteome Res. 2019;18:664–77.

47. Hu H, Mao Y, Huang Y, Lin C, Zaia J. Bioinformatics of glycosaminoglycans. Perspect Sci. 2017;11:40–4.

48. Woodin CL, Maxon M, Desaire H. Software for automated interpretation of mass spectrometry data from glycans and glycopeptides. Analyst. 2013;138:2793–803.

49. Li F, Glinskii OV, Glinsky VV. Glycobioinformatics: current strategies and tools for data mining in MS-based glycoproteomics. Proteomics. 2013;13:341–54.

50. Tang H, Mechref Y, Novotny MV. Automated interpretation of MS/MS spectra of oligosaccharides. Bioinformatics. 2005;21:i431–9.
51. Apte A, Meitei NS. Bioinformatics in glycomics: glycan characterization with mass spectrometric data using SimGlycan. Methods Mol Biol. 2010;600:269–81.
52. Hong P, Sun H, Sha L, Pu Y, Khatri K, Yu X, Tang Y, Lin C. GlycoDeNovo—an efficient algorithm for accurate de novo glycan topology reconstruction from tandem mass spectra. J Am Soc Mass Spectrom. 2017;11:2288–301.
53. Dallas DC, Martin WF, Hua S, German JB. Automated glycopeptide analysis—review of current state and future direction. Brief Bioinform. 2013;14:361–74.
54. Wu SW, Liang SY, Pu TH, Chang FY, Khoo KH. Sweet-Heart—an integrated suite of enabling computational tools for automated MS2/MS3 sequencing and identification of glycopeptides. J Proteome. 2013;84:1–16.
55. Mayampurath A, Yu CY, Song E, Balan J, Mechref Y, Tang H. Computational framework for identification of intact glycopeptides in complex samples. Anal Chem. 2014;86:453–63.
56. Bern M, Kil YJ, Becker C. Byonic: advanced peptide and protein identification software. Curr Protoc Bioinformatics. 2012;13:13–20.
57. Toghi ES, Shah P, Yang W, Li X, Zhang H. GP-Quest: a spectral library matching algorithm for site-specific assignment of tandem mass spectra to intact N-glycopeptides. Anal Chem. 2015;87:5181–8.
58. Eavenson M, Kochut KJ, Miller JA, Ranzinger R, Tiemeyer M, Aoki K, et al. Qrator: a web-based curation tool for glycan structures. Glycobiology. 2015;25:66–73.
59. Toukach P, Joshi HJ, Ranzinger R, Knirel YA, Von der Lieth CW. Sharing of worldwide distributed carbohydrate-related digital resources: online connection of the bacterial carbohydrate structure database and GLYCOSCIENCES.de. Nucleic Acids Res. 2007;35:D280–6.
60. Toukach PV. Bacterial carbohydrate structure database version 3. Glycoconj J. 2009;26:856–63.
61. Toukach PV, Egorova KS. Carbohydrate structure database merged from bacterial, archaeal, plant and fungal parts. Nucleic Acids Res. 2016;44:D1229–36.
62. Loss A, Stenutz R, Schwarzer E, Von der Lieth C-W. GlyNest and CASPER: two independent approaches to estimate ^{1}H and ^{13}C NMR shifts of glycans available through a common web-interface. Nucleic Acids Res. 2006;34:W733–7.
63. Kang X, Zhao W, Dickwella Widanage MC, Kirui A, Ozdenvar U, Wang T. CCMRD: a solid-state NMR database for complex carbohydrates. J Biomol NMR. 2020;74:239–45.
64. Aoki-Kinoshita KF, Lisacek F, Mazumder R, York WS, Packer NH. The GlySpace alliance: toward a collaborative global glycoinformatics community. Glycobiology. 2020;30:70–1.

Chapter 14
Concluding Remarks

Abstract Carbohydrates (sugars) are the major components of our food, providing us with energy and calories. In our body, carbohydrates also form parts of proteins and lipids. There is no doubt that carbohydrates play fundamental roles in several biological processes in living systems. To understand the functions of carbohydrates (glycans), their detailed structures must be known. The novel developments in analytical methodologies during the last two decades were beneficial to glycoscience. Carbohydrate analysis is an unavoidable item in life science, biomedicine and carbohydrate biotechnology.

Keywords Glycoscience · HMOs · Xenotransplantation · Disease biomarkers · Biologics · Vaccins · glycoRNAs

The past 20 years have shown tremendous progress in many fields throughout the biosciences. The ability to investigate and define cellular processes at the molecular level has dramatically increased due to the development and improvement in sophisticated, sensitive instruments, including mass spectrometry and NMR spectroscopy, next to advanced automatic separation instrumentation and on-line bioinformatics databases. Consequently, glycoscience has rapidly expanded in the last two decades and plays an integral role in the academic biology study and many of the world's industrial processes.

Nowadays, it is evident that carbohydrates are extremely important. They are involved in many biological processes in the living world. The structure of glycans from glycoproteins and other glycoconjugates is highly relevant for their function. Accurate structural analysis of glycans has opened the way for the meaningful study of their biological and pathophysiological functions and effects. In the human Genome-Wide Association Studies (GWAS), the biomolecular glycosylation is referred to as the third language of biology. Because sugars are not encoded directly in genomes, much information concerning the roles of glycans is still derived from structural analysis. Many biological roles of glycans are known now. Carbohydrates not only are dietary sources of energy but also participate in cellular functions such as cell growth, recognition, adhesion, signaling, protection, and fertilization/reproduction and in extracellular matrix organization. Furthermore, carbohydrates are responsible for protein folding, protein trafficking, triggering of endocytosis and

G. J. Gerwig, *The Art of Carbohydrate Analysis*, Techniques in Life Science and Biomedicine for the Non-Expert, https://doi.org/10.1007/978-3-030-77791-3_14

phagocytosis, clearance of damaged glycoconjugates and cells, and also tissue elasticity. Many new functions of carbohydrates will continue to be discovered.

As mentioned earlier, the importance of carbohydrates is clearly demonstrated for human breast milk. Milk oligosaccharides act as antimicrobial defense factors against pathogenic bacteria [1]. The constituting oligosaccharides are natural antagonists of intestinal infections in infants and promote the growth of beneficial gut flora, including bifidobacteria. Commercially available baby nutritional products are prepared from cow milk by removing bad carbohydrates (e.g., bovine milk oligosaccharides can contain N-glycolylneuraminic acid) and adding new oligosaccharides (galacto- and fructo-oligosaccharides, and fucose-containing oligosaccharides) to impart the beneficial properties provided by the natural human milk oligosaccharides (HMOs).

Reliable analytical methods are needed for characterizing and quantifying potentially immunogenic glycans, such as those bearing the non-human Gal(α1→3) Gal(β1→4)GlcNAc motifs, which may cause fatal anaphylaxis in certain patients' populations [2, 3]. This Gal(α1→3)Gal epitope (glycotope), present on glycoproteins and glycolipids of cells and tissues of many non-primate animal species (e.g., pigs), but absent in human, still is one of the main barriers to organ xenotransplantation (hyperacute rejection, HAR). Besides, other carbohydrate epitopes, non-human sialic acid Neu5Gc and the risk of infections by animal viruses play a role [4].

As said before, so far, molecular characterization of the oligosaccharides linked to proteins and lipids has contributed to the knowledge of the multiple functions that these biomolecules serve. However, the picture of the role of glycans in life science is still scattered. The full extent of glycan distribution and exact functions remain unclear. This means that decoding the functions of complex carbohydrates still is a very exciting scientific challenge. The glycan profiling of normal and diseased forms of a glycoprotein has already provided new insights for future research in rheumatoid arthritis, prion disease, and congenital disorders of glycosylation. The important role of glycans is demonstrated by the growing discovery list of human inborn errors in metabolism and diseases that are the result of defects in glycan biosynthesis or glycan breakdown (lysosomal storage disorders) [5]. Furthermore, altered glycosylation has been observed as a feature of Alzheimer's disease, cystic fibrosis, and multiple sclerosis. Ongoing glycobiology studies in these areas are challenging. For assessing the role of glycans, it is a prerequisite to understanding their structures and distribution in biologic systems. In this regard, sensitive and quantitative methodologies are required and the search for more sophisticated techniques continues.

It has clearly been demonstrated that carbohydrates are implicated in various diseases such as cancer, immune diseases, infection diseases, muscle-degenerative diseases, neurodegenerative diseases, and diabetes. The connection between glycosylation and disease includes mild as well as severe syndromes and can span from early neonatal to adult life. Nowadays, certain types of glycan structures (e.g., with increased levels of fucosylation) are used as markers for tumor progression. Glycobiology presents a fertile ground for disease biomarker discovery. Understanding the role of glycan structure and function in fundamental biology will enable the development of new-age diagnostics and therapeutics with higher efficacy. The clinical application of glycobiology is growing each year and the glycosciences are fully integrated with the life sciences. Several carbohydrate-based vaccines are under development to treat cancer based on ganglioside immunogens (G_{M2} and G_{D2}) present on certain types of cancer cells (melanoma and breast cancer) [6].

In addition, the pharmaceutical industry shows an increased production of therapeutic glycoproteins (biologics), prepared as recombinant secretory products in cell culture systems or in transgenic animals. They are used for the treatment of various diseases including cancer, cardiovascular diseases, diabetes, infections, inflammatory, and autoimmune disorders, and more. The control of glycosylation takes on major importance during the development of these drugs, since the carbohydrate chains have dramatic effects on their properties, such as stability, bioactivity, action, and pharmacodynamics in intact organisms. For example, core fucosylation influences the effector function of mAbs, and mannose-6-phosphate moieties on glycans of therapeutic enzymes are essential for trafficking to the lysosomes where the enzymes need to be catalytically active. Sialic acids are important for the lifetime of the glycoprotein. In-depth characterization of the glycan structures is therefore an essential requirement before these carbohydrate-containing recombinant biotherapeutics can be used in clinical trials [7, 8]. Besides, the glycosylation must be controlled during the production of these biopharmaceuticals to satisfy regulatory requirements for batch-to-batch product consistency.

Bacteria (including some pathogenic ones) express highly specific glycans on their cell surface and produce extracellular polysaccharides, which are involved in bacterial infection. Bacterial vaccines are interesting examples of carbohydrate-based therapeutics. These vaccines contain the concerning bacterial-derived oligosaccharide coupled to a protein carrier to trigger the immune system. Furthermore, parasitic protozoa GPI-based vaccines are in development.

Curative drugs, containing sialic acid analogs, battle the influenza virus by inhibiting the action of viral sialidases, which allow newly produced viruses to expand the cell surface binding sites to infect other cells. Coronaviridae contains spike (S) glycoproteins and hemagglutinin esterases that utilize O-acetyl forms of sialic acids for attachment. Desialylation of cells, as a result of the action of viral sialidases, may result in autoimmune disease. Bronchopulmonary disease results from desialylation of lung epithelia after an influenza virus infection. The recent pandemic of the COVID-19 virus illustrates that much research is still needed concerning the involvement of sialic acids/glycoproteins and related enzymes (e.g., Coronavirus haemagglutinin-esterase) in this area.

In this book, we have looked at some of the major methodologies used nowadays in the (structural) analysis of carbohydrates, including polysaccharides and glycoconjugates. There has been significant progress made in glycan analysis. The availability of commercial kits that contain the necessary reagents for the release of N- and O-glycans and labeling of monosaccharides, sialic acids, and glycans have made it easier for laboratories to adopt these technologies.

The progress in mass spectrometry and its data analysis by computer programs have brought about big changes in carbohydrate analysis. Glycan profiling has almost reached the status of automatization. However, at this moment, still there is no single method that permits complete complex carbohydrate analysis, and therefore, a panel of methodologies is often required to fully characterize the glycans. This means that the development of new carbohydrate analysis techniques will continue. The extreme complexity and diversity of glycoconjugates structures continue to demand new processes for their elucidation. Eventually, new techniques may emerge that allow total glycan analysis to be performed automatically and also be capable of rapid throughput, together with computerized data interpretation. To get a complete understanding of the functions of glycans in glycoproteins, the detailed study of site-specific glycosylation (heterogeneity) is a prerequisite. Identification of the three-dimensional structures of the glycans will also aid in understanding the functions of glycoproteins. Despite all the progress in our knowledge on protein glycosylation and many practical achievements in this area, our understanding of the molecular code translating the effect of protein attached glycans into the functional protein space remains far from complete, in the context of the ~500 genes involved in glycosylation in the cell.

The surfaces of all cells in nature are covered with a dense and complex coating of glycans, which is taxon-, species-, and cell-type specific. These glycans are the first point of contact for cellular interactions. It is important to recognize that the full range of types and distributions of glycans in nature is still largely unexplored [9]. Efforts to elucidate the molecular mechanisms that underlie glycan functions and to understand how carbohydrates are assembled and disassembled can reap benefits in fields ranging from bioenergy to human medicine. This all is an intriguing scientific challenge. Further progress in understanding may also be expected about the involvement of carbohydrates in xenotransplantation and stem cell glycobiology. There remains much more to be understood. There are still a lot of questions as to how bacteria, viruses, and toxins evade the body's immune system using carbohydrates and initiate infective processes.

It is clear that carbohydrates are an integral constituent of all living organisms and are associated with numerous vital functions that sustain life. It is now accepted that complex glycans play major roles in biology, such as the development of the embryo, the function of the immune system, microbial and viral pathogenesis, and cellular communication. It is the ultimate goal of glycoscience research to be able to link glycan structures with their biological functions. It is expected that the field of glycomics will continue to thrive in the future.

That glycobiology is still open for further interesting research is demonstrated by the next example. It is known that (m/t) RNA can have post-transcriptional modifications (PTMs). Recently, it has been discovered that RNA can also be target of glycosylation [10]. There is evidence that multiple mammalian species use RNA as a third scaffold for glycosylation, next to proteins and lipids. Specific small noncoding RNAs (e.g., Y RNA) bear protein-associated N-type glycans that are fucosylated and sialylated. They are called "glycoRNAs" and the majority is present on the cell surface, where they can interact with sialic acid-binding immunoglobulin lectin receptors (Siglecs). This feature expands the role for RNA in extracellular biology.

New discoveries are awaiting. Profit will be obtained by the accessibility of curated glycobioinformatics databases and efficient software for glycan analysis. The advancement of glycoscience will be stimulated by expanding the integration of data describing glycoproteins, glycolipids, polysaccharides, and the genome-coded enzymatic machinery that builds or breaks down these glycans, together with the ever-increasing information about the interactions of these carbohydrate compounds with other components in the living cell. Furthermore, a better understanding of the glycosylation of proteins will facilitate the development of the next generation of glycoprotein-based therapeutics. Undoubtedly, many important medicines will come from the field of glycoscience in the future. Glycoscience has become an essential part of modern innovative biotechnology. There is an increasing understanding that specific glycans can improve health and prevent disease.

Of course, it was impossible to discuss all existing carbohydrate analytical methods in this book due to scope and space limitations. More (specific) analytical methods used in carbohydrate chemistry can be found in the more specialized scientific carbohydrate literature as regularly referenced in the text. At present, glycomics and glycoproteomics are still limited to relatively specialized research groups. Hopefully, this book adds a few steps in transforming glycoscience from a field dominated by specialists to a widely studied and integrated discipline, which could lead to a more complete understanding of carbohydrates by a broader research community.

References

1. Peterson R, Cheah WY, Grinyer J, Packer N. Glycoconjugates in human milk: protecting infants from disease. Glycobiology. 2013;23:1425–38.
2. Macher BA, Galili U. The Galα1,3Galβ1,4GlcNAc-R (α-Gal) epitope: a carbohydrate of unique evolution and clinical relevance. Biochim Biophys Acta. 2008;1780:75–88.
3. Steinke JW, Platts-Mills TA, Commins SP. The alpha gal story: lessons learned from connecting the dots. J Allergy Clin Immunol. 2015;135:589–97.
4. Jang K-S, Kim Y-G, Adhya M, Park H-M, Kim B-G. The sweets standing at the borderline between allo- and xenotransplantation. Xenotransplantation. 2013;20:199–208.
5. Sun A. Lysosomal storage disease overview. Ann Transl Med. 2018;6(24):476.
6. Yin Z, Huang X. Recent development in carbohydrate-based anti-cancer vaccines. J Carbohydr Chem. 2012;31:143–86.
7. Jones A. N-glycan analysis of biotherapeutic proteins. BioPharm Int. 2017;30:20–5.
8. O'Flaherty RM, Trbojevic-Akmacic I, Greville G, Rudd PM, Lauc G. The sweet spot for biologics: recent advances in characterization of biotherapeutic glycoproteins. Expert Rev Proteomics. 2018;15:13–29.
9. Planinc A, Bones J, Dejaegher B, van Antwerpen P, Delporte C. Glycan characterization of biopharmaceuticals: updates and perspectives. Anal Chim Acta. 2016;921:13–27.
10. Flynn RA, Pedram K, Malaker SA, Bastista PJ, Smith BAH, Johnson AG, George BM, Majzoub K, Villata PW, Carette JE, Bertozzi CR. Small RNAs are modified with N-glycans and displayed on the surface of living cells. Cell. 2021;184:3109–24.

Appendix A

EI-MS of Partially Methylated Alditol Acetates (PMAAs)

Electron Impact-Mass Spectra (EI-MS) of Partially Methylated Alditol Acetates (PMAAs) derived from: (1) hexopyranoses (2) hexofuranoses (3) pentopyranoses (4) pentofuranoses (5) N-acetyl hexosamines

© Springer Nature Switzerland AG 2021   319
G. J. Gerwig, *The Art of Carbohydrate Analysis*, Techniques in Life Science and
Biomedicine for the Non-Expert, https://doi.org/10.1007/978-3-030-77791-3

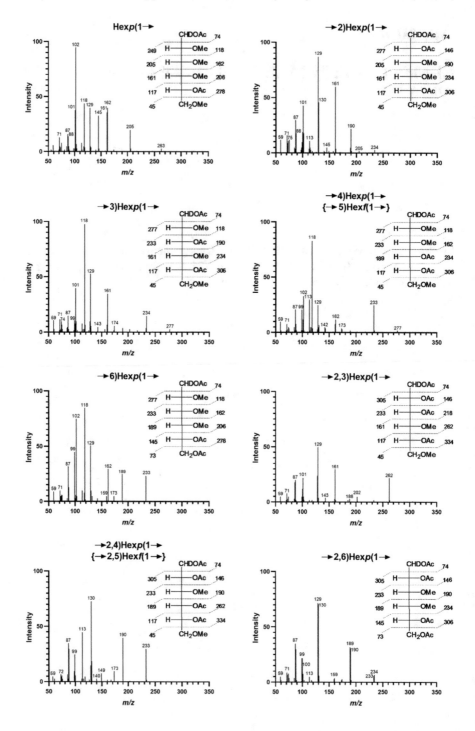

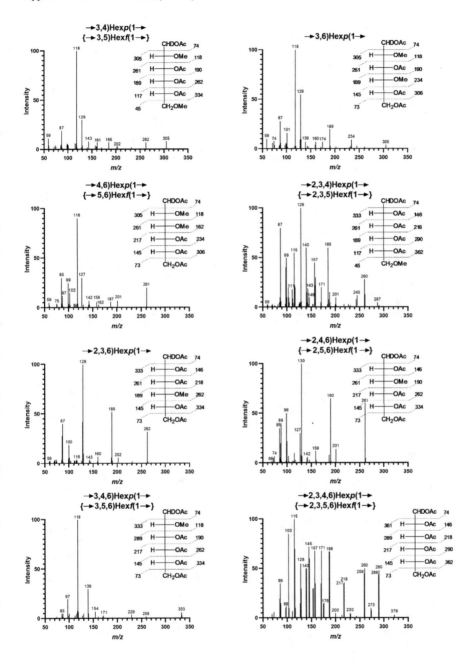

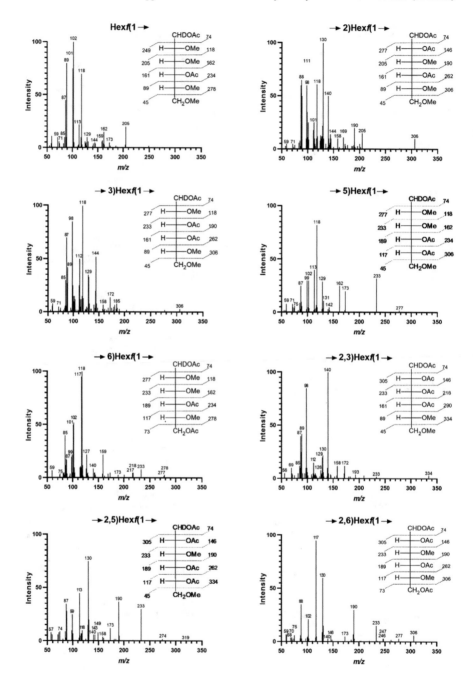

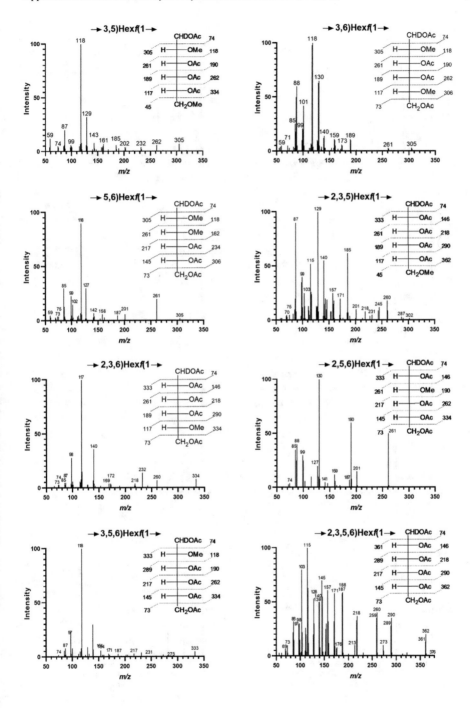

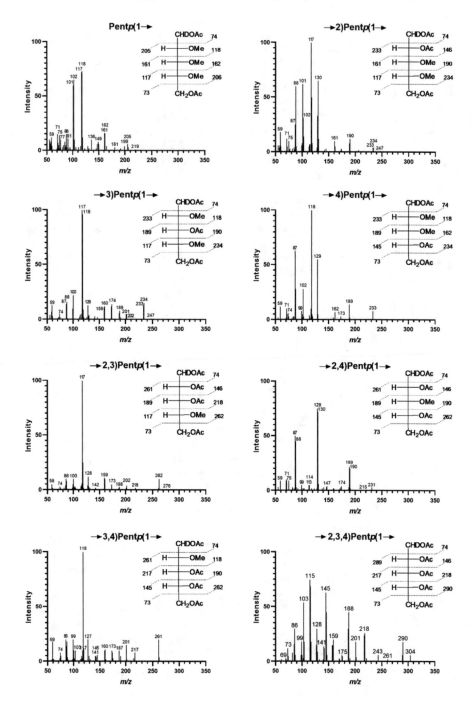

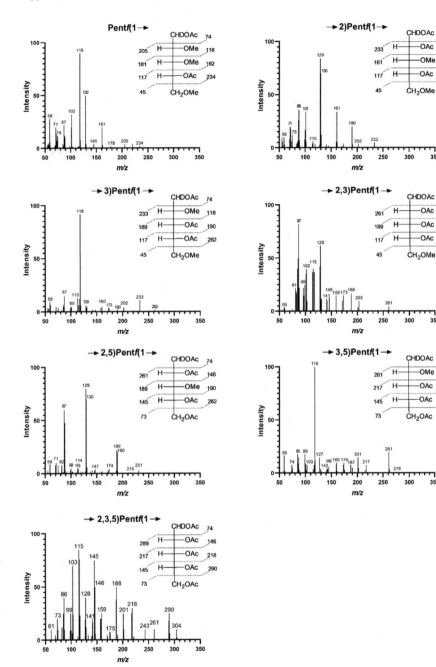

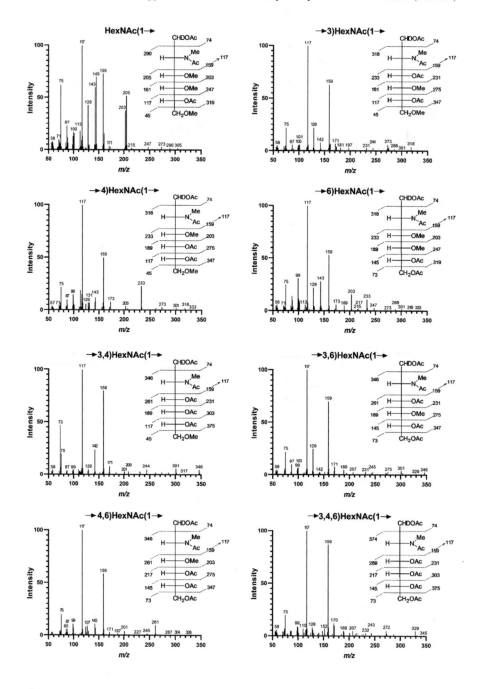

Appendix B

MALDI-TOF-MS of Permethylated N-glycans

Pseudomolecular signal ions [M+Na]$^+$ in positive ion mode MALDI-TOF-MS of permethylated derivatives of N-Glycans, which are mostly found in glycoproteins. Glycan structures are drawn using GlycoWorkBench 2.14.

Signal (*m/z*) [M+Na]$^+$	Composition				Tentative structure assignment
	Hex ⚪⚪	HexNAc ◼	Fuc ▼	Neu5Ac ◆	
1171.5	3	2	–	–	
1345.6	3	2	1	–	
1375.6	4	2	–	–	
1416.7	3	3	–	–	
1549.5	4	2	1	–	
1579.9	5	2	–	–	
1590.7	3	3	1	–	
1661.7	3	4	–	–	
1784.0	6	2	–	–	
1824.8	5	3	–	–	

(continued)

© Springer Nature Switzerland AG 2021

G. J. Gerwig, *The Art of Carbohydrate Analysis*, Techniques in Life Science and Biomedicine for the Non-Expert, https://doi.org/10.1007/978-3-030-77791-3

Signal (*m/z*) [M+Na]⁺	Composition Hex ⚪	HexNAc ■	Fuc ▼	Neu5Ac ◆	Tentative structure assignment
1835.8	3	4	1	–	
1865.6	4	4	–	–	
1987.9	7	2	–	–	
2040.0	4	4	1	–	
2069.8	5	4	–	–	
2080.5	3	5	1	–	
2155.8	4	3	1	1	
2192.1	8	2	–	–	
2226.7	4	4	–	1	
2243.9	5	4	1	–	
2285.0	4	5	1	–	
2395.9	9	2	–	–	
2400.5	4	4	1	1	
2418.0	5	4	2	–	
2430.7	5	4	–	1	
2519.4	6	5	–	–	
2592.1	5	4	3	–	
2605.1	5	4	1	1	

(continued)

Signal (m/z) [M+Na]⁺	Composition				Tentative structure assignment
	Hex ⊙◯	HexNAc ■	Fuc ▼	Neu5Ac ◆	
2693.5	6	5	1	–	
2779.2	5	4	2	1	
2792.0	5	4	–	2	
2850.2	5	5	1	1	
2867.5	6	5	2	–	
2880.2	6	5	–	1	
2966.3	5	4	1	2	
2968.3	7	6	–	–	
3024.1	5	5	2	1	
3037.0	5	5	–	2	
3054.3	6	5	1	1	
3142.4	7	6	1	–	
3211.1	5	5	1	2	
3213.3	7	7	–	–	
3241.4	6	5	–	2	

(continued)

Signal (m/z) [M+Na]+	Composition				Tentative structure assignment
	Hex ⚪⚪	HexNAc ⬛	Fuc ▼	Neu5Ac ◆	
3316.5	7	6	2	–	
3329.5	7	6	–	1	
3415.4	6	5	1	2	
3490.8	7	6	3	–	
3503.5	7	6	1	1	
3602.6	6	5	–	3	
3690.5	7	6	–	2	
3766.0	8	7	2	–	
3776.7	6	5	1	3	
3864.6	7	6	1	2	
3940.1	8	7	3	–	
3950.7	6	5	2	3	

(continued)

Signal (m/z) [M+Na]+	Composition				Tentative structure assignment
	Hex ⬤⭕	HexNAc ◼	Fuc ▼	Neu5Ac ◆	
4051.8	7	6	–	3	
4226.1	7	6	1	3	
4412.8	7	6	–	4	
4586.8	7	6	1	4	

Appendix C

ES-MS [M-H]⁻ Signals of Some O-glycans (as alditols)

Pseudomolecular signal ions [M-H]⁻ of some O-Glycans (as alditols) regularly found in mucin glycoproteins.

| Signal (m/z) [M-H]⁻ | Composition | | | | Tentative structure assignment |
	Hex ○	HexNAc ▢▮	Fuc ▼	Neu5Ac ◆	
384.1	1	1	–	–	
425.2	–	2	–	–	
513.2	–	1	–	1	
530.2	1	1	1	–	
587.2	1	2	–	–	
675.3	1	1	–	1	
716.3	–	2	–	1	
733.3	1	2	1	–	
749.3	2	2	–	–	
787.3	1	2	–	–	
790.3	1	3	–	–	
821.3	1	1	1	1	
878.3	1	2	–	1	
879.3	1	2	2	–	
895.3	2	2	1	–	
936.4	1	3	1	–	
952.4	2	3	–	–	

(continued)

© Springer Nature Switzerland AG 2021

G. J. Gerwig, *The Art of Carbohydrate Analysis*, Techniques in Life Science and Biomedicine for the Non-Expert, https://doi.org/10.1007/978-3-030-77791-3

Signal (m/z) [M-H]⁻	Composition				Tentative structure assignment
	Hex ○	HexNAc ▢▣	Fuc ▼	Neu5Ac ◆	
966.3	1	1	–	2	
1024.4	1	2	1	1	
1040.4	2	2	–	1	
1041.4	2	2	2	–	
1081.4	1	3	–	1	
1098.4	2	3	1	–	
1114.4	3	3	–	–	
1155.4	2	4	–	–	
1169.4	1	2	–	2	
1170.4	1	2	2	1	
1186.4	2	2	1	1	
1187.5	2	2	3	–	
1243.5	2	3	–	1	
1244.5	2	3	2	–	
1301.5	2	4	1	–	
1331.5	2	2	–	2	
1332.5	2	2	2	1	
1364.5	4	2	–	1	
1389.5	2	3	1	1	
1390.5	2	3	3	–	
1405.5	3	3	–	1	
1406.5	3	3	2	–	
1552.6	3	3	3	–	

(continued)

Signal (*m/z*) [M-H]⁻	Composition				Tentative structure assignment
	Hex ◯	HexNAc ▢▦	Fuc ▼	Neu5Ac ◆	
1593.6	2	4	3	–	
1609.6	3	4	2	–	
1666.6	3	5	1	–	
1770.6	4	4	–	1	
1812.7	3	5	2	–	

Glossary

Adhesins Proteins on the surface of bacteria, viruses, or parasites that interacts with glycans of eukaryotic cell membranes.

Aglycone Noncarbohydrate part of a glycoconjugate or glycoside.

Aldose Monosaccharide with an aldehyde group or potential aldehydic carbonyl group.

Amino Sugar Monosaccharide in which a hydroxyl group (-OH) is replaced by an amino group (-NH$_2$).

Anomeric carbon C1 for aldoses and C2 for ketoses.

Anomericity α and β configuration of the glycosidic bond of a monosaccharide to another monosaccharide or an aglycone.

Anomers Stereoisomers of a monosaccharide that differ only in configuration at the anomeric carbon.

Antibody Serum protein produced by plasma cells in response to immunogens and that binds specifically to (glycan)-antigens.

Antigens Foreign molecules that specifically binds to lymphocytes or antibodies.

Apoptosis Programmed cell death.

Biosimilar A biopharmaceutical which is highly similar to the original biological drug (medicine) in terms of safety, purity, and efficacy.

Carbohydrate A generic term used interchangeably with sugar, saccharides, monosaccharides, oligosaccharides, polysaccharides or glycans.

Cartilage Smooth elastic tissue, rubber-like padding that covers and protects the ends of long bones at the joints and nerves.

Ceramide Lipid part (with fatty acid) of glycosphingolipids.

Cerebroside A glycolipid composed of ceramide with an attached galactose (galactosylceramide) or glucose (glucosylceramide).

Chondroitin Sulfate A type of glycosaminoglycan defined by the disaccharide unit [GalNAc($\beta 1 \rightarrow 4$)GlcA($\beta 1 \rightarrow 3$)]$_n$, modified with ester-linked sulfate at certain positions and typically found covalently linked to a proteoglycan core protein.

Collagen Collection of fibrous proteins of the extracellular matrix.

© Springer Nature Switzerland AG 2021
G. J. Gerwig, *The Art of Carbohydrate Analysis*, Techniques in Life Science and Biomedicine for the Non-Expert, https://doi.org/10.1007/978-3-030-77791-3

Deoxy Sugar   Monosaccharide in which a hydroxyl group (-OH) is replaced by a hydrogen atom (-H).

Dermatan Sulfate   A modified form of chondroitin sulfate in which a portion of the β-glucuronate residues is epimerized to α-iduronates.

Diastereoisomers   Monosaccharides with identical formulas but having a different spatial distribution of atoms, e.g., glucose vs galactose.

β-Elimination   Base-catalyzed, non-hydrolytic cleavage of O-linked glycan attached to the hydroxyl group of Serine or Threonine within a protein/peptide.

Enantiomers   Structural mirror images, e.g., D-glucose versus L-glucose.

Endoglycosidase   An enzyme that catalyzes the cleavage of an internal glycosidic linkage in an oligo/polysaccharide or releases oligosaccharides from glycoconjugates.

Epimers   Monosaccharides differing only in the configuration of a single chiral carbon. For example, galactose is the C4 epimer of glucose, mannose is the C2 epimer of glucose.

Epithelium   Cells that form the epidermis of the respiratory tract.

Exoglycosidase   An enzyme that catalyzes the cleavage of a terminal monosaccharide from the non-reducing end of an oligo/polysaccharide.

Furanose   Five-membered (four carbons and one oxygen) ring form of a monosaccharide.

Ganglioside   Anionic glycosphingolipid containing one or more sialic acid residues.

Glycan   A generic term for an assembly of sugars, polysaccharide, or the carbohydrate part of a glycoprotein or glycolipid.

Glycemic index   (GI) is a relative ranking of carbohydrate in foods according to how they affect blood glucose levels.

Glycobiology   Study of the structure, chemistry, biosynthesis, and biological functions of carbohydrates and glycoconjugates.

Glycocalyx   Glycan pericellular matrix environing the plasma membrane.

Glycoconjugate   A molecule in which one or more glycan units are covalently linked to a noncarbohydrate moiety (e.g., lipid or protein).

Glycoforms   Variants of a glycoconjugate in which the glycan moieties exhibit heterogeneity.

Glycogene   A family of genes that codes glycosyltransferases, sulfotransferases, nucleotide-sugar transporters.

Glycolipid   General term denoting a molecule containing a saccharide linked to a lipid aglycone. In higher organisms, most glycolipids are glycosphingolipids and in plants are glycerol-glycolipids.

Glycolysis   Metabolic pathway that converts glucose into pyruvate to extract energy.

Glycomics   Study of the biological role of glycans in cells, tissue, and organisms.

Glycopeptide   An oligopeptide having one or more covalently attached glycan units.

Glycoprotein   Protein having one or more covalently attached glycan units.

Glycoproteome   Description of the total complement of glycosylated proteins and their glycans in a time- and space-defined environment/system.

Glycoproteomics   Study of the system-wide site-specific analysis of protein glycosylation.

Glycosaminoglycans   GAGs (previously known as mucopolysaccharides) are polysaccharide side-chains of proteoglycans composed of disaccharide repeating units, consisting of a hexosamine and uronic acid (usually containing N-or O-sulfate groups.

Glycosidase   Enzyme that catalyzes the hydrolysis of glycosidic bonds.

Glycoside   Sugar having an anomeric glycosidic linkage to an aglycone.

Glycosidic linkage   Covalent bond between a monosaccharide and another monosaccharide via the hydroxyl group of the anomeric carbon.

Glycosphingolipid   Glycolipid containing a glycan glycosidically attached to the primary hydroxyl group of a ceramide (Sphingosine).

Glycosylation   Covalent attachment of a carbohydrate to a polypeptide, lipid, polynucleotide, carbohydrate, or other organic compounds, generally catalyzed by glycosyltransferases, utilizing sugar nucleotide donor substrates.

Glycosyltransferase   Enzyme that catalyzes the transfer of sugar from a sugar nucleotide donor to a substrate or from carbohydrate to another carbohydrate.

Golgi apparatus   Organelle involved in glycosylation in most eukaryotic cells, bearing various glycosyltransferases and related proteins.

Hemiacetal   Compound formed by the reaction of an aldehyde (carbonyl group) with an alcohol group, as in ring closure of an aldose.

Hemiketal   Compound formed by the reaction of a ketone with an alcohol group, as in ring closure of a ketose.

Heparan sulfate   Glycosaminoglycan (linear polysaccharide) defined by the disaccharide unit $[GlcNAc(\alpha1\rightarrow4)GlcA(\beta1\rightarrow4)/IdoA(\alpha1\rightarrow4)]_n$, containing N- and O-sulfate esters at various positions.

Heparin   Highly sulfated GAG made by mast cells containing iduronic acid and N-, 2-O- and 6-O-sulfate residues.

Heteropolysaccharide   Polysaccharide containing more than one type of monosaccharide.

Hexose   A 6-carbon monosaccharide typically with an aldehyde at the C1 position (aldohexose) and hydroxyl groups at all other positions.

Hexosamine   Hexose with an amino group in place of the hydroxyl group at the C2 position (e.g., $GlcNH_2$, $GalNH_2$).

Homeostasis   State of steady internal, physical, and chemical conditions maintained by living systems.

Homopolysaccharide   A polysaccharide composed of only one type of monosaccharide (e.g., glucan).

Hyaluronan   A glycosaminoglycan (also called hyaluronic acid) consisting of a polymer of the disaccharide unit $[\rightarrow3)GlcNAc(\beta1\rightarrow4)GlcA(\beta1\rightarrow]_n$ that is neither sulfated nor covalently linked to protein.

Immunoglobulin   Y-shaped protein (antibody) used by the immune system to identify and neutralize foreign objects such as pathogenic bacteria and viruses.

Keratan Sulfate   A proteoglycan-polylactosamine $[\rightarrow3)Gal(\beta1\rightarrow4)GlcNAc(\beta1\rightarrow]_n$ with sulfate esters at C6 of GlcNAc and Gal residues.

Ketose Monosaccharide with a ketone group or a potential ketonic carbonyl group (typically at the C2 position).

Lectin A protein of non-immune origin with a specific sugar-binding activity. They bear at least two sugar-binding sites, which make them able to agglutinate cells and/or precipitate glycoconjugates.

Lipopolysaccharide (LPS) polysaccharide linked to a lipid moiety, expressed on the cell wall of Gram-negative bacteria, also called endotoxins.

Macroheterogeneity Variable occupancy of different glycosylation sites on glycoproteins.

Microheterogeneity Structural variations (glycoforms) of a glycan attached to a single glycosylation site of a glycoprotein.

Monoclonal antibodies (mAbs) identical immunoglobulin molecules that are produced by cloned antibody-producing cells.

Monosaccharide Carbohydrate that cannot be hydrolyzed into a simpler carbohydrate. Simple monosaccharides are polyhydroxy aldehydes or polyhydroxy ketones with three to nine carbon atoms.

Mucins Large glycoproteins of mucus with high content of Ser, Thr, and Pro and numerous oligosaccharides O-linked via GalNAc, often occurring in clusters on the polypeptide.

Mutarotation Coming to an equilibrium mixture of α/β furanose, α/β pyranose, and open-chain form of a monosaccharide (non-reducing sugars will not mutarotate).

N-glycosylation Carbohydrates (glycans) linked to a protein via asparagine residues.

Non-reducing terminus (Non-reducing end) Outermost end of an oligosaccharide or polysaccharide chain, opposite to that of the reducing end.

Oligosaccharide Linear or branched chain of monosaccharides attached to one another via glycosidic linkages. (Disaccharide, Trisaccharide, Tetrasaccharide, etc.).

Peptidoglycan GAG of MurNAc(β1-4)GlcNAc(β1-4) repeating units covalently linked to a short peptide.

Polysaccharide Linear or branched polymer of more than 10 monosaccharides (often called glycan).

Proteoglycans Special class of glycoproteins with more than one glycosaminoglycan (GAG) covalently O-linked (via Xyl) to the polypeptide core.

Pyranose Six-membered (five carbons and one oxygen) ring form of a monosaccharide; the most common form found for hexoses and pentoses.

Reducing terminus (Reducing end) Monosaccharide at the end of an oligosaccharide chain that has reducing power because it is unattached (shows α/β configuration).

Saccharide A generic term for any carbohydrate or assembly of carbohydrates.

Sialic acids A general term for a family of 9-carbon monosaccharides derived from neuraminic acid, usually attached to the nonreducing end of the sugar chain.

Sialidase Enzyme that removes a terminal sialic acid residue from a glycan.

Sphingosine An 18-carbon amino alcohol with an unsaturated hydrocarbon chain (2-amino-4-octadecene-1,3-diol), which forms the primary part of sphingolipids.

Sugar A generic term for carbohydrates, usually of low molecular mass and sweet taste, in the general language used for Sucrose.

Sugar Nucleotide Activated form of a monosaccharide, such as UDP-Gal, GDP-Fuc, and CMP-Sialic acid, typically used as donor substrates by glycosyltransferases.

Index

© Springer Nature Switzerland AG 2021
G. J. Gerwig, *The Art of Carbohydrate Analysis*, Techniques in Life Science and
Biomedicine for the Non-Expert, https://doi.org/10.1007/978-3-030-77791-3